Shi Jin • Lorenzo Pareschi
Editors

Uncertainty Quantification for Hyperbolic and Kinetic Equations

 Springer

Editors
Shi Jin
Department of Mathematics
University of Wisconsin
Madison, WI, USA

Lorenzo Pareschi
Dipartimento di Matematica e Informatica
Università degli Studi di Ferrara
Ferrara, Italy

ISSN 2199-3041   ISSN 2199-305X   (electronic)
SEMA SIMAI Springer Series
ISBN 978-3-319-67109-3   ISBN 978-3-319-67110-9   (eBook)
https://doi.org/10.1007/978-3-319-67110-9

Library of Congress Control Number: 2018934713

Printed on acid-free paper

This Springer imprint is published by the registered company Springer International Publishing AG part
of Springer Nature.
The registered company address is: Gewerbestrasse 11, 6330 Cham, Switzerland

SEMA SIMAI Springer Series

Series Editors: Luca Formaggia • Pablo Pedregal (Editor-in-Chief)
Mats G. Larson • Tere Martínez-Seara Alonso • Carlos Parés • Lorenzo Pareschi •
Andrea Tosin • Elena Vazquez • Jorge P. Zubelli • Paolo Zunino

Volume 14

More information about this series at http://www.springer.com/series/10532

Preface

Kinetic equations describe the probability density distribution of interacting particles, and often are also subject to the influence of external fields. The equations consist of transport, collision, source and forcing terms. Although kinetic equations, most notably the Boltzmann equation, are often derived from the mean-field approximations of the first principle Newton's Second Law for N-particles by taking the $N \to \infty$ limit, one can hardly exactly obtain the collision kernel, which is instead determined empirically, giving rise to *uncertainty*. More recently, kinetic equations have found applications in other fields, like socio-economy and life sciences. In all these new emerging fields, the derivation from first principles is not possible and the derivation of the kinetic models is based on empirical observations and contains uncertain parameters.

Hyperbolic equations usually consist of some conservation laws which need to be closed by equations of state or constitutive relations. These closures are empirical, and thus will necessarily contain uncertainties.

Due to difficulty of measurement, in these problems uncertainty can also arise from initial or boundary data, the source and forcing terms, or even geometry.

Uncertainty quantification (UQ) is important to study the propagation of uncertainty, to identify sensitive input parameters, to determine the likelihood of certain outcomes, and to help to validate and improve these equations (models). While UQ has been a popular field in science, engineering and industry in the last two decades, the study of UQ for hyperbolic and kinetic equations has encountered some major difficulties, and relevant literature is scarce.

One of the main difficulties for (nonlinear) hyperbolic equations is the formation of singular solutions, known as shocks and contact discontinuities, in the physical space. When the uncertainty is modeled by random variables, the singularity in the physical space will propagate into the random space, preventing many modern UQ methods, for example, the polynomial chaos method—which boosts spectral convergence (in contrast to the half-th order convergence of the classical Monte Carlo method) given sufficient regularities in the random space—, from convergence at a high order or from convergence at all. The singular solution also gives rise to burdens for the theoretical investigation of the UQ method. For example, the stochastic Galerkin method is based on the L_2 theory, while a discontinuous solution is better described mathematically by the L_1 norm. Furthermore, some expansion methods, such as the polynomial chaos expansion, could lead to the loss of hyperbolicity, triggering potential instability of the problem although the original problem is well posed.

Kinetic equations, defined in the phase space, already suffer from the "dimension curse". The extra dependence on the random uncertainties—which are often high dimensional—will only make the computation prohibitively expansive. The nonlinearities in the kinetic equations provide further challenges for the theoretical study of the UQ problems. Finally, possible multiple time and spatial scales in kinetic equations make the computation of uncertain kinetic problems more difficult.

The various chapters collected in this book, from leading experts in the relevant field, provide introductions to this exciting field, as well as surveys of state-of-the-art computational tools for some of the aforementioned challenges. Among the topics covered are stochastic finite element and Galerkin methods for hyperbolic and kinetic equations, multi-level Monte Carlo methods for hyperbolic conservation laws, kinetic formulation for conservation laws in the context of UQ, computational methods to tackle high dimensional uncertain kinetic equations, kinetic formulation of high-frequency wave propagations in random media, and applications of uncertain kinetic equations to socio-economic and life sciences.

We hope, through this book, to attract more researchers into this important and interesting, yet relatively unexplored field of applied mathematics and scientific computing.

Madison, WI, USA   Shi Jin
Ferrara, Italy   Lorenzo Pareschi
July 2017

Contents

About the Editors

Shi Jin is a Vilas Distinguished Achievement Professor of Mathematics at the University of Wisconsin-Madison. He earned his B.S. from Peking University and his Ph.D. from the University of Arizona. His research fields include computational fluid dynamics, kinetic equations, hyperbolic conservation laws, high frequency waves, quantum dynamics, and uncertainty quantification—fields in which he has published over 140 papers. He has been honored with the Feng Kang Prize in Scientific Computing and the Morningside Silver Medal of Mathematics at the Fourth International Congress of Chinese Mathematicians, and is a Fellow of both the American Mathematical Society and the Society for Industrial and Applied Mathematics (SIAM). He is an Invited Speaker at the International Congress of Mathematicians in 2018.

Lorenzo Pareschi is a Full Professor of Numerical Analysis at the Department of Mathematics and Computer Science, University of Ferrara, Italy. He received his Ph.D. in Mathematics from the University of Bologna, Italy and subsequently held visiting professor appointments at the University of Wisconsin-Madison, the University of Orleans and University of Toulouse, France, and the Imperial College, London, UK. His research interests include multiscale modeling and numerical methods for phenomena described by time dependent nonlinear partial differential equations, in particular by means of hyperbolic balance laws and kinetic equations. He is the author/editor of nine books and more than 110 papers in peer-reviewed journals.

The Stochastic Finite Volume Method

Rémi Abgrall and Svetlana Tokareva

Abstract We give the general principle of the Stochastic Finite Volume method and show its versatility by many examples from standard ODE to fluid problems. We derive the error estimates for the mean and variance resulting from the SFVM and show that the convergence rates of the statistical quantities are equivalent to the convergence rates of the deterministic solution. We propose the anisotropic choice of the mesh nodes for high-dimensional stochastic parameter spaces and analyze the efficiency of the anisotropic stochastic mesh adaptation algorithm.

We finally generalize the SFVM approach and apply the DG discretization on the unstructured triangular grids in the physical space. We demonstrate the efficiency and the scaling of the implemented methods on various numerical tests.

1 Introduction

1.1 Deterministic Scalar Hyperbolic Conservation Laws

Many problems in physics and engineering are modeled by hyperbolic systems of conservation or balance laws. As examples for these equations, we mention only the Shallow Water Equations of hydrology, the Euler Equations for inviscid, compressible flow and the Magnetohydrodynamic (MHD) equations of plasma physics, see, e.g. [10, 15].

The simplest example for a system of hyperbolic conservation laws is the Cauchy problem for the scalar (single) conservation law:

$$\frac{\partial u}{\partial t} + \sum_{j=1}^{d} \frac{\partial}{\partial x_j}\left(f_j(u)\right) = 0, \quad x = (x_1, \ldots, x_d) \in \mathbb{R}^d, \ t > 0 \tag{1}$$

R. Abgrall (✉) · S. Tokareva
Institute of Mathematics, University of Zurich, Zurich, Switzerland
e-mail: remi.abgrall@math.uzh.ch; svetlana.tokareva@math.uzh.ch

© Springer International Publishing AG, part of Springer Nature 2017
S. Jin, L. Pareschi (eds.), *Uncertainty Quantification for Hyperbolic and Kinetic Equations*, SEMA SIMAI Springer Series 14,
https://doi.org/10.1007/978-3-319-67110-9_1

augmented by the initial data

$$u(x, 0) = u_0(x), \quad x \in \mathbb{R}^d. \tag{2}$$

Here the unknown is $u : \mathbb{R}^d \mapsto \mathbb{R}$ and f_j is the flux function in the j-th dimension.

Solutions of (1)–(2) develop discontinuities in finite time even when the initial data $u_0(x)$ is smooth and must be interpreted in the weak sense (e.g. [10, 15, 16, 32]).

1.2 Stochastic Conservation Laws

Many efficient numerical methods have been developed to approximate the entropy solutions of systems of conservation laws [15, 21], e.g. finite volume or discontinuous Galerkin methods. The classical assumption in designing efficient numerical methods is that the initial data $\mathbf{U}_0$ is known *exactly*. However, in many situations of practical interest, these data are not known exactly due to inherent uncertainty in modelling and measurements of physical parameters. In the present work, we follow [25] and describe incomplete information in the uncertain data mathematically as random fields. Such data are described in terms of statistical quantities of interest like the mean, variance, higher statistical moments; in some cases the distribution law of the stochastic data is also assumed to be known.

A mathematical framework of *random entropy solutions* for scalar conservation laws with random initial data has been developed in [25]. There, existence and uniqueness of random entropy solutions has been shown for scalar hyperbolic conservation laws, also in multiple dimensions. Furthermore, the existence of the statistical quantities of the random entropy solution such as the statistical mean and k-point spatio-temporal correlation functions under suitable assumptions on the random initial data have been proven. The existence and uniqueness of the random entropy solutions for scalar conservation laws with random fluxes has been proven in [27].

Numerical methods for uncertainty quantification in hyperbolic conservation laws have been proposed and studied recently in e.g. [2, 8, 17, 22, 23, 25, 26, 29, 35, 36].

1.3 Random Fields and Probability Spaces

We introduce a probability space $(\Omega, \mathscr{F}, \mathbb{P})$, with Ω being the set of all elementary events, or space of outcomes, and $\mathscr{F}$ a σ-algebra of all possible events, equipped with a probability measure $\mathbb{P}$. Random entropy solutions are random functions taking values in a function space; to this end, let $(E, \mathscr{G}, \mathbb{G})$ denote any measurable space. Then an E-valued random variable is any mapping $Y : \Omega \rightarrow E$ such that

$\forall A \in \mathscr{G}$ the preimage $Y^{-1}(A) = \{\omega \in \Omega : Y(\omega) \in A\} \in \mathscr{F}$, i.e. such that Y is a $\mathscr{G}$-measurable mapping from Ω into E.

We confine ourselves to the case that E is a complete metric space; then $(E, \mathscr{B}(E))$ equipped with a Borel σ-algebra $\mathscr{B}(E)$ is a measurable space. By definition, E-valued random variables $Y : \Omega \rightarrow E$ are $(E, \mathscr{B}(E))$ measurable. Furthermore, if E is a separable Banach space with norm $\| \circ \|_E$ and with topological dual E^*, then $\mathscr{B}(E)$ is the smallest σ-algebra of subsets of E containing all sets

$$\{x \in E : \varphi(x) < \alpha\}, \varphi \in E^*, \alpha \in \mathbb{R}.$$

Hence, if E is a separable Banach space, $Y : \Omega \rightarrow E$ is an E-valued random variable if and only if for every $\varphi \in E^*$, $\omega \mapsto \varphi(Y(\omega)) \in \mathbb{R}$ is an $\mathbb{R}$-valued random variable. Moreover, there hold the following results on existence and uniqueness [25].

For a simple E-valued random variable Y and for any $B \in \mathscr{F}$ we set

$$\int_B Y(\omega) \, \mathbb{P}(d\omega) = \int_B Y \, d\mathbb{P} = \sum_{i=1}^{N} x_i \mathbb{P}(A_i \cap B). \tag{3}$$

For such $Y(\omega)$ and all $B \in \mathscr{F}$ holds

$$\left\| \int_B Y(\omega) \, \mathbb{P}(d\omega) \right\|_E \leq \int_B \| Y(\omega) \|_E \, \mathbb{P}(d\omega). \tag{4}$$

For any random variable $Y : \Omega \rightarrow E$ which is Bochner integrable, there exists a sequence $\{Y_m\}_{m \in \mathbb{N}}$ of simple random variables such that, for all $\omega \in \Omega$, $\| Y(\omega) - Y_m(\omega) \|_E \rightarrow 0$ as $m \rightarrow \infty$. Therefore (3) and (4) can be extended to any E-valued random variable. We denote the expectation of Y by

$$\mathbb{E}[Y] = \int_\Omega Y(\omega) \, \mathbb{P}(d\omega) = \lim_{m \rightarrow \infty} \int_\Omega Y_m(\omega) \mathbb{P}(d\omega) \in E.$$

Denote by $L^p(\Omega, \mathscr{F}, \mathbb{P}; E)$ for $1 \leq p \leq \infty$ the Bochner space of all p-summable, E-valued random variables Y and equip it with the norm

$$\| Y \|_{L^p(\Omega; E)} = \left(\mathbb{E}[\| Y \|_E^p] \right)^{1/p} = \left(\int_\Omega \| Y(\omega) \|_E^p \, \mathbb{P}(d\omega) \right)^{1/p}.$$

For $p = \infty$ we can denote by $L^\infty(\Omega, \mathscr{F}, \mathbb{P}; E)$ the set of all E-valued random variables which are essentially bounded and equip this space with the norm

$$\| Y \|_{L^\infty(\Omega; E)} = \underset{\omega \in \Omega}{\text{ess sup}} \, \| Y(\omega) \|_E.$$

2   General Framework

2.1   General Principles

Assume a deterministic problem is written as

$$\mathscr{L}(u) = 0, \tag{5}$$

defined in a domain $K \subset \mathbb{R}^d$ with boundary conditions, and if needed initial conditions. Since the discussion of this section is formal, we include different initial and boundary conditions of the problem in the symbol $\mathscr{L}$. Let the operator $\mathscr{L}$ depend in some way on parameters (for example, considering fluids, in the equation of state, or the parameter of a turbulent model), that in many cases are not known exactly. Hence we assume that they are random variables defined on some random space Ω, and that these random variables are measurable with respect to a measure $d\mu$ defined on Ω. Hence our problem can formally be seen as a "stochastic" PDE of the type

$$\mathscr{L}(u, X) = 0, \tag{6}$$

defined in a domain $K \subset \mathbb{R}^d$, subject to initial and boundary conditions, and where X is a random variable defined on Ω. For simplicity, we use the same notation $\mathscr{L}$ for the problem. The operator $\mathscr{L}$ depends on $u \equiv u(x, t, X)$ and $X \equiv X(\omega)$ where $x \in \mathbb{R}^s$ for $s \in \{1, 2, 3\}$ and $t \in \mathbb{R}^+$ are respectively the space coordinate and time, and the random event (or random parameter) ω belongs to Ω. In the case of steady problems, the time is omitted. The random variable may also depend on space and time, as well as the measure μ, and the technique in principle can be extended to this case but the discussion is beyond the scope of this chapter for simplicity of exposure.

We will identify Ω to some subset of $\mathbb{R}^s$, s being the number of random parameters to define X. Thus we can also see (6) as a problem defined on a subset $K \subset \mathbb{R}^d$ of dimension $d = s + p$.

For any realization of Ω, we are able to solve the following deterministic form of (6) in space and time, by some numerical method:

$$\mathscr{L}_h(u_h, X(\omega)) = 0. \tag{7}$$

In order to approximate a solution of (6), the first step is to discretize the probability space Ω. We construct a partition of Ω, i.e. a set of $\Omega_j, j = 1, \ldots, N$ that are mutually independent

$$\mathbb{P}(\Omega_i \cap \Omega_j) = 0 \text{ for any } i \neq j \tag{8}$$

and that cover Ω

$$\Omega = \cup_{i=1}^{N} \Omega_i. \tag{9}$$

We assume $\mu(\Omega_i) = \int_{\Omega_i} d\mu > 0$ for any i. We wish to approximate the solution of (6) by the average conditional expectancies $E(u_h|\Omega_j)$

$$\mathbb{E}\left(u_h \mid \Omega_j\right) = \frac{\int_{\Omega_j} u_h d\mathbb{P}}{\int_{\Omega_j} d\mathbb{P}} \tag{10}$$

from the knowledge of the operator $\mathcal{L}_h$ and thanks to a reconstruction procedure inspired from the methods for high order finite volume methods: ENO, WENO, etc. This idea, initially developed in [2], will be detailed in Sects. 3 and 4 for the finite volume method on Cartesian and unstructured meshes, but can be used for ODEs as in [3] as we show now by an example.

Remark Depending on the context, the described method will either be named as semi intrusive (SI) (since very few modifications of an existing code need to be done), or stochastic finite volume (SFV) method since in essence the conditional expectancies can also be seen as integrals over a stochastic finite volume.

2.2 A First Example: The Kraichnan-Orszag Three-Mode Problem

The Kraichnan-Orszag three-mode problem has been introduced by Kraichnan [19] and Orszag [28]. It has been intensively studied to demonstrate that gPC expansion could suffer from accuracy loss for problems involving long time integration. In [37], the exact solution is given, and different computations have been performed in [7, 11, 13, 24, 37, 38]. This problem is defined by the following system of nonlinear ordinary differential equations

$$\frac{dy_1}{dt} = y_1 y_3,$$

$$\frac{dy_2}{dt} = -y_2 y_3, \tag{11}$$

$$\frac{dy_3}{dt} = -y_1^2 + y_2^2$$

subject to stochastic initial conditions

$$y_1(0) = y_1(0;\omega), \qquad y_2(0) = y_2(0;\omega), \qquad y_3(0) = y_3(0;\omega). \tag{12}$$

In the literature, generally uniform distributions are considered, except in [38] where beta and Gaussian distributions are also taken into account. The computational cost of our SI/SFV method for the Kraichnan-Orszag problem is compared to that of other methods, a quasi-random Sobol (MC-SOBOL) sequence with 8×10^6

iterations, and a Polynomial Chaos Method (PC) with Clenshaw-Curtis sparse grid. The error in variance of y_1 is considered at a final time t_f of 50. We define the error between two numerically integrated functions $f_1(t_j)$ and $f_2(t_j), j = 1, \cdots, n_t$, as:

$$\varepsilon_{L^2} = \frac{\frac{1}{n_t}\sqrt{\sum_{j=1}^{n_t}\left(f_1(t_j) - f_2(t_j)\right)^2}}{\frac{1}{n_t}\sqrt{\sum_{j=1}^{n_t}\left(f_1(t_j)\right)^2}}, \tag{13}$$

where f_1 is considered the Monte Carlo converged solution. For different error levels, corresponding computational cost is computed.

2.2.1 One Random Variable

First, we will study the 1D problem corresponding to initial conditions of the form

$$y_1(0) = 1.0, \qquad y_2(0) = 0.1\omega, \qquad y_3(0) = 0.0, \tag{14}$$

where ω is a uniformly distributed random variable varying in $[-1, 1]$. We use SI, MC-SOBOL and PC method to compute the variance of y_1. In Table 1, we show the results in terms of number of samples required to reach a prescribed error ε_{L^2}. Performances of SI methods are comparable and even better than PC methods.

Then, the same problem described previously but with a different probability distribution for $y_2(0)$ has been considered. In particular, ω is discontinuous on $[a, b] = [-1, 1]$ with probability density function (pdf) defined by:

$$f(\gamma) = \frac{1}{M} \times \begin{cases} \dfrac{1 + \cos(\pi x)}{2} & \text{if } x \in [-1, 0] \\ 10 + \dfrac{1 + \cos(\pi x)}{2} & \text{if } x \in [0, 1] \\ 0 & \text{else} \end{cases} \tag{15}$$

and $M = \frac{11}{2}$ to ensure normalization. Because of the discontinuous pdf, only MC-SOBOL and SI solutions can be compared, showing the great flexibility given by SI method with respect to the form of the pdf. In Fig. 1, variance of $y_1(t)$ is reported for the converged solutions obtained with MC-SOBOL and SI. The SI method permits to reproduce exactly MC-SOBOL solution. In Fig. 2, a convergence study for SI method is reported by using an increasing number of points in the stochastic space. In Table 2, we reported number of samples required to reach a prescribed error ε_{L^2}. SI method shows to be very competitive in terms of efficiency and computational

Table 1 Number of samples required for the 1D K-O problem for time $t \in [0, 10]$

Error level ε_{L^2}	MC-SOBOL	PC	SI
10^{-1}	20	12	5
10^{-2}	240	19	10
10^{-3}	2200	23	20

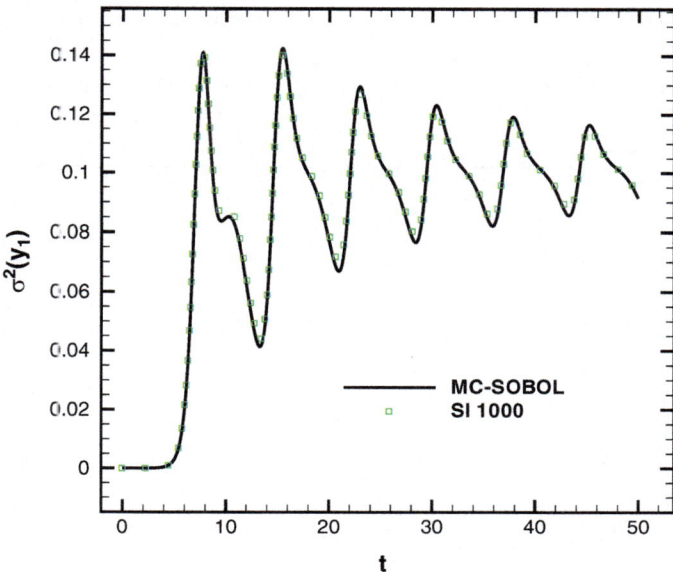

Fig. 1 Variance of y_1 computed by means of SI and MC-SOBOL methods (Reproduced with permission from [3])

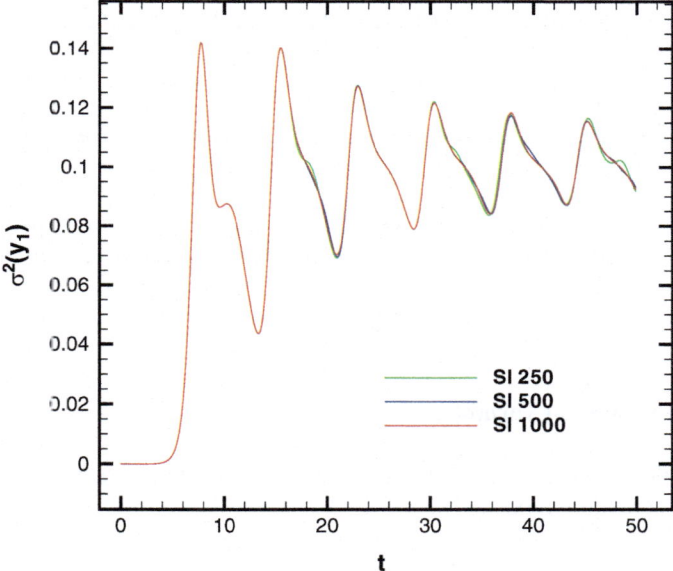

Fig. 2 Variance of y_1 computed by means of SI for different meshes in the stochastic space (Reproduced with permission from [3])

Table 2 Number of samples required for the 1D-discontinuous K-O problem for time $t \in [0, 50]$

Error level ε_{L^2}	MC-SOBOL	SI
10^{-1}	35	7
10^{-2}	250	160
10^{-3}	2500	900

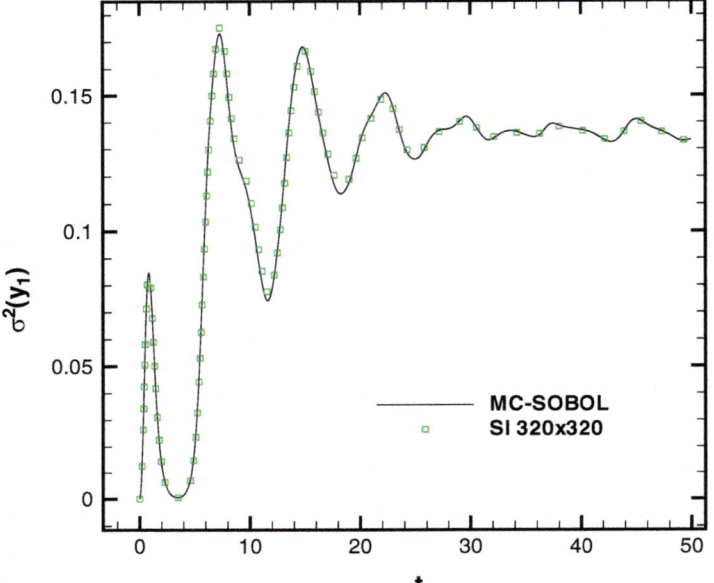

Fig. 3 Variance of y_1 computed by means of SI and MC-SOBOL methods (Reproduced with permission from [3])

cost with respect to MC-SOBOL method when whatever form of pdf is used (a discontinuous pdf in this case). We remark that a uniform grid is used in the stochastic plan without any type of adaptation. This displays the great potentiality of this method if coupled with an adaptive method.

2.2.2 Two Random Variables

In this section, we use SI method to study the Kraichnan-Orszag problem with two-dimensional random inputs:

$$y_1(0) = 1.0, \qquad y_2(0) = 0.1\omega_1, \qquad y_3(0) = \omega_2, \tag{16}$$

where ω_1 is discontinuous on $[a, b] = [-1, 1]$ with a density defined by Eq. (15) and ω_2 is a uniform random variable in $[-1, 1]$.

In Fig. 3, the SI capability to reproduce exactly MC-SOBOL solution is represented. SI and MC-SOBOL solutions are nearly coincident also for long time

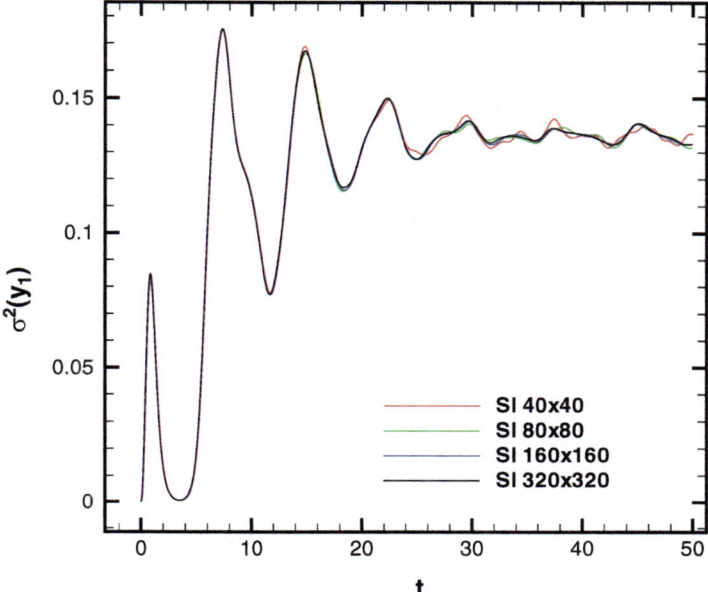

Fig. 4 Variance of y_1 computed by means of SI for different meshes in the stochastic space (Reproduced with permission from [3])

Table 3 Number of samples required for the 2D-discontinuous K-O problem for time $t \in [0, 50]$

Error level ε_{L^2}	MC-SOBOL	SI
10^{-1}	160	81
10^{-2}	10,000	2500
10^{-3}	300,000	102,400

($t = 50$). The mesh convergence study in the stochastic space for SI is reported in Fig. 4 showing that the solution obtained with a mesh of 320×320 is well converged. In Table 3 computational cost required to reach a prescribed error of ε_{L^2} is reported. Reductions from 50 to 66% are obtained using SI with respect to MC-SOBOL solutions.

3   Stochastic Finite Volume Method on Cartesian Grids

In this chapter, we concentrate on the analysis of the stochastic hyperbolic conservation laws with random initial data and flux coefficients. Many efficient numerical methods have been developed to approximate the entropy solutions of systems of conservation laws [15, 21], however, in many practical applications it is not always possible to obtain exact data due to, for example, measurement or modeling errors. We describe incomplete information in the conservation law mathematically as random fields. Such data are described in terms of statistical quantities of interest like the mean, variance, higher statistical moments; in some cases the distribution law of the stochastic initial data is also assumed to be known. There exist several techniques to quantify the uncertainty (i.e. determine the mean flow and its statistical moments), such as the Monte-Carlo (MC), the Multi-Level Monte Carlo (MLMC) and Stochastic Galerkin method, see [17, 22, 23, 25, 26, 29, 35, 36]. Here we analyse a different approach to the uncertainty quantification in the conservation laws, the Stochastic Finite Volume Method (SFVM), which is based on the finite volume framework and was first introduced in [2, 8]. The SFVM is formulated to solve numerically the system of conservation laws with sources of randomness in both flux coefficients and initial data.

Consider the hyperbolic system of conservation laws with random flux coefficients

$$\frac{\partial \mathbf{U}}{\partial t} + \nabla_x \cdot \mathbf{F}(\mathbf{U}, \omega) = \mathbf{0}, \ t > 0; \tag{17}$$

$\mathbf{x} = (x_1, x_2, x_3) \in D_x \subset \mathbb{R}^3$, $\mathbf{U} = [u_1, \dots, u_p]^\top$, $\mathbf{F} = [\mathbf{F}_1, \mathbf{F}_2, \mathbf{F}_3]$, $\mathbf{F}_k = [f_1, \dots, f_p]^\top$, $k = 1, 2, 3$, and random initial data

$$\mathbf{U}(\mathbf{x}, 0, \omega) = \mathbf{U}_0(\mathbf{x}, \omega), \ \omega \in \Omega. \tag{18}$$

A mathematical framework of *random entropy solutions* for scalar conservation laws has been developed in [25]. There, existence and uniqueness of random entropy solutions to (17)–(18) has been shown for scalar conservation laws, also in multiple dimensions. Furthermore, the existence of the statistical quantities of the random entropy solution such as the statistical mean and k-point spatio-temporal correlation functions under suitable assumptions on the random initial data have been proven.

3.1 Stochastic Finite Volume Method

We parametrize all the random inputs in Eqs. (17)–(18) using the random variable $\mathbf{y} = \mathbf{Y}(\omega)$ which takes values in $D_y \subset \mathbb{R}^q$ and rewrite the stochastic conservation law in the parametric form:

$$\partial_t \mathbf{U} + \nabla_x \cdot \mathbf{F}(\mathbf{U}, \mathbf{y}) = \mathbf{0}, \ \mathbf{x} \in D_x \subset \mathbb{R}^3, \ \mathbf{y} \in D_y \subset \mathbb{R}^q, \ t > 0; \tag{19}$$

$$\mathbf{U}(\mathbf{x}, 0, \mathbf{y}) = \mathbf{U}_0(\mathbf{x}, \mathbf{y}), \ \mathbf{x} \in D_x \subset \mathbb{R}^3, \ \mathbf{y} \in D_y \subset \mathbb{R}^q. \tag{20}$$

Let $\mathcal{T}_x = \cup_{i=1}^{N_x} K_x^i$ be the triangulation of the computational domain D_x in the physical space and $\mathcal{C}_y = \cup_{j=1}^{N_y} K_y^j$ be the Cartesian grid in the domain D_y of the parametrized probability space.

We further assume the existence of the probability density function $\mu(\mathbf{y})$ and compute the expectation of the n-th solution component of the conservation law (19)–(20) as follows:

$$\mathbb{E}[u_n] = \int\limits_{D_y} u_n \mu(\mathbf{y}) \, d\mathbf{y}, \; n = 1, \ldots, p$$

The scheme of the Stochastic Finite Volume method (SFVM), see e.g. [34], can be obtained from the integral form of Eqs. (19)–(20):

$$\iint\limits_{K_y^j K_x^i} \partial_t \mathbf{U} \, \mu(\mathbf{y}) \, dxdy + \iint\limits_{K_y^j K_x^i} \nabla_x \cdot \mathbf{F}(\mathbf{U}, \mathbf{y}) \, \mu(\mathbf{y}) \, dxdy = 0.$$

Introducing the cell average

$$\bar{\mathbf{U}}_{ij}(t) = \frac{1}{|K_x^i||K_y^j|} \iint\limits_{K_y^j K_x^i} \mathbf{U}(\mathbf{x}, t, \mathbf{y})\mu(\mathbf{y}) \, dxdy$$

with the cell volumes

$$|K_x^i| = \int\limits_{K_x^i} d\mathbf{x}, \quad |K_y^j| = \int\limits_{K_y^j} \mu(\mathbf{y}) \, d\mathbf{y}$$

and performing the partial integration over K_x^i we get

$$\frac{d\bar{\mathbf{U}}_{ij}}{dt} + \frac{1}{|K_x^i||K_y^j|} \int\limits_{K_y^j} \left[\int\limits_{K_x^i} \mathbf{F}(\mathbf{U}, \mathbf{y}) \cdot \mathbf{n} \, dS \right] \mu(\mathbf{y}) \, d\mathbf{y} = 0$$

Next, we use any standard numerical flux approximation $\hat{\mathbf{F}}\big(\tilde{\mathbf{U}}_L(\mathbf{x}, t, \mathbf{y}), \tilde{\mathbf{U}}_R(\mathbf{x}, t, \mathbf{y}), \mathbf{y}\big)$ to replace the discontinuous flux through the element interface $\mathbf{F}(\mathbf{U}, \mathbf{y}) \cdot \mathbf{n}$. Here $\tilde{\mathbf{U}}_{L,R}$ denote the boundary extrapolated solution values at the edge of the cell K_x^i, obtained by the high order reconstruction from the cell averages. The complete numerical flux is then approximated by a suitable quadrature rule as

$$\bar{\mathbf{F}}_{ij}(t) = \frac{1}{|K_y^j|} \int\limits_{K_y^j} \left[\int\limits_{K_x^i} \hat{\mathbf{F}}(\tilde{\mathbf{U}}_L, \tilde{\mathbf{U}}_R, \mathbf{y}) \right] \mu(\mathbf{y}) \, d\mathbf{y} \approx \frac{1}{|K_y^j|} \sum\limits_{\mathbf{m}} \hat{\mathscr{F}}(t, \mathbf{y_m})\mu(\mathbf{y_m})w_{\mathbf{m}},$$

$$(21)$$

where we have denoted the flux integral over the physical cell as $\hat{\mathscr{F}}$, $\mathbf{m} = (m_1, \ldots, m_q)$ is the multi-index, $\mathbf{y_m}$ and $w_{\mathbf{m}}$ are quadrature nodes and weights, respectively.

The SFV method then results in the solution of the following ODE system:

$$\frac{d\bar{\mathbf{U}}_{ij}}{dt} + \frac{1}{|K_x^i|}\bar{\mathbf{F}}_{ij}(t) = \mathbf{0}, \tag{22}$$

for all $i = 1, \ldots, N_x$, $j = 1, \ldots, N_y$. Therefore, to obtain the high-order scheme we first need to provide the high-order flux approximation based, for example, on the ENO/WENO reconstruction in the physical space. Second, we have to guarantee the high-order integration in (21) also by applying the ENO/WENO reconstruction in the stochastic space and choosing the suitable quadrature rule. Finally, we need the high-order time-stepping algorithm to solve the ODE system (22), such as Runge-Kutta method.

3.2 Numerical Convergence Analysis

We perform the convergence analysis of the SFVM for a simple linear advection equation with uncertain phase initial condition

$$u_t + au_x = 0, \quad x \in (0, 1),$$

$$u(x, 0) = \sin\left(2\pi\left(x + 0.1Y(\omega)\right)\right).$$

The random variable $y = Y(\omega)$ is assumed to be distributed uniformly on $[0, 1]$.

The reference solution in this and other experiments of this chapter involving convergence analysis has been computed exactly using the method of characteristics.

In Figs. 5 and 6, we plot the $L^1(0, 1)$ error for the expectation and the variance of u with respect to the mesh size and the computational time. We investigate the influence of different reconstruction orders in spatial and stochastic variables on the convergence rates and therefore present the convergence plots for the SFVM based on different combinations of ENO/WENO reconstruction in x and y. We compare the SFVM with 1st, 2nd and 3rd order of accuracy in physical space combined with 3rd and 5th order reconstruction in stochastic variable. The type of reconstruction is indicated in Figs. 5 and 6 as follows: for example, the line "SFV-x2y5" corresponds to the 2nd order piecewise-linear ENO reconstruction in x and 5th order piecewise-quadratic WENO reconstruction in y, the line "SFV-x3y5" stands for 3rd order piecewise-linear WENO reconstruction in x with 5th order WENO reconstruction in y, etc. The numerical flux used in all the numerical experiments of this paper is the Rusanov flux. The results show that, while the convergence rate is dominated by the order of accuracy in x, the algorithms with higher order reconstruction in y are

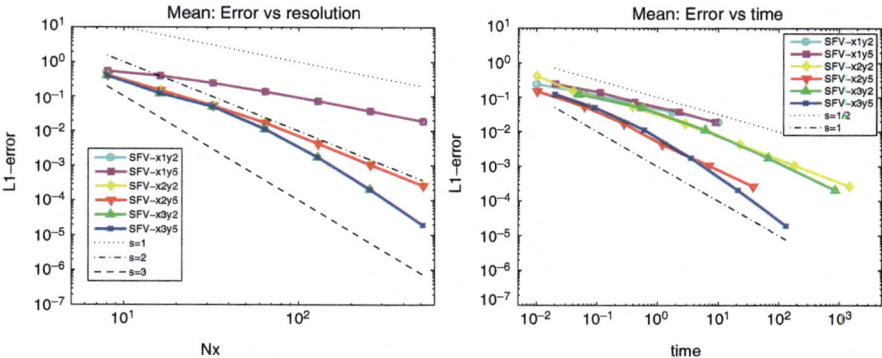

Fig. 5 Mean: dependence of the error on the mesh resolution and computational time (Reproduced with permission from [34])

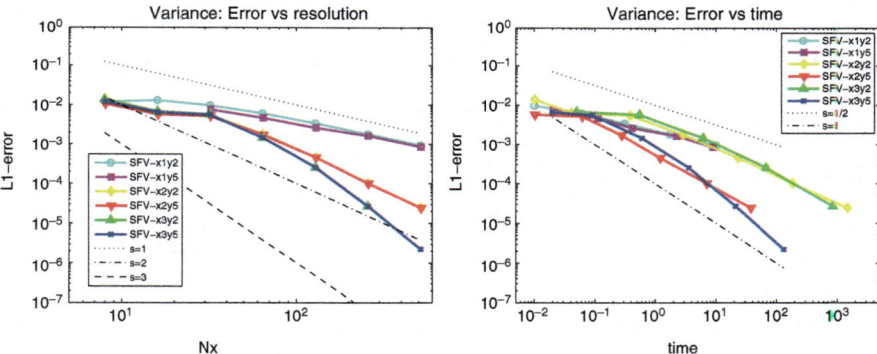

Fig. 6 Variance: dependence of the error on the mesh resolution and computational time (Reproduced with permission from [34])

more efficient computationally since the same error can be reached with less overall computational time as compared to the lower order reconstruction in y.

3.3 Numerical Results

3.3.1 Buckley-Leverett Equation

As a second example for a scalar conservation law, we consider the Buckley-Leverette equation with random flux:

$$\frac{\partial u}{\partial t} + \frac{\partial f(\omega; u)}{\partial x} = 0, \; x \in (0, L), \; t > 0; \tag{23}$$

$$u(x, 0) = u_0(x), \tag{24}$$

where

$$f(\omega; u) = \frac{u^2}{u^2 + \alpha(\omega)(1 - u)^2}$$

and $\alpha = \alpha(\omega)$ is the random variable with known distribution. Assume further that $u_0(x)$ is the Riemann initial data, that is

$$u_0(x) = \begin{cases} u_L, & \text{if } x < x_0, \\ u_R, & \text{if } x > x_0. \end{cases}$$

Note that the Buckley-Leverette equation models water flooding in a one-dimensional petroleum reservoir and the above introduction of uncertainty reflects the inherent uncertainty in measuring the relative permeability. We apply the Stochastic Finite Volume method to solve (23)–(24) with $L = 2.5$, $x_0 = 1.0$, $u_L = 0.8$, $u_R = 0.3$ and uniformly distributed $\alpha(\omega)$. The computational results for two different distributions of $\alpha(\omega)$ are presented in Figs. 7 and 8. The solution mean

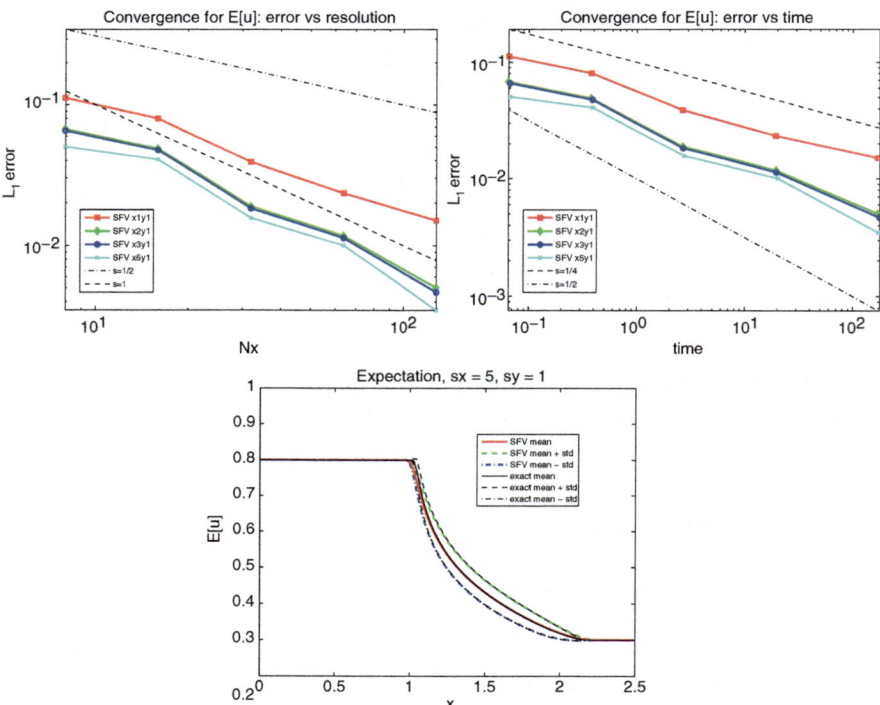

Fig. 7 Convergence w.r.t. resolution (upper left) and computational time (upper right) of the Stochastic Finite Volume method and solution mean (lower) for Buckley-Leverett equation with random flux, $\alpha(\omega) \sim \mathcal{U}[0.05, 0.15]$

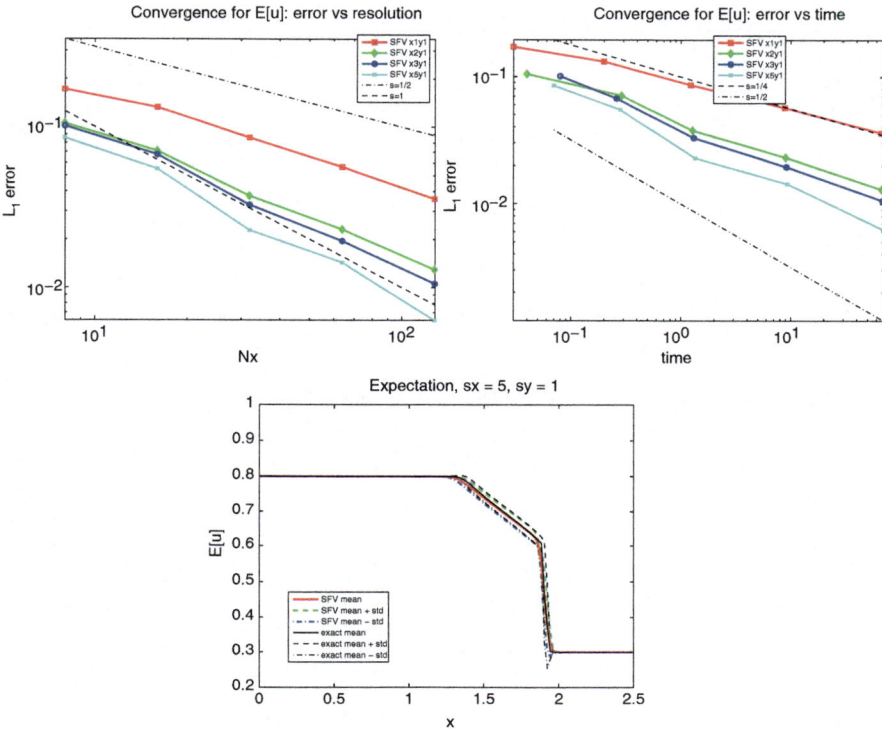

Fig. 8 Convergence w.r.t. resolution (upper left) and computational time (upper right) of the Stochastic Finite Volume method and solution mean (lower) for Buckley-Leverett equation with random flux, $\alpha(\omega) \sim \mathscr{U}[0.8, 1.2]$

is complicated on account of the formation of a compound shock. Furthermore, increasing the order of the spatio-temporal discretization does lead to a better approximation of the solution. Note that increasing the order does not imply an increase in the convergence rate as the solution is discontinuous.

3.3.2 Stochastic Sod's Shock Tube Problem with Random Initial Data

Consider the Riemann problem for the Euler equations

$$\frac{\partial \mathbf{U}}{\partial t} + \frac{\partial \mathbf{F(U)}}{\partial x} = \mathbf{0}, \quad x \in (0, 2), \tag{25}$$

$$\mathbf{U}(x, 0, y) = \mathbf{U}_0(x, y) = \begin{cases} \mathbf{U}_L, & x < Y(\omega); \\ \mathbf{U}_R, & x > Y(\omega), \end{cases} \tag{26}$$

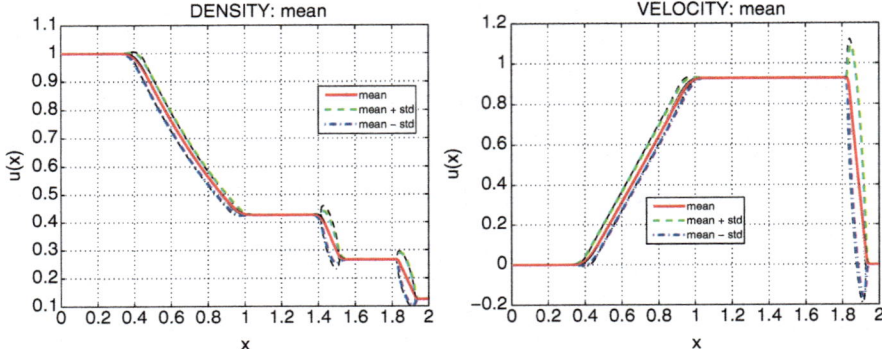

Fig. 9 Sod's shock tube problem with random shock location: density (left) and velocity (right)

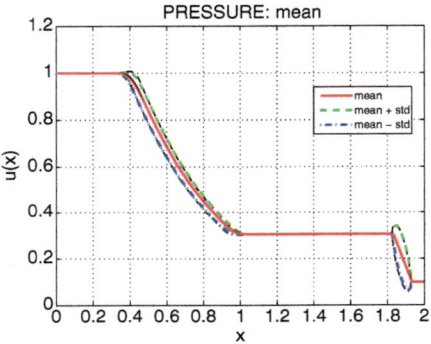

Fig. 10 Sod's shock tube problem with random shock location: pressure

with $y = Y(\omega)$, $\omega \in \Omega$ and

$$\mathbf{U} = [\rho, \rho u, E]^\top, \quad \mathbf{F} = [\rho u, \rho u^2 + p, \rho u(E + p)]^\top.$$

The initial data is set in primitive variables as

$$\mathbf{W}_0(x, \omega) = [\rho_0(x, \omega), u_0(x, \omega), p_0(x, \omega)]^\top = \begin{cases} [1.0, 0.0, 1.0] & \text{if} \quad x < Y(\omega), \\ [0.125, 0.0, 0.1] & \text{if} \quad x > Y(\omega). \end{cases}$$

We apply the SFV method to solve the system (25)–(26) with $Y(\omega)$ uniformly distributed on [0.95, 1.05]. We have used the 3rd order WENO reconstruction in both x and y variables. The results are presented in Figs. 9 and 10, in which the solution mean (solid line) as well as mean plus/minus standard deviation (dashed lines) are shown. The typical deterministic solution of the Sod's shock tube problem with the given initial conditions consists of the left-traveling rarefaction wave and the right-traveling shock wave separated by the contact discontinuity. However, a continuous transition between the intermediate states instead of the discontinuities is observed in the mean flow. This effect is unrelated to the diffusion of the numerical scheme and is due to the smoothing properties of the probabilistic shock profile [31].

3.3.3 Stochastic Sod's Shock Tube Problem with Random Flux and Initial Data

Consider the Riemann problem for the one-dimensional Euler equations with randomness in both flux and initial data

$$\frac{\partial \mathbf{U}}{\partial t} + \frac{\partial \mathbf{F}(\mathbf{U}, \omega)}{\partial x} = \mathbf{0}, \quad x \in (0, 2), \tag{27}$$

$$\mathbf{U}(x, 0, \omega) = \mathbf{U}_0\big(x, Y_1(\omega), Y_2(\omega)\big) = \begin{cases} \mathbf{U}_L\big(Y_2(\omega)\big), & x < Y_1(\omega); \\ \mathbf{U}_R, & x > Y_1(\omega), \end{cases} \tag{28}$$

with $y_j = Y_j(\omega)$, $j = 1, 2, 3$, $\omega \in \Omega$ and

$$\mathbf{U} = [\rho, \rho u, E]^\top, \quad \mathbf{F} = [\rho u, \rho u^2 + p, \rho u(E + p)]^\top,$$

$$p = (\gamma - 1)\Big(E - \frac{1}{2}\rho u^2\Big).$$

We also assume the randomness in the adiabatic constant, $\gamma = \gamma\big(Y_3(\omega)\big)$, and therefore

$$\mathbf{F}(\mathbf{U}, \omega) = \mathbf{F}\big(\mathbf{U}, Y_3(\omega)\big).$$

The initial data is set in primitive variables as

$$\mathbf{W}_0(x, \omega) = [\rho_0(x, \omega), u_0(x, \omega), p_0(x, \omega)]^\top$$

$$= \begin{cases} [1.0, 0.0, 1.0] & \text{if } x < Y_1(\omega), \\ [0.125 + 0.5\, Y_2, 0.0, 0.1] & \text{if } x > Y_1(\omega). \end{cases}$$

We apply the SFVM to solve the system (27)–(28) with $Y_1(\omega) \sim \mathscr{U}[0.95, 1.05]$, $Y_2(\omega) \sim \mathscr{U}[-0.1, 0.1]$, $Y_3(\omega) \sim \mathscr{U}[1.2, 1.6]$ using the 3rd order WENO reconstruction in both physical and stochastic variables. The results are presented in Figs. 11 and 12, in which the solution mean (solid line) as well as mean plus/minus standard deviation (dashed lines) are plotted.

The convergence results (dependence of the error on the number of mesh points and on the computational time) for the solution mean are presented in Fig. 13. Due

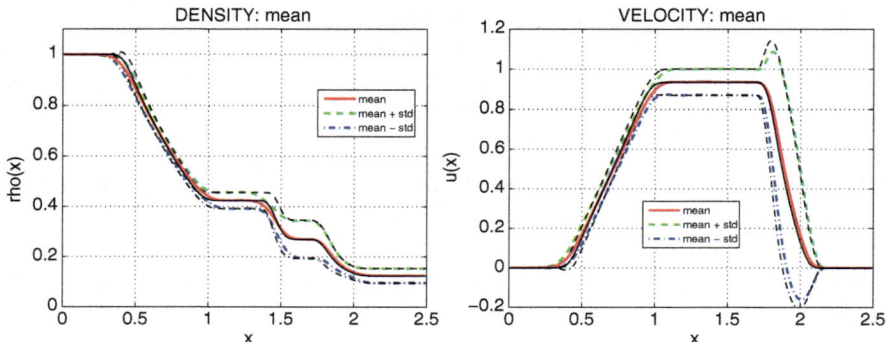

Fig. 11 Sod's shock tube problem with random flux and initial data: density (left) and velocity (right) (Reproduced with permission from [34])

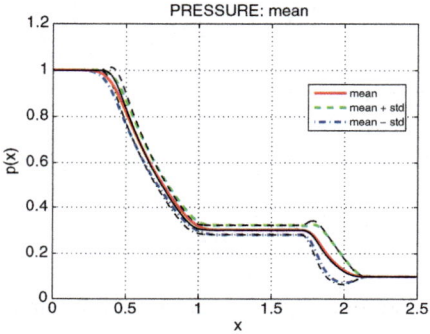

Fig. 12 Sod's shock tube problem with random flux and initial data: pressure (Reproduced with permission from [34])

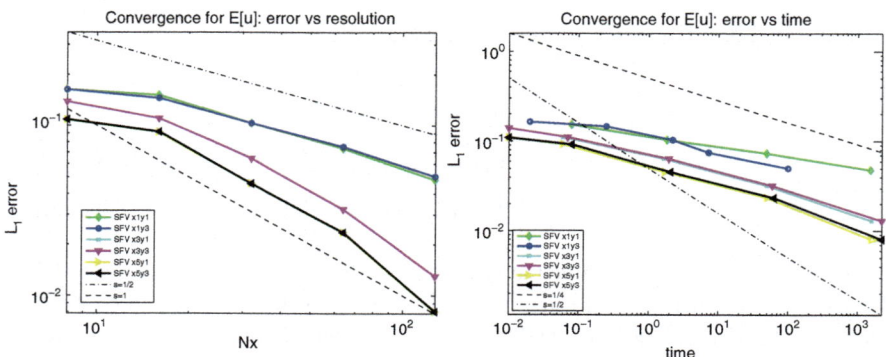

Fig. 13 Sod's shock tube problem with random flux and initial data: convergence of the mean w.r.t. physical mesh resolution (left) and computational time (right)

to the shock formation in the path-wise solution the maximum order of convergence for the mean is limited to 1.

3.4 Adaptive Parametrization of the Stochastic Space for SFVM

In this section we introduce the mesh adaptation technique for the Stochastic Collocation FVM and apply it to reduce the computational cost of the Stochastic Finite Volume method. To this end, we consider the following model problem:

$$\frac{\partial u}{\partial t} + \frac{\partial f(u, \omega)}{\partial x} = 0, \; x \in D = [0, L] \subset \mathbb{R}, \; t > 0; \tag{29}$$

$$u(x, 0, \omega) = u_0(x, \omega), \; x \in D, \; \omega \in \Omega. \tag{30}$$

Assume that the initial data is given as the Karhunen-Loève expansion:

$$u_0(x, \omega) = \bar{u}_0(x) + \sum_{j \geq 1} Y_j(\omega) \sqrt{\lambda_j} \Phi_j(x), \tag{31}$$

where $\Phi_j(x)$ and λ_j are the eigenfunctions and eigenvalues of the integral operator with covariance kernel:

$$\int_D C_Y(x_1, x_2) \Phi(x_1) \, dx_1 = \lambda \Phi(x_2).$$

We can therefore choose the random variable to parametrize the stochastic conservation law as $y = (y_1, y_2, \ldots) = \mathbf{Y}(\omega) = (Y_1(\omega), Y_2(\omega), \ldots)$, then

$$u_0(x, \omega) = u_0(x, y)\Big|_{y = \mathbf{Y}(\omega)} = \bar{u}_0(x) + \sum_{j \geq 1} y_j \sqrt{\lambda_j} \Phi_j(x),$$

The adaptation technique remains absolutely the same in the case of SCL with random fluxes, e.g. when the flux has the form

$$f(u; \omega) = \bar{f}(u) + \sum_{j \geq 1} Y_j(\omega) \sqrt{\lambda_j} \Phi_j(u),$$

where $\Phi_j(u)$ and λ_j are the eigenfunctions and eigenvalues of the integral operator with covariance kernel:

$$\int_D C_Y(u_1, u_2) \Phi(u_1) \, du_1 = \lambda \Phi(u_2).$$

Let $u_0(x, \omega)$ be the Gaussian process with exponential covariance [14]

$$C_Y(x_1, x_2) = \sigma_Y^2 e^{-|x_1 - x_2|/\eta},$$

then

$$\lambda_j = \frac{2\eta\sigma_Y^2}{\eta^2 w_j^2 + 1}, \quad \Phi_j(x) = \frac{1}{\sqrt{(\eta^2 w_j^2 + 1)L/2 + \eta}}[\eta w_j \cos(w_j x) + \sin(w_j x)],$$

where w_j are the roots of

$$(\eta^2 w^2 - 1)\sin(wL) = 2\eta w \cos(wL)$$

and

$$Y_j \sim \mathcal{N}(0, 1), \quad \mathbb{E}[Y_j \, Y_k] = \delta_{jk}$$

We next consider the Burgers' equation with following initial conditions:

$$u_0(x, y) = \sin(\pi x) + 0.1x(x - L)\left(\sum_{j=1}^{q} y_j \sqrt{\lambda_j}\Phi_j(x)\right)$$

In order to reduce the computational cost of the SFV method, we propose the mesh adaptation in the stochastic space based on the choice of the number of nodes in each of the stochastic coordinates according to

$$N_y^j = C \, N_x \sqrt{\lambda_j}. \tag{32}$$

Table 4 lists the convergence rates of the SFVM with adaptive meshing algorithm, for the s-th order in the physical variable and p_j-th order in the stochastic variables.

Table 4 Convergence rates of the SFVM with anisotropic stochastic mesh

Nx	$s = 1, p_j = 1$	Nx	$s = 1, p_j = 5$	Nx	$s = 2, p_j = 2$
4	–	4	–	4	–
8	1.100440	8	1.212217	8	1.408558
16	3.056992	16	3.024343	16	2.638068
32	0.741016	32	0.724594	32	2.723459
Nx	$s = 3, p_j = 3$	Nx	$s = 3, p_j = 5$	Nx	$s = 5, p_j = 5$
4	–	4	–	4	–
8	1.458768	8	1.558384	8	1.746105
16	2.553230	16	2.478849	16	2.204640
32	2.900257	32	2.817779	32	3.619253

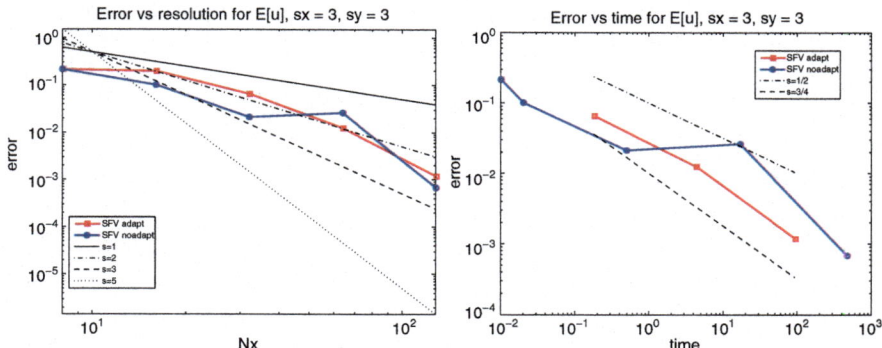

Fig. 14 Convergence: error vs space resolution (left) and error vs runtime (right)

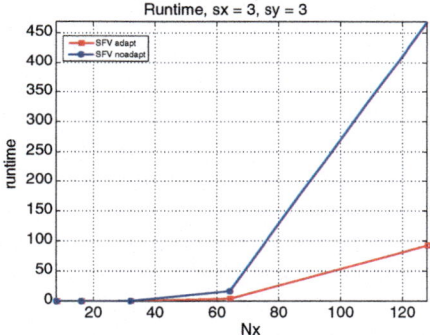

Fig. 15 Runtime (seconds) (Reproduced with permission from [34])

Figure 14 shows the convergence of the adaptive SFVM algorithm ("SFV adapt") and the SFVM without stochastic mesh adaptation ("SFV noadapt"). The non-adaptive version of the SFVM simply uses equal number of cells in each stochastic coordinate, while the adaptive version chooses the number of cells in each y_j according to (32). The computational time needed to perform both algorithms is shown in Fig. 15. Clearly, the proposed adaptation of the algorithm improves the convergence properties of the SFV method.

3.5 Efficiency of the SFVM

We compare the efficiencies of the SFV and MLMC methods [25, 26] for the solution of the one-dimensional stochastic Sod's problem for the Euler equations described in Sect. 3.3.3. Figure 16 illustrates the convergence of SFVM and MLMC based on 1st and 2nd order FV ENO/WENO solvers.

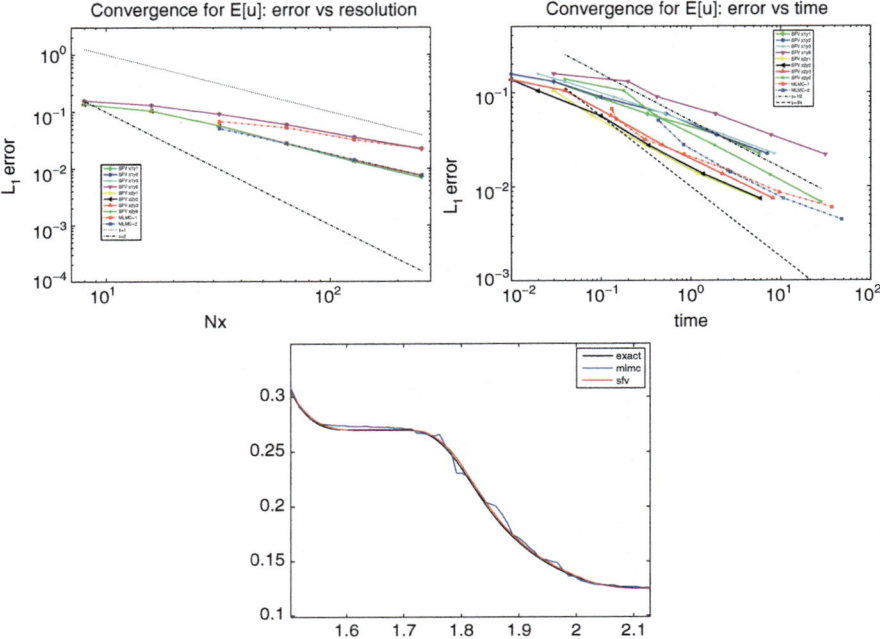

Fig. 16 Convergence of the SFVM and MLMC (upper plots) and detail of the solution mean (lower plot) (Reproduced with permission from [34])

Figure 16 demonstrates that both approaches lead to the same orders of convergence in space while SFVM with properly chosen reconstruction orders appears to be more efficient in terms of error-to-work estimates. Moreover, the solution mean generated by the MLMC method contains spurious oscillations which do not disappear by physical mesh refinement, while the SFVM produces monotone statistical solution at the same level of stochastic resolution.

3.6   SFVM Error Estimates for the Statistical Solution

3.7   Estimates in L_∞-Norm

Consider the stochastic scalar conservation law in the parametric form

$$\frac{\partial u}{\partial t} + \frac{\partial f(u; y)}{\partial x} = 0, \quad x \in D, \ y \in Y, \ t > 0, \tag{33}$$

$$u(x, 0; y) = u_0(x; y). \tag{34}$$

Let $Y = \bigcup\limits_{k=1}^{N_y} Y_k$, where Y_k is the mesh cell in the stochastic variable y. Denote the probability density function by $\mu(y)$.

Assume $u = u(x, t; y)$ is the exact solution to (33)–(34) and u_{ik}^n is its approximation at a fixed time moment $t = t^n$, resulting from SFVM. Denote $u_h = \{u_{ik}^n\}$. *Assume* further that the following error estimate holds:

$$\|u - u_h\|_{L^\infty(D\times Y)} \leq C_1 \Delta x^p + C_2 \Delta y^r, \tag{35}$$

where Δx and Δy are typical mesh sizes in the physical and stochastic coordinates, respectively. Note that the error analysis for the L_∞-norm applies to the case of smooth solutions; in the presence of shocks one should consider L_1-norm instead.

3.7.1 Error Estimate for the Mean $E_h[u_h]$

We have:

Lemma 1 *The mean value of the exact solution at the point (x_i, t^n) is*

$$\mathbb{E}[u](x_i, t^n) = \int\limits_Y u(x_i, t^n; y)\mu(y)\, dy, \tag{36}$$

and the corresponding SFVM approximation is computed as follows:

$$E_h[u_h]_i^n = \sum_{k=1}^{N_y} u_{ik}^n \omega_k, \quad \omega_k = \int\limits_{Y_k} \mu(y)\, dy. \tag{37}$$

Then

$$\left\| \mathbb{E}[u] - E_h[u_h] \right\|_{L^\infty(D)} \leq \|u - u_h\|_{L^\infty(D\times Y)}. \tag{38}$$

Proof

$$\left| \mathbb{E}[u](x_i, t^n) - E_h[u_h]_i^n \right| = \left| \int\limits_Y u(x_i, t^n; y)\mu(y)\, dy - \sum_k u_{ik}^n \omega_k \right|$$

$$= \left| \sum_k \int\limits_{Y_k} u(x_i, t^n; y)\mu(y)dy - \sum_k u_{ik}^n \int\limits_{Y_k} \mu(y)dy \right| = \left| \sum_k \int\limits_{Y_k} [u(x_i, t^n; y) - u_{ik}^n]\mu(y)dy \right|$$

$$\leq \sum_k \int_{Y_k} |u(x_i, t^n; y) - u_{ik}^n| \mu(y) dy \leq \sup_{Y_k} |u(x_i, t^n; y) - u_{ik}^n| \sum_k \int_{Y_k} \mu(y)\, dy$$

$$= \sup_{Y_k} |u(x_i, t^n; y) - u_{ik}^n| \int_Y \mu(y)\, dy = \sup_{Y_k} |u(x_i, t^n; y) - u_{ik}^n|, \qquad (39)$$

and

$$\big\| \mathbb{E}[u] - E_h[u_h] \big\|_{L^\infty(D)} \leq \| u - u_h \|_{L^\infty(D \times Y)}.$$

3.7.2   Error Estimate for the Variance $V_h[u_h]$

We have

Lemma 2 *The variance of the exact solution at (x_i, t^n) is equal to*

$$\mathbb{V}[u](x_i, t^n) = \mathbb{E}\big[\big(u(x_i, t^n) - \mathbb{E}[u](x_i, t^n) \big)^2 \big] = \mathbb{E}\big[u^2(x_i, t^n) \big] - \big(\mathbb{E}[u](x_i, t^n) \big)^2 \qquad (40)$$

and is approximated by

$$V_h[u_h]_i^n = E_h\big[u_h^2 \big]_i^n - \big(E_h[u_h]_i^n \big)^2. \qquad (41)$$

Then

$$\big\| \mathbb{V}[u] - V_h[u_h] \big\|_{L^\infty(D)} \leq C \| u - u_h \|_{L^\infty(D \times Y)}. \qquad (42)$$

Proof The approximation error for the variance can be computed as

$$\big\| \mathbb{V}[u] - V_h[u_h] \big\|_{L^\infty(D)} = \big\| \mathbb{E}[u^2] - (\mathbb{E}[u])^2 - E_h[u_h^2] + (E_h[u_h])^2 \big\|_{L^\infty(D)}$$

$$= \big\| \big(\mathbb{E}[u^2] - E_h[u_h^2] \big) - \big((\mathbb{E}[u])^2 - (E_h[u_h])^2 \big) \big\|_{L^\infty(D)}$$

$$\leq \big\| \mathbb{E}[u^2] - E_h[u_h^2] \big\|_{L^\infty(D)} + \big\| (\mathbb{E}[u])^2 - (E_h[u_h])^2 \big\|_{L^\infty(D)}. \qquad (43)$$

We can estimate the first term as

$$\big| \mathbb{E}[u^2(x_i, t^n)] - E_h[u_h^2]_i^n \big| = \bigg| \int_Y u^2(x_i, t^n; y) \mu(y)\, dy - \sum_k (u_{ik}^n)^2 \omega_k \bigg|$$

$$= \bigg| \sum_k \int_{Y_k} u^2(x_i, t^n; y) \mu(y)\, dy - \sum_k (u_{ik}^n)^2 \int_{Y_k} \mu(y)\, dy \bigg|$$

$$= \left| \sum_k \int_{Y_k} \left[u^2(x_i, t^n; y) - \left(u_{ik}^n \right)^2 \right] \mu(y) \, dy \right|$$

$$= \left| \sum_k \int_{Y_k} \left[u(x_i, t^n; y) - u_{ik}^n \right] \sqrt{\mu(y)} \left[u(x_i, t^n; y) + u_{ik}^n \right] \sqrt{\mu(y)} \, dy \right|$$

$$\leq \left| \sum_k \left(\int_{Y_k} \left[u(x_i, t^n; y) - u_{ik}^n \right]^2 \mu(y) \, dy \right)^{1/2} \cdot \left(\int_{Y_k} \left[u(x_i, t^n; y) + u_{ik}^n \right]^2 \mu(y) \, dy \right)^{1/2} \right|$$

$$\leq C \left| \sum_k \left(\int_{Y_k} \left[u(x_i, t^n; y) - u_{ik}^n \right]^2 \mu(y) \, dy \right)^{1/2} \right|$$

$$\leq C \sum_k \left(\int_{Y_k} \left| u(x_i, t^n; y) - u_{ik}^n \right|^2 \mu(y) \, dy \right)^{1/2}$$

$$\leq C \sup_{Y_k} \left| u(x_i, t^n; y) - u_{ik}^n \right| \left(\int_Y \mu(y) \, dy \right)^{1/2} = C \sup_{Y_k} \left| u(x_i, t^n; y) - u_{ik}^n \right|, \tag{44}$$

and hence

$$\left\| \mathbb{E}[u^2] - E_h[u_h^2] \right\|_{L^\infty(D)} \leq C \| u - u_h \|_{L^\infty(D \times Y)}. \tag{45}$$

For the second term we have

$$\left\| \left(\mathbb{E}[u](x_i, t^n) \right)^2 - \left(E_h[u_h]_i^n \right)^2 \right\|_{L^\infty(D)}$$

$$= \left\| \left(\mathbb{E}[u](x_i, t^n) - E_h[u_h]_i^n \right) \left(\mathbb{E}[u](x_i, t^n) + E_h[u_h]_i^n \right) \right\|_{L^\infty(D)}$$

$$\leq \left\| \mathbb{E}[u](x_i, t^n) - E_h[u_h]_i^n \right\|_{L^\infty(D)} \left\| \mathbb{E}[u](x_i, t^n) + E_h[u_h]_i^n \right\|_{L^\infty(D)}$$

$$\leq C \left\| \mathbb{E}[u](x_i, t^n) - E_h[u_h]_i^n \right\|_{L^\infty(D)} \leq C \| u - u_h \|_{L^\infty(D \times Y)}. \tag{46}$$

Finally,

$$\left\| \mathbb{V}[u] - V_h[u_h] \right\|_{L^\infty(D)} \leq C \| u - u_h \|_{L^\infty(D \times Y)}.$$

Analogous estimates can be obtained for higher moments.

3.8   Estimates in L^1-Norm

Denote by u the exact solution of (33), by u_h^y the numerical solution which is exact in x variable and discretized in y and by u_h^{xy} the numerical discretized in both variables. Assume that the numerical solution converges with rate p in x variable and rate r in y variable, that is

$$\|u_h^y - u_h^{xy}\|_{L^1(D)} \leqslant C_1 \Delta x^p \ \forall y \in Y. \tag{47}$$

$$\|u - u_h^y\|_{L^1(Y)} \leqslant C_2 \Delta y^r \ \forall x \in D, \tag{48}$$

The next estimate follows immediately from this assumption:

$$\|u - u_h^{xy}\|_{L^1(D \times Y)} \leqslant C_1 \Delta x^p + C_2 \Delta y^r. \tag{49}$$

3.8.1   Convergence of $E_h[u_h^{xy}]$ in L^1-Norm

We have

Lemma 3 *The expected value of the exact solution is a deterministic function*

$$\mathbb{E}[u](x_i, t^n) = \int_Y u(x_i, t^n; y)\mu(y)\, dy, \tag{50}$$

and the approximation of the expectation of the numerical solution is, as before, equal to

$$E_h[u_h^{xy}]_i^n = \sum_{k=1}^{N_y} u_{ik}^n \omega_k = \sum_{k=1}^{N_y} u_{ik}^n \int_{Y_k} \mu(y)\, dy = \sum_{k=1}^{N_y} \int_{Y_k} u_{ik}^n \mu(y)\, dy$$

$$= \int_Y u_{ik}^n \mu(y)\, dy = \mathbb{E}[u_h^{xy}](x_i, t^n). \tag{51}$$

Then

$$\left\|\mathbb{E}[u] - \mathbb{E}[u_h^{xy}]\right\|_{L^1(D)} \leqslant C_1 \Delta x^p + C_2 \Delta y^r. \tag{52}$$

Proof

$$\left\|\mathbb{E}[u] - \mathbb{E}[u_h^{xy}]\right\|_{L^1(D)} = \left\|\mathbb{E}[u] - \mathbb{E}[u_h^y] + \mathbb{E}[u_h^y] - \mathbb{E}[u_h^{xy}]\right\|_{L^1(D)}$$

$$\leqslant \left\|\mathbb{E}[u] - \mathbb{E}[u_h^y]\right\|_{L^1(D)} + \left\|\mathbb{E}[u_h^y] - \mathbb{E}[u_h^{xy}]\right\|_{L^1(D)}$$

$$= \int_D \left| \mathbb{E}[u] - \mathbb{E}[u_h^y] \right| dx + \int_D \left| \mathbb{E}[u_h^y] - \mathbb{E}[u_h^{xy}] \right| dx$$

$$= \int_D \left| \int_Y (u - u_h^y) \mu(y) \, dy \right| dx + \int_D \left| \int_Y (u_h^y - u_h^{xy}) \mu(y) \, dy \right| dx$$

$$\leq \int_D \int_Y |u - u_h^y| \mu(y) \, dy dx + \int_D \int_Y |u_h^y - u_h^{xy}| \mu(y) \, dy dx. \qquad (53)$$

The first integral in (53) can be estimated as follows:

$$\int_D \int_Y |u - u_h^y| \mu(y) \, dy dx \leq \int_D \sup_Y \mu(y) \int_Y |u - u_h^y| \, dy dx$$

$$= C \|u - u_h^y\|_{L^1(Y)} \leq C \Delta y^r, \qquad (54)$$

and for the second integral we have

$$\int_D \int_Y |u_h^y - u_h^{xy}| \mu(y) \, dy dx = \int_Y \left[\int_D |u_h^y - u_h^{xy}| \, dx \right] \mu(y) \, dy$$

$$= \|u_h^y - u_h^{xy}\|_{L^1(D)} \int_Y \mu(y) \, dy = \|u_h^y - u_h^{xy}\|_{L^1(D)} \leq C \Delta x^p. \qquad (55)$$

Hence, the convergence rate of the expectation in L^1-norm can be estimated as

$$\left\| \mathbb{E}[u] - \mathbb{E}[u_h^{xy}] \right\|_{L^1(D)} \leq C_1 \Delta x^p + C_2 \Delta y^r.$$

3.8.2 Convergence of $V_h[u_h^{xy}]$ in L^1-Norm

We have:

Lemma 4 *The variance of the exact solution at (x_i, t^n) is equal to*

$$\mathbb{V}[u](x_i, t^n) = \mathbb{E}\left[(u(x_i, t^n) - \mathbb{E}[u](x_i, t^n))^2 \right] = \mathbb{E}[u^2(x_i, t^n)] - (\mathbb{E}[u](x_i, t^n))^2, \qquad (56)$$

and can be approximated as

$$V_h[u_h^{xy}]_i^n = E_h\left[(u_h^{xy})^2 \right]_i^n - \left(E_h[u_h^{xy}]_i^n \right)^2 = \mathbb{E}\left[(u_h^{xy})^2 \right]_i^n - \left(\mathbb{E}[u_h^{xy}]_i^n \right)^2 = \mathbb{V}[u_h^{xy}]_i^n. \qquad (57)$$

Then

$$\left\| \mathbb{V}[u] - \mathbb{V}[u_h^{xy}] \right\|_{L^1(D)} \leq C_1 \Delta x^p + C_2 \Delta y^r. \tag{58}$$

Proof

$$\left\| \mathbb{V}[u] - \mathbb{V}[u_h^{xy}] \right\|_{L^1(D)} = \left\| \mathbb{E}[u^2] - \left(\mathbb{E}[u]\right)^2 - \mathbb{E}[(u_h^{xy})^2] + \left(\mathbb{E}[u_h^{xy}]\right)^2 \right\|_{L^1(D)}$$

$$\leq \left\| \mathbb{E}[u^2] - \mathbb{E}[(u_h^{xy})^2] \right\|_{L^1(D)} + \left\| \left(\mathbb{E}[u]\right)^2 - \left(\mathbb{E}[u_h^{xy}]\right)^2 \right\|_{L^1(D)}. \tag{59}$$

The following estimate holds for the first integral in (59):

$$\left\| \mathbb{E}[u^2] - \mathbb{E}[(u_h^{xy})^2] \right\|_{L^1(D)} = \int_D \left| \mathbb{E}[u^2] - \mathbb{E}[(u_h^{xy})^2] \right| dx =$$

$$= \int_D \left| \int_Y \left[u^2 - (u_h^{xy})^2 \right] \mu(y)\, dy \right| dx \leq \int_D \int_Y \left| u^2 - (u_h^{xy})^2 \right| \mu(y)\, dy dx =$$

$$= \int_D \int_Y \left| u - u_h^{xy} \right| \left| u + u_h^{xy} \right| \mu(y)\, dy dx \leq C \int_D \int_Y \left| u - u_h^{xy} \right| dy dx =$$

$$= C \| u - u_h^{xy} \|_{L^1(D \times Y)} \leq C \Delta x^p + C_7 \Delta y^r. \tag{60}$$

For the second integral in (59) we get

$$\left\| \left(\mathbb{E}[u]\right)^2 - \left(\mathbb{E}[u_h^{xy}]\right)^2 \right\|_{L^1(D)} = \int_D \left| \left(\mathbb{E}[u]\right)^2 - \left(\mathbb{E}[u_h^{xy}]\right)^2 \right| dx =$$

$$= \int_D \left| \mathbb{E}[u] - \mathbb{E}[u_h^{xy}] \right| \left| \mathbb{E}[u] + \mathbb{E}[u_h^{xy}] \right| dx \leq$$

$$\leq C \left\| \mathbb{E}[u] - \mathbb{E}[u_h^{xy}] \right\|_{L^1(D)} \leq C_1 \Delta x^p + C_2 \Delta y^r. \tag{61}$$

Finally, from (60)–(61) we get

$$\left\| \mathbb{V}[u] - \mathbb{V}[u_h^{xy}] \right\|_{L^1(D)} \leq C_1 \Delta x^p + C_2 \Delta y^r.$$

Similar estimates are also valid for higher moments of u.

3.9   Error vs Work Estimates for SFVM

In the previous section it has been shown that the error of the expectation approximation is given by

$$E = \left\| \mathbb{E}[u] - \mathbb{E}[u_h^{xy}] \right\|_{L^1(D)} \leqslant C_1 \Delta x^p + C_2 \Delta y^r, \tag{62}$$

where p and r are the convergence rates of the SFVM solver in physical and stochastic variables, respectively. Based on this result, we derive the error vs work estimates for SFVM.

Let $x \in \mathbb{R}^n$, $y \in \mathbb{R}^m$. Assume that the CFL condition is satisfied, such that $\Delta t = O(\Delta x)$. The total work W (or total time) required to compute the solution of the stochastic scalar conservation law using SFVM is proportional to the total numbers of grid points in x, y and t axes, denoted respectively by N_x, N_y and N_t, i.e.

$$W = C N_x N_y N_t = C \frac{1}{\Delta x^n} \frac{1}{\Delta y^m} \frac{1}{\Delta t} = \frac{C}{\Delta x^{n+1} \Delta y^m} = C \Delta x^{-(n+1)} \Delta y^{-m}. \tag{63}$$

Further derivation of the estimate depends on the choice of the mesh sizes equilibration, that is, on the relation between Δx and Δy.

1. Assume that the mesh sizes are equilibrated according to the expected orders of convergence p and r: $\Delta y = \Delta x^{p/r}$. Then $E = C \Delta x^p$ and $\Delta x = C E^{1/p}$. Substituting these relations into Eq. (63) we get

$$W = C \Delta x^{-(n+1)} \Delta x^{-pm/r} = C \Delta x^{-(n+1+pm/r)} = C E^{-\frac{n+1+pm/r}{p}} \tag{64}$$

and hence

$$E = C W^{-\frac{p}{n+1+pm/r}}, \tag{65}$$

which is the desired error vs work estimated.

2. Assume now that the mesh size Δy is obtained by the following scaling: $\Delta y = \eta \Delta x$, where η is the constant scaling factor, meaning that the stochastic mesh is isotropic (same Δy for all random variables). Define $q = \min(p, r)$. Then $E = C \Delta x^q$ and $\Delta x = C E^{1/q}$, and the total work is defined as

$$W = C \Delta x^{-(n+1)} \Delta y^{-m} = C \Delta x^{-(n+m+1)} = C E^{-\frac{n+m+1}{q}}, \tag{66}$$

which finally gives

$$E = C W^{-\frac{q}{n+m+1}}. \tag{67}$$

Note that the estimate (67) is equivalent to the complexity result for the deterministic finite-volume method in the $(n + m)$-dimensional space, which sets strict limitations on the number of random variables that can be handled by the SFVM if the scaling factor η is close to 1. However, computational practice shows that it is sufficient to use few computational cells to discretize the equations in the stochastic space to obtain a good quality approximation of the statistical quantities and therefore the SFVM is essentially much more efficient as deterministic FVM. Another significant simplification of the approach is the absence of the fluxes in the stochastic variables y, which also contributes to the efficiency of the SFVM.

3. Assume that the stochastic mesh is anisotropic, that is the mesh sizes Δy_k are different for $k = 1, \ldots, m$: $\Delta y_k = \eta_k \Delta x$. Applying the same technique as above we obtain

$$E = C_1 \Delta x + C_2 \sum_{k=1}^{m} \Delta y_k^r = C_1 \Delta x + C_2 \Delta x^r \sum_{k=1}^{m} \eta_k^r \leqslant C \Delta x^q \Big(1 + \sum_{k=1}^{m} \eta_k^r\Big), \quad (68)$$

where $q = \min(p, r)$ as before. We have also assumed that $\Delta x << 1$ such that $\Delta x^p < \Delta x^q$ and $\Delta x^r < \Delta x^q$. Then the mesh size Δx can be represented as

$$\Delta x = \left(\frac{E}{1 + \sum\limits_{k=1}^{m} \eta_k^r} \right)^{1/q}. \quad (69)$$

The total work is

$$W = C\Big(\prod_{k=1}^{m} \eta_k^{-1}\Big) \Delta x^{-(n+m+1)} = C\Big(\prod_{k=1}^{m} \eta_k^{-1}\Big) \left(\frac{E}{1 + \sum\limits_{k=1}^{m} \eta_k^r} \right)^{-(n+m+1)/q}, \quad (70)$$

and the resulting error vs work estimate is

$$E = C\Big(1 + \sum_{k=1}^{m} \eta_k^r\Big)\Big(\prod_{k=1}^{m} \eta_k^{-\frac{q}{n+m+1}}\Big) W^{-\frac{q}{n+m+1}}. \quad (71)$$

Note that in the isotropic case, when all $\eta_k = \eta = $ const, formula (71) results in

$$E = C(1 + m\eta^r)\, \eta^{-\frac{qm}{n+m+1}}\, W^{-\frac{q}{n+m+1}}. \quad (72)$$

Comparing (71) and (72) we notice that the proper choice of scaling factors η_k in the anisotropic stochastic mesh construction, while not affecting the convergence *rates*, can reduce the convergence constant, which means increasing computational efficiency. The choice of η_k should be based on the sensitivity analysis of the random entropy solution to each of the m random variables.

Let us demonstrate the efficiency provided by the anisotropic mesh adaptation. We compare the convergence constants:

$$C_i = (1 + m\eta^r) \, \eta^{-\frac{qm}{n+m+1}} \tag{73}$$

for the *isotropic* mesh with equal mesh sizes in all stochastic coordinates, $\Delta y_k = \eta \Delta x$, $k = 1, \ldots, m$, and

$$C_a = \left(1 + \sum_{k=1}^{m} \eta_k^r\right) \prod_{k=1}^{m} \eta_k^{-\frac{q}{n+m+1}} \tag{74}$$

for the *anisotropic* stochastic mesh with mesh size scaling according to $\Delta y_k = \eta_k \Delta x$, $k = 1, \ldots, m$. Assume further that $\eta_1 > \eta_2 > \ldots \eta_m$ and $\eta_k > 1$ for all k, such that $\Delta y_k > \Delta x$.

Our goal is to show that the convergence constants ratio $\delta_m = \dfrac{C_i}{C_a} > 1$ as $m \to \infty$ if $\eta_1 < \eta$ and $r > q$, that is, the anisotropic stochastic mesh increases the algorithm efficiency as the number of random variables grows if the convergence rate r in the stochastic space is higher than q, the minimum of the convergence rates in physical and stochastic coordinates.

We start by noting that under the assumption $\eta_1 < \eta$ the following inequality is valid:

$$C_a = \left(1 + \sum_{k=1}^{m} \eta_k^r\right) \prod_{k=1}^{m} \eta_k^{-\frac{q}{n+m+1}} < (1 + m\eta_1^r)\eta_1^{-\frac{qm}{n+m+1}} = C_a^1, \tag{75}$$

and therefore

$$\delta_m = \frac{C_i}{C_a} > \frac{C_i}{C_a^1} = \frac{(1 + m\eta^r) \, \eta^{-\frac{qm}{n+m+1}}}{(1 + m\eta_1^r)\eta_1^{-\frac{qm}{n+m+1}}} = \left(\frac{1 + m\eta^r}{1 + m\eta_1^r}\right)\left(\frac{\eta}{\eta_1}\right)^{-\frac{qm}{n+m+1}}. \tag{76}$$

Hence, the limit of the constants ratio is

$$\delta = \lim_{m \to \infty} \delta_m = \left(\frac{\eta}{\eta_1}\right)^{r-q}, \tag{77}$$

and clearly $\delta > 1$ if $r > q$.

Let's analyse in more detail the possible values of δ in dependence on the convergence rates p and r in x and y variables, respectively.

Smooth Solution If the solution is smooth in x and y, then the convergence rate of the SFVM is the expected one, therefore by applying high-order finite-volume approximations in both variables one can obtain the full convergence rates p and r.

- If $p < r$, then $q = \min(p, r) = p$ and $r - q = r - p > 0$, $\delta > 1$ and hence the SFVM will converge faster on anisotropic stochastic mesh.

- If $p > r$, then $q = \min(p, r) = p$ and $r - q = r - p < 0, \delta < 1$, therefore the anisotropic mesh doesn't improve the convergence.

Shock Solution Recall that if the shock wave appears in the physical space, then it also propagates into the stochastic space, so that the solution becomes discontinuous in both x and y. In this case one typically has $p = 1/2$ according to the Kuznetsov's result [20] and $r = 1$ as shown in [27], therefore $q = \min(p, r) = 1/2$ and $\delta = \sqrt{\frac{\eta}{\eta_1}} > 1$. This means that the SFVM on anisotropic mesh in the stochastic space is more efficient than SFVM on the uniform mesh even if the solution has a shock.

3.10 Anisotropic Mesh Adaptation for Euler Equations

We reconsider the stochastic Sod's shock tube problem and apply the anisotropic stochastic mesh adaptation which is similar to the one proposed for the scalar conservation laws with Karhunen-Loève flux (or initial data) expansion. Clearly, such an expansion is not available for the realistic systems of conservation laws like the Euler equations since the flux function and the random variables are pre-defined. However, it is possible to scale the random variables according to their influence on the random solution based on *empirical* considerations.

For the stochastic version of Sod's shock tube problem studied above, one can see that the uncertainty in the γ flux coefficient is practically unimportant (but not negligible) for the statistical solution, while the uncertain initial discontinuity location as well as random density amplitude being most important. Therefore we propose the following scaling for the number of cells in the stochastic coordinates: $N_y^i = C N_x \lambda_i$, where we take $C = 1/32$ and $\lambda_1 = 3$ (random shock location), $\lambda_2 = 2$ (random density amplitude) and $\lambda_3 = 1$ (random γ).

Figures 17, 18 and 19 demonstrate the convergence for the density for the 1st, 3rd and 5th order WENO reconstruction in stochastic coordinates $y_k, k = 1, 2, 3$, respectively. Each of the plots contains the results for 1st, 3rd and 5th order WENO reconstruction in the physical coordinate x. The results are presented for both adaptive and non-adaptive meshes in the stochastic space and clearly show the superior efficiency of the adaptive SFVM algorithm.

3.11 Numerical Approximation of the Probability Density Function for the Random Solution of Euler Equations

The advantage of the SFV method is the possibility to construct the empirical probability density functions (statistical histograms) after only one run since the *complete* information about the random solution is generated (that is, its approximation as a function of x and y_k is provided).

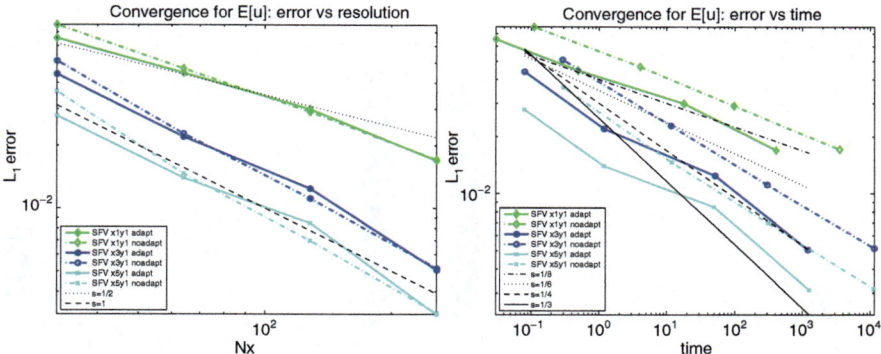

Fig. 17 Adaptive vs non-adaptive mesh, 1st order WENO in y_k, $k = 1, 2, 3$

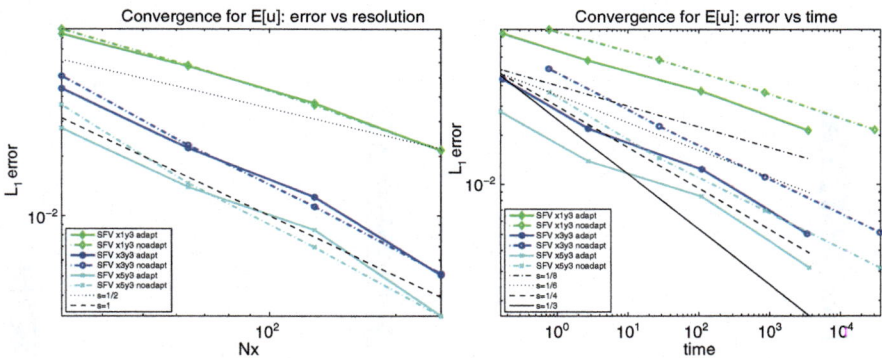

Fig. 18 Adaptive vs non-adaptive mesh, 3rd order WENO in y_k, $k = 1, 2, 3$

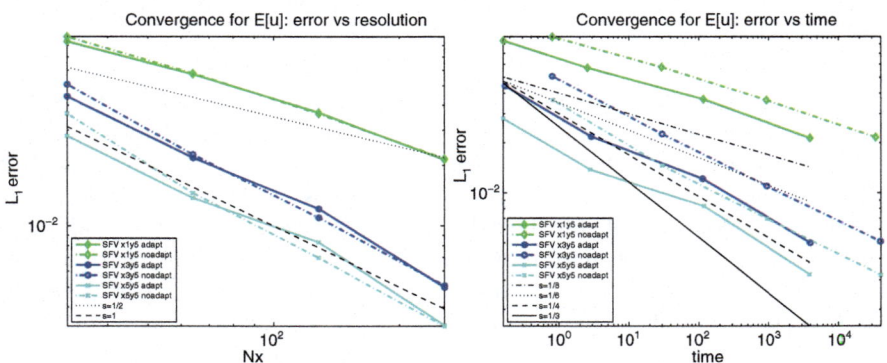

Fig. 19 Adaptive vs non-adaptive mesh, 5th order WENO in y_k, $k = 1, 2, 3$

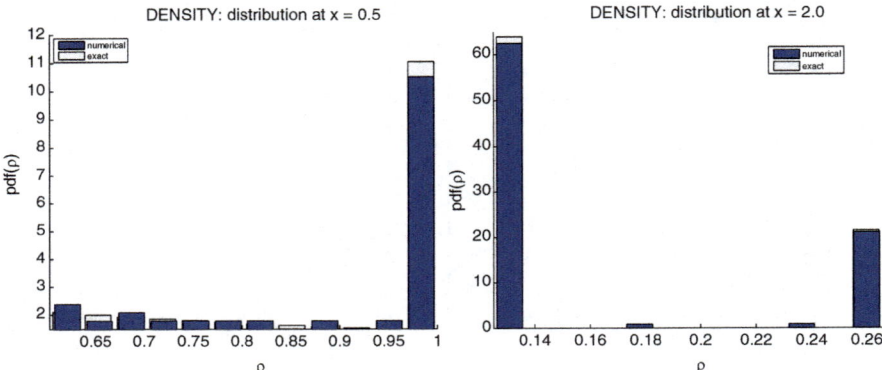

Fig. 20 Density distribution at $x = 0.5$ (left) and $x = 2$ (right). Random initial discontinuity location

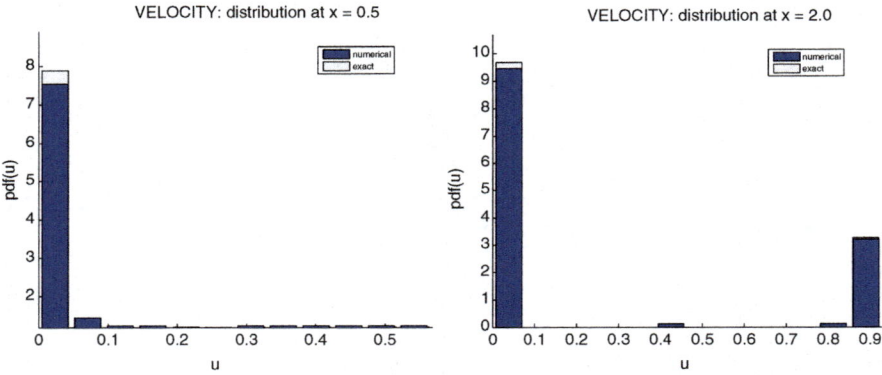

Fig. 21 Velocity distribution at $x = 0.5$ (left) and $x = 2$ (right). Random initial discontinuity location

We demonstrate the performance of SFVM for the approximation of the probability density functions for the solution of the stochastic Sod's shock tube problem. To this end, we solve the problem using the SFVM with 5th order WENO reconstruction in x and y_k, $k = 1, 2, 3$ for two cases: (1) with one uniformly distributed random variable for shock location, $x_0 \sim \mathcal{U}[0.75, 1.25]$; (2) with three random variables on the anisotropic mesh (see previous section), and plot the distribution histograms at $x = 0.5$ (rarefaction wave) and at $x = 2$ (shock wave). The number of bins to plot the diagrams is chosen according to

- square-root choice: $k = [\sqrt{n}]$, if the total number of grid points is small (practically, less than 30, like in case (1)),
- Sturges' formula [33]: $k = [\log_2 n + 1]$, otherwise.

The corresponding histograms are presented in Figs. 20, 21, and 22 for one random variable and in Figs. 23, 24, and 25 for three random variables. For

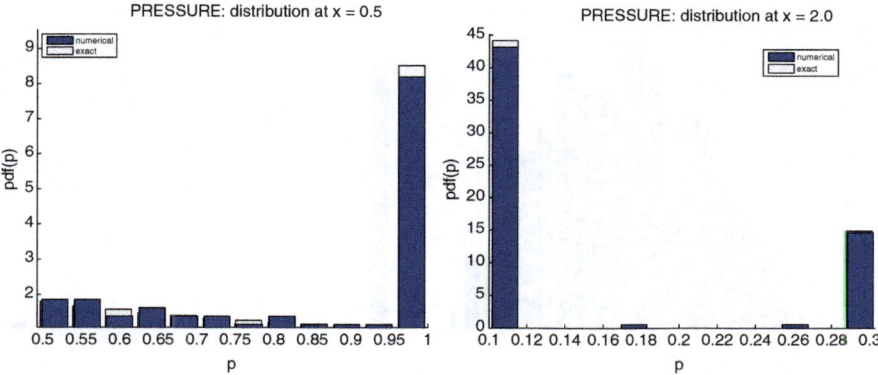

Fig. 22 Pressure distribution at $x = 0.5$ (left) and $x = 2$ (right). Random initial discontinuity location

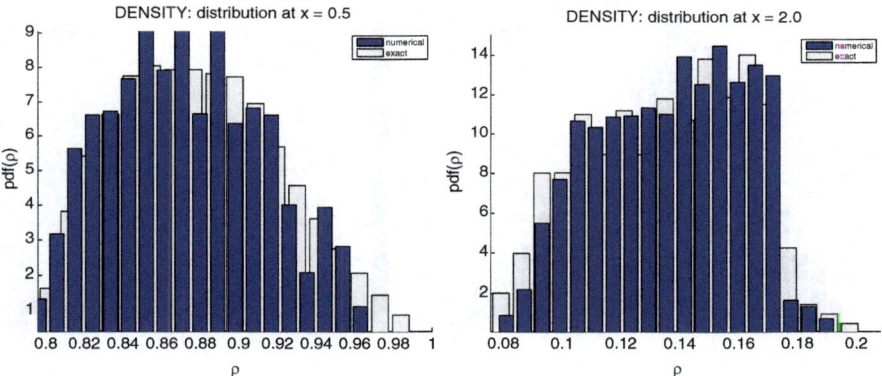

Fig. 23 Density distribution at $x = 0.5$ (left) and $x = 2$ (right). Random initial discontinuity location, density amplitude and γ

comparison, each of the plots contains the histograms for the exact solution of the problem with fine resolution in the stochastic space. The computed probability density functions indicate the bimodal character of gas parameter distributions in the stochastic Sod's problem.

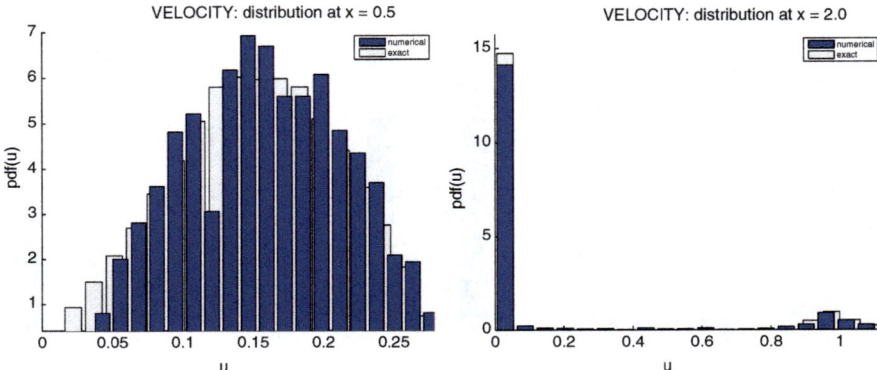

Fig. 24 Velocity distribution at $x = 0.5$ (left) and $x = 2$ (right). Random initial discontinuity location, density amplitude and γ

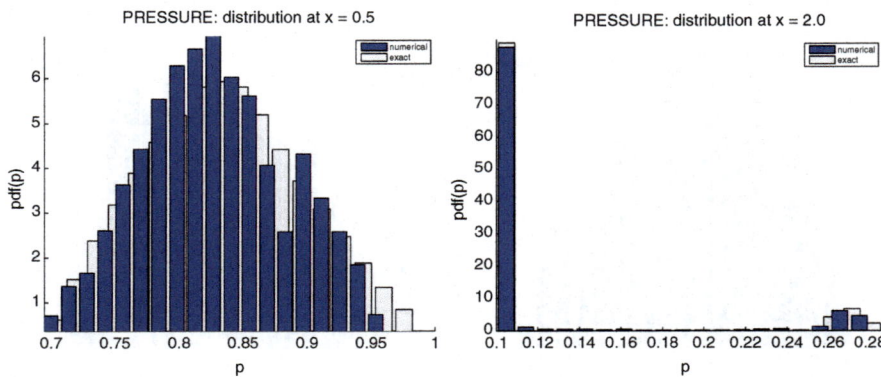

Fig. 25 Pressure distribution at $x = 0.5$ (left) and $x = 2$ (right). Random initial discontinuity location, density amplitude and γ

4 Stochastic Finite Volume Method on Unstructured Grids

4.1 Mixed DG/FV Formulation for the Stochastic Conservation Law in Multiple Dimensions

In this section we generalize the approach to uncertainty quantification described previously in order to efficiently apply high-order approximation techniques on unstructured grid in physical domains with complicated geometry. To this end, we use the Discontinuous Galerkin (DG) method to discretize the equations in the physical space and combine it with the finite-volume discretization in the stochastic variables as described in Sect. 3. Note that we can still use Cartesian grids in the stochastic space since the computational domain in this space is a q-dimensional rectangle.

As before, we start with the parametric form of the stochastic conservation law:

$$\frac{\partial \mathbf{U}}{\partial t} + \nabla_x \cdot \mathbf{F}(\mathbf{U}, \mathbf{y}) = \mathbf{0}, \ \mathbf{x} \in D_x \subset \mathbb{R}^3, \ \mathbf{y} \in D_y \subset \mathbb{R}^q, \ t > 0; \tag{78}$$

$$\mathbf{U}(\mathbf{x}, 0, \mathbf{y}) = \mathbf{U}_0(\mathbf{x}, \mathbf{y}), \ \mathbf{x} \in D_x \subset \mathbb{R}^3, \ \mathbf{y} \in D_y \subset \mathbb{R}^q. \tag{79}$$

Let $\mathcal{T}_x = \cup_{i=1}^{N_x} K_x^i$ be the triangulation of the computational domain D_x in the physical space and $\mathcal{C}_y = \cup_{j=1}^{N_y} K_y^j$ be the Cartesian grid in the domain D_y of the parametrized probability space. On each element K_x^i of the physical domain triangulation we apply the DG discretization of solution in the physical variable $\mathbf{x}$, that is, on each element of the triangulation we choose a system of basis functions $\{\varphi_l(\mathbf{x})\}$, $l = 1, \ldots, p$, and represent the numerical solution as the decomposition over the chosen basis

$$\mathbf{U}_h(\mathbf{x}, t, \mathbf{y}) = \sum_{l=1}^{p} \mathbf{U}_l^i(t, \mathbf{y}) \varphi_l(\mathbf{x}), \ x \in K_x^i, \tag{80}$$

with the coefficients $\mathbf{U}_l(t, \mathbf{y})$ to be determined. Next, according to the DG discretization procedure, we multiply the governing Eqs. (78)–(79) to each of the basis functions φ_k, $k = 1, \ldots, p$ and integrate the result over the element K_x^i. Application of the Gauss' theorem to the volume integral yields the following semi-discrete DG formulation $\forall k = 1, \ldots, p$:

$$\sum_{l=1}^{p} \partial_t \mathbf{U}_l^i(t, \mathbf{y}) \int_{K_x^i} \varphi_l(\mathbf{x}) \varphi_k(\mathbf{x}) \, d\mathbf{x} + \int_{\partial K_x^i} \mathbf{F}(\mathbf{U}_h, \mathbf{y}) \cdot \mathbf{n} \, \varphi_k(\mathbf{x}) \, d\mathbf{x} -$$

$$- \int_{K_x^i} \mathbf{F}(\mathbf{U}_h, \mathbf{y}) \nabla \varphi_k(\mathbf{x}) \, d\mathbf{x} = \mathbf{0}. \tag{81}$$

$$\sum_{l=1}^{p} \mathbf{U}_l^i(0, \mathbf{y}) \int_{K_x^i} \varphi_l(\mathbf{x}) \varphi_k(\mathbf{x}) \, d\mathbf{x} = \int_{K_x^i} \mathbf{U}_0(\mathbf{x}, \mathbf{y}) \varphi_k(\mathbf{x}) \, d\mathbf{x}. \tag{82}$$

The physical flux $\mathbf{F}(\mathbf{U}_h, \mathbf{y}) \cdot \mathbf{n}$ is in general discontinuous across the cell boundary and therefore needs to be replaced by any standard numerical flux approximation $\hat{\mathbf{F}}(\mathbf{U}_h^{int}, \mathbf{U}_h^{ext}, \mathbf{y})$ depending on two boundary extrapolated solution values $\mathbf{U}_h^{int}$ and $\mathbf{U}_h^{ext}$ (inside and outside of the cell, respectively) Note that at this stage the DG coefficients $\mathbf{U}_l^i(t, \mathbf{y})$ are still functions of the random variable $\mathbf{y}$ and time t and thus to get rid of this dependence we introduce the DG coefficients averaged over an element of the stochastic grid

$$\mathbf{U}_l^{ij}(t) = \frac{1}{|K_y^j|} \int\limits_{K_y^j} \mathbf{U}_l^i(t, \mathbf{y})\, \mu(\mathbf{y})\, d\mathbf{y}$$

and apply the finite-volume discretization over each cell K_y^j in the random variable, which leads to

$$\sum_{l=1}^{p} \int\limits_{K_y^j} \left[\partial_t \mathbf{U}_l^i(t, \mathbf{y}) \int\limits_{K_x^i} \varphi_l(\mathbf{x})\varphi_k(\mathbf{x})\, d\mathbf{x} \right] \mu(\mathbf{y})\, d\mathbf{y} +$$

$$+ \int\limits_{K_y^j} \left[\int\limits_{\partial K_x^i} \hat{\mathbf{F}}(\mathbf{U}_h^{int}, \mathbf{U}_h^{ext}, \mathbf{y})\, \varphi_k(\mathbf{x})\, d\mathbf{x} \right] \mu(\mathbf{y})\, d\mathbf{y} -$$

$$- \int\limits_{K_y^j} \left[\int\limits_{K_x^i} \mathbf{F}(\mathbf{U}_h, \mathbf{y}) \nabla \varphi_k(\mathbf{x})\, d\mathbf{x} \right] \mu(\mathbf{y})\, d\mathbf{y} = 0, \quad k = 1, \dots, p. \quad (83)$$

Finally, the resulting scheme becomes

$$\sum_{l=1}^{p} \frac{d\mathbf{U}_l^{ij}(t)}{dt} \int\limits_{K_x^i} \varphi_l(\mathbf{x})\varphi_k(\mathbf{x})\, d\mathbf{x} + \frac{1}{|K_y^j|} \iint\limits_{K_y^j \partial K_x^i} \hat{\mathbf{F}}(\mathbf{U}_h^{int}, \mathbf{U}_h^{ext}, \mathbf{y})\, \varphi_k(\mathbf{x})\, \mu(\mathbf{y}) d\mathbf{x} d\mathbf{y} -$$

$$- \frac{1}{|K_y^j|} \iint\limits_{K_y^j K_x^i} \mathbf{F}(\mathbf{U}_h, \mathbf{y}) \nabla \varphi_k(\mathbf{x})\, \mu(\mathbf{y}) d\mathbf{x} d\mathbf{y} = 0, \quad k = 1, \dots, p. \quad (84)$$

The initial data for $\mathbf{U}_l^{ij}(t)$ is obtained similarly: for $k = 1, \dots, p$

$$\sum_{l=1}^{p} \mathbf{U}_l^{ij}(0) \int\limits_{K_x^i} \varphi_l(\mathbf{x})\varphi_k(\mathbf{x})\, d\mathbf{x} = \frac{1}{|K_y^j|} \iint\limits_{K_y^j K_x^i} \mathbf{U}_0(\mathbf{x}, \mathbf{y})\varphi_k(\mathbf{x})\, \mu(\mathbf{y}) d\mathbf{x} d\mathbf{y}. \quad (85)$$

Equations (84)–(85) form an ODE system with respect to the coefficients $\mathbf{U}_l^{ij}(t)$ which can be solved using the Runge-Kutta method of the appropriate order. The slope limiting procedure has to be applied at each intermediate stage of the Runge-Kutta method in order to ensure the stability of the resulting DG scheme. This is done using the algorithm proposed in [9].

4.2 Numerical Results

4.2.1 Stochastic Cloud-Shock Interaction Problem (Random Flux)

Consider the two-dimensional Euler equations with deterministic initial data

$$[\rho_0, u_0, v_0, p_0] = \begin{cases} [3.86859, 11.2536, 0, 167.345], & \text{if } x_1 < 0.05, \\ [1, 0, 0, 1], & \text{if } x_1 > 0.05, \end{cases}$$

and a high-density cloud lying to the right of the shock:

$$\rho_0 = 10, \text{ if } \sqrt{(x_1 - 0.25)^2 + (x_2 - 0.5)^2} \leqslant 0.15.$$

Assume the random $\gamma = \gamma(\omega)$ in the equation of state (EOS)

$$p = (\gamma(\omega) - 1)\Big(E - \frac{1}{2}\rho(u^2 + v^2)\Big),$$

$$\gamma(\omega) \sim \mathscr{U}\big(5/3 - \epsilon, 5/3 + \epsilon\big), \ \epsilon = 0.1$$

The results of the simulation are presented in Fig. 26. In our computations we have used the 2nd order DG method in **x** variable and 3rd order WENO method in **y** variable, triangular mesh in **x** consisting of about 170,000 cells and Cartesian mesh in **y** consisting of 16 cells. Note that no symmetry conditions have been imposed on the mesh. The results are plotted at $T = 0.06$.

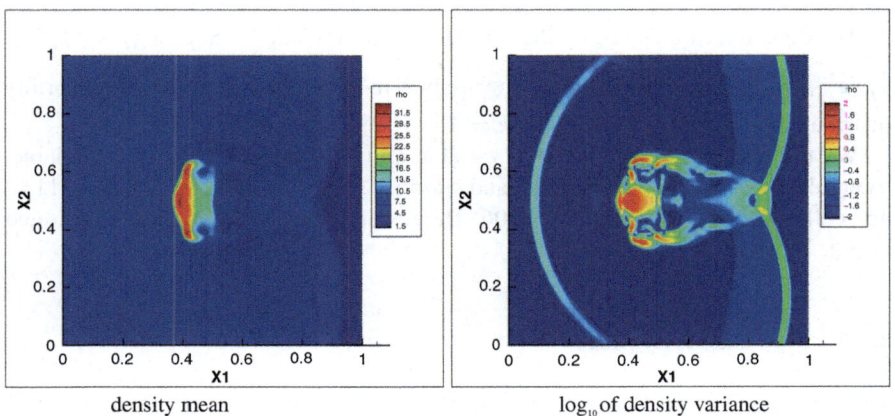

density mean $\log_{10}$ of density variance

Fig. 26 Stochastic cloud-shock interaction problem (Reproduced with permission from [34])

4.2.2 Forward-Facing Step Channel

Consider the stochastic flow in the channel with the forward facing step with random Mach number of the inflowing gas: M $\sim \mathscr{U}(2.9, 3.1)$. We have used the mesh of about 13,000 triangular cells in the physical space and 15 equally-sized cells in the stochastic space, the methods used are 2nd order DG and 3rd order WENO in physical and random variables, respectively. The results of the simulation are given in Fig. 27, indicating that the uncertainty in the Mach number influences the position and intensity of shock in front of the step, while having little effect on the shocks reflected from the channel walls.

4.2.3 Stochastic Cloud-Shock Interaction Problem (Random IC)

We use the mesh adaptation approach similar to the one described in Sect. 3.4 to solve the stochastic cloud-shock interaction problem with initial data depending on *four* random variables. Note that the usage of non-adaptive algorithm for such simulation would lead to excessive computational cost of SFVM.

Consider the two-dimensional Euler equations with deterministic initial data

$$
\mathbf{W}_0 = \begin{cases} [3.86859 + 0.1Y_2(\omega), 11.2536, 0, 167.345], & \text{if } x_1 < 0.04 + 0.01Y_1(\omega), \\ [1, 0, 0, 1], & \text{if } x_1 > 0.04 + 0.01Y_1(\omega), \end{cases}
$$

with a high-density cloud to the right of the shock:

$$
\rho_0 = 10 + 0.5Y_3(\omega), \text{ if } \sqrt{(x_1 - 0.25)^2 + (x_2 - 0.5)^2} \leqslant 0.15 + 0.02Y_4(\omega).
$$

The equations are closed by the following deterministic EOS: $p = (\gamma - 1)\Big(E - \frac{1}{2}\rho(u^2 + v^2)\Big)$, $\gamma = 5/3$. The random variables in the initial condition are uniformly distributed on [0, 1]: $Y_k \sim \mathscr{U}[0, 1]$, $k = 1, \ldots, 4$.

We use the 2nd order DG in **x** variable and 3rd order WENO in **y** variable, triangular mesh in **x** (170,000 cells) and *adaptive* Cartesian mesh in **y** ($3 \cdot 2 \cdot 7 \cdot 11 = 462$ cells), the output time is $T = 0.06$. The results of this simulation are illustrated in Fig. 28.

4.2.4 Flow Past a Cylinder

We have applied the SDGFV method for the simulation of the stochastic flow around a cylinder which is modeled by the Navier-Stokes equations. For this study we have chosen one random variable random variables: $Y_0 \sim \mathscr{U}[-1, 1]$

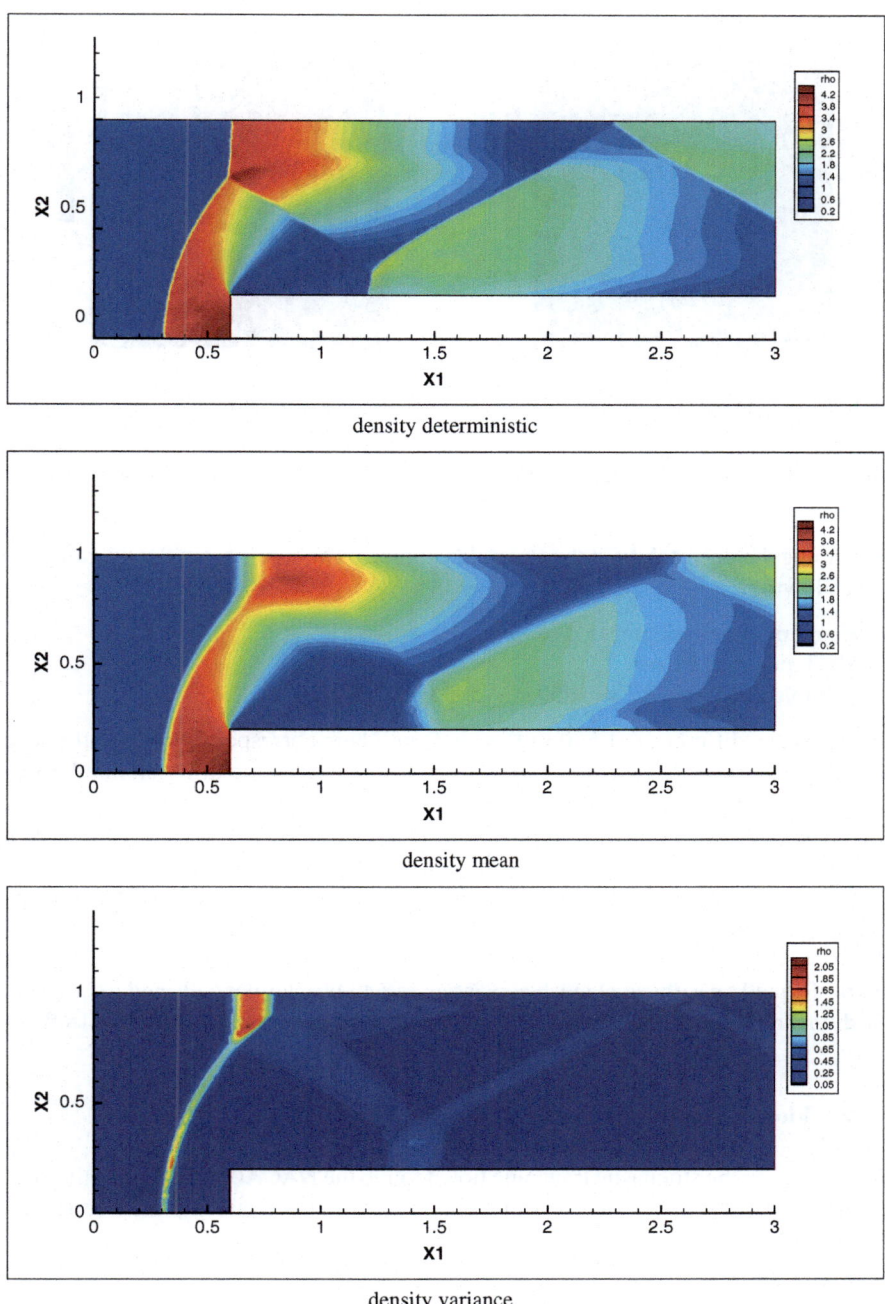

density deterministic

density mean

density variance

Fig. 27 Stochastic flow in a forward-facing step channel (Reproduced with permission from [34])

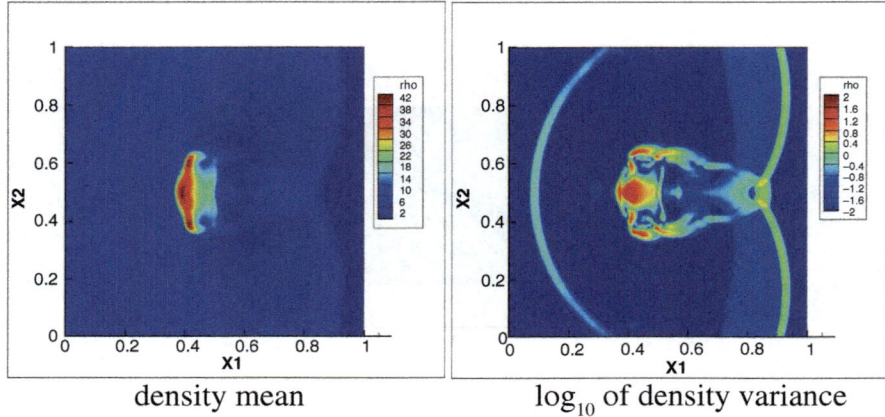

density mean $\log_{10}$ of density variance

Fig. 28 Stochastic cloud-shock interaction problem (Reproduced with permission from [34])

describing the uncertainty in the Reynolds number. The free-stream flow parameters are following:

- Angle of attack (degrees): $\alpha = 0$
- Mach number: M = 0.1
- Reynolds number: Re $= 2000.0 + 500.0\, Y_0(\omega)$

Note that the difference of 500 in Reynolds numbers corresponds to the difference in speeds of about 5 m/s which in turn leads to the fluctuation of Mach numbers of only 5% which is negligible compared to 25% fluctuation of the Reynolds number. Therefore the reduction of the number of random variables to one in this simulation appears to be reasonable.

The results of the computations (mean, variance and deterministic distribution of Mach numbers) are presented in Fig. 29. For this simulation we again use the 2nd order DG in **x** variable and 3rd order WENO method in **y** variable, the computations are performed on a physical mesh consisting of 6434 triangular cells and a Cartesian mesh in stochastic space consisting of 16 elements, the output time is $T = 18.0$.

4.2.5 Flow Around NACA0012 Airfoil

We next study the stochastic transonic flow around the NACA0012 wing profile. The flow is modeled by the system of Euler equations with uncertainty in the free-stream parameters:

- Angle of attack (degrees): $\alpha = 1.25 + 0.05\, Y_0(\omega)$
- Mach number: M $= 0.8 + 0.05\, Y_1(\omega)$

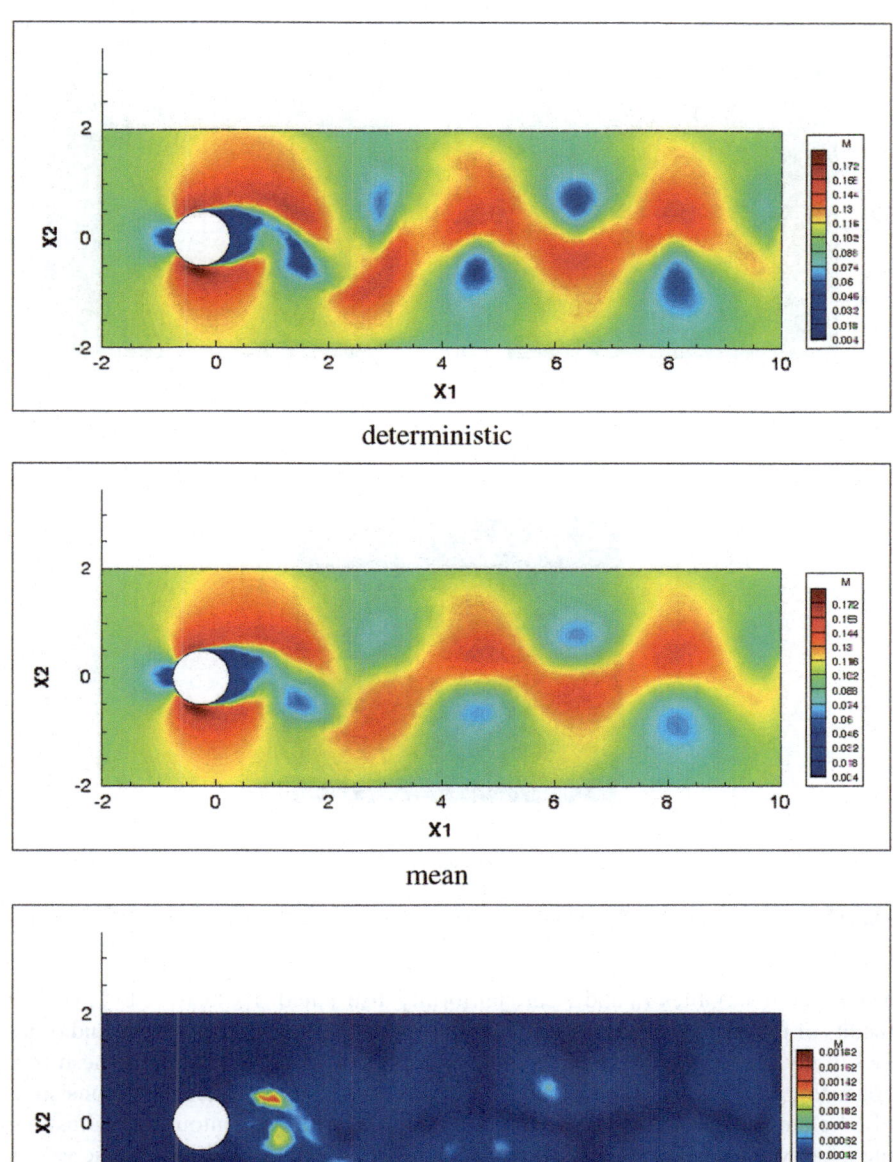

deterministic

mean

variance

Fig. 29 Flow around a cylinder. Mach number contour lines

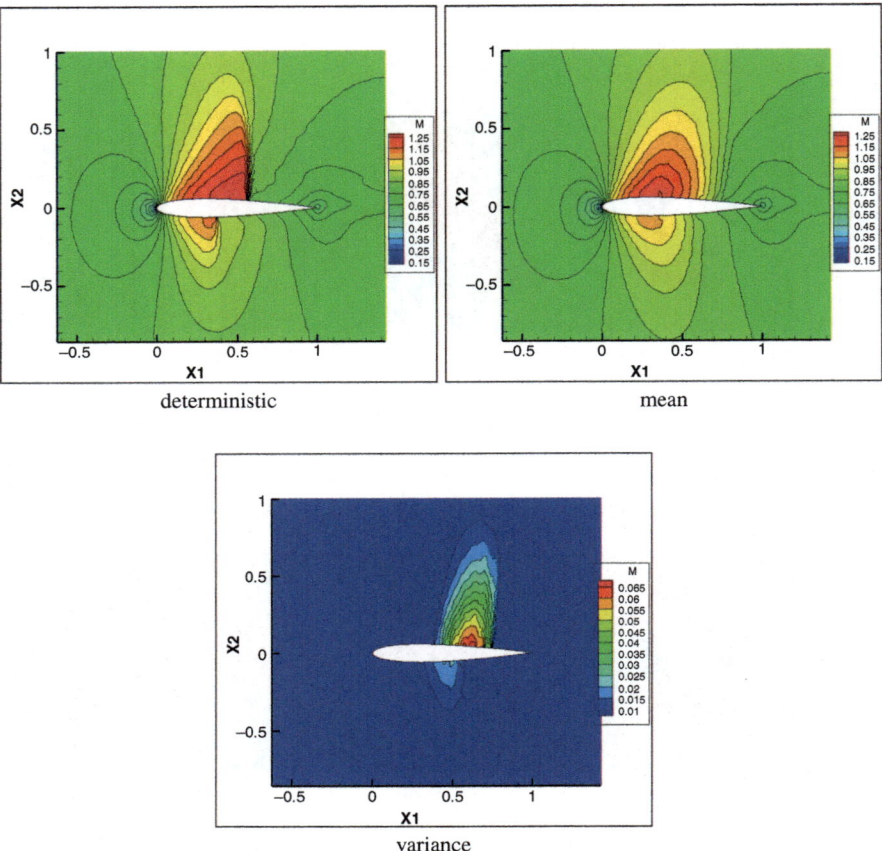

deterministic   mean

variance

Fig. 30 Inviscid transonic flow around NACA0012 airfoil. Mach number contour lines

The random variables Y_0 and Y_1 are uniformly distributed: $Y_0, Y_1 \sim \mathcal{U}[-1, 1]$. The results of this simulation are presented in Figs. 30, 31, and 32. We use the 2nd order DG in $\mathbf{x}$ variable and 3rd order WENO method in $\mathbf{y}$ variable, triangular mesh in $\mathbf{x}$ consisting of 92,023 elements and two-dimensional Cartesian mesh in $\mathbf{y}$ consisting of 64 elements, the output time is $T = 10.0$. In Fig. 30, the contour lines illustrate the mean value and variance of the Mach number as well as its deterministic values around the wing profile. Clearly, two shock waves are present in the deterministic run: one on the lower and one on the upper surface of the profile (see Fig. 31 for the distribution of the pressure coefficients). These shock waves however are smoothed in the mean flow, which is in accordance with the results of [31]. Finally, the approximations of the probability density functions for the distribution of the drag and lift coefficients are given in Fig. 32.

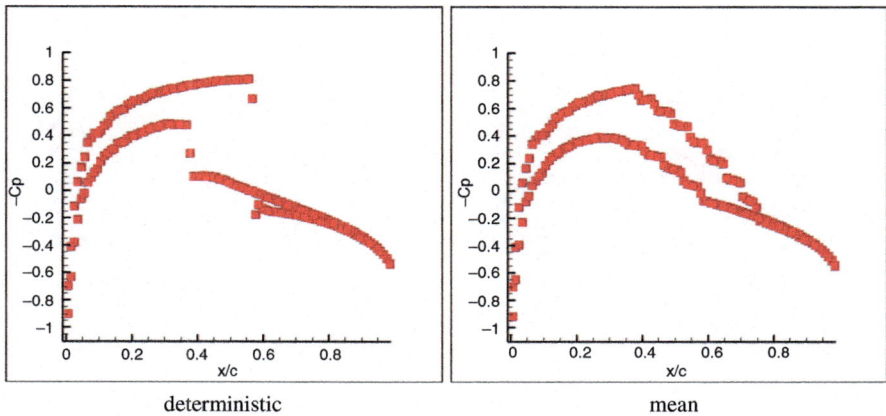

deterministic   mean

Fig. 31 Inviscid transonic flow around NACA0012 airfoil. Pressure coefficients

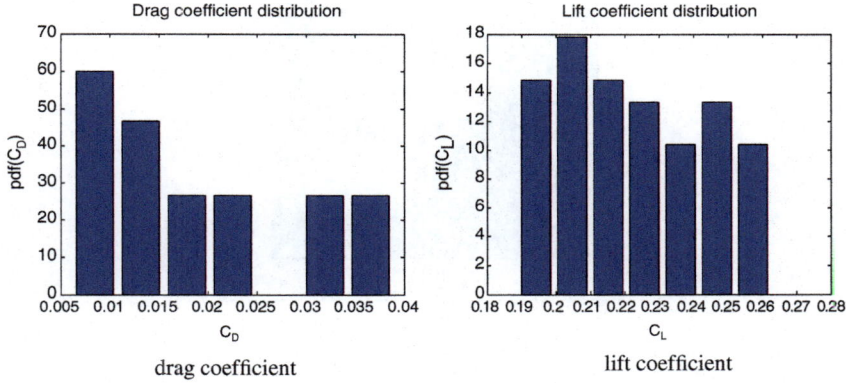

drag coefficient   lift coefficient

Fig. 32 Inviscid transonic flow around NACA0012 airfoil. Distribution of drag (left) and lift (right) coefficients

4.2.6 Flow Around NACA23012 Airfoil with Flap

We run the NACA23012 airfoil simulation with the following parameters:
Random variables: $Y_0, Y_1 \sim \mathscr{U}[-1, 1]$.
Free-stream flow parameters:

- Angle of attack (degrees): $\alpha = 8.0 + 0.5\, Y_0(\omega)$
- Flap deflection angle (degrees): $\alpha_f = 30$
- Mach number: $M = 0.1 + 0.015\, Y_1(\omega)$
- Reynolds number: $Re = 2100000.0 + 300000.0\, Y_1(\omega)$
- Prandtl number: $Pr = 0.72$

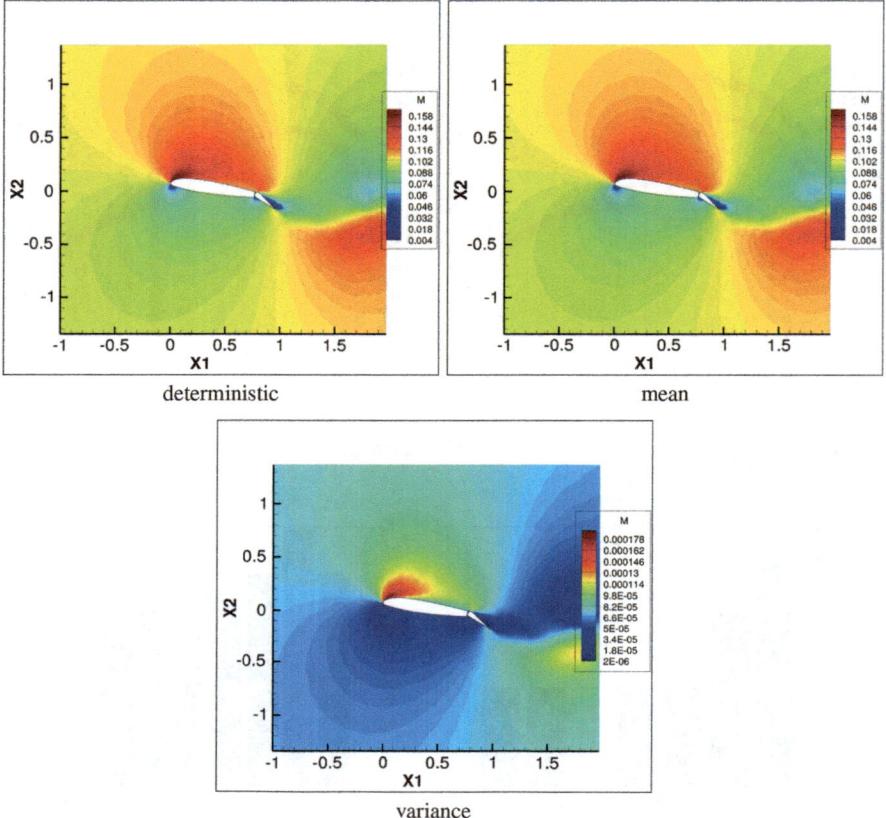

Fig. 33 Flow around NACA23012 airfoil with flap. Mach number contour lines

In order to run this simulation we adapt the stochastic mesh as follows: we take $N_y^0 = 3$ cells in the stochastic coordinate Y_0 and $N_y^1 = 6$ cells in the coordinate Y_1. The adaptation is based on a simple argument that the range of random angles of attack makes about 6% of its mean value ($0.5/8 = 0.0625$) and the range of both random Mach and Reynolds numbers makes 15% of the mean value, which results in double number of cells. This simulation has been performed on a triangular physical mesh of 17,418 cells, the numerical methods used are 2nd order DG in physical variables and 1st order FV in stochastic variables. The results of the simulation at time is $T = 1.0$ are presented in Fig. 33.

4.2.7   Flow Around RAE2822 Airfoil

Finally, we perform the stochastic simulation of the flow around the RAE2822 airfoil. The setup of the deterministic problem is the following.

- Angle of attack (degrees): $\alpha_\infty = 2.31$
- Mach number: $M_\infty = 0.729$
- Reynolds number: $Re_\infty = 6.5 \times 10^6$

The random parameters are modeled by means of the Beta distribution on $[y_L, y_R]$:

$$B(y, a, b) = \frac{1}{B(a, b)}(y - y_L)^{a-1}(y_R - y)^{b-1}(y_R - y_L)^{-(a+b-1)}$$

In this simulation, we assume that the angle of attach and the Mach numbers are random variables defined by

- $\alpha_\infty(\omega) \sim B(y, a, b)$ with $a = b = 4$, $y_L = 0.98\alpha_\infty$, $y_R = 1.02\alpha_\infty$
- $M_\infty(\omega) \sim B(y, a, b)$ with $a = b = 4$, $y_L = 0.95\,M_\infty$, $y_R = 1.05\,M_\infty$

We use the 2nd order DG method in **x** and 3rd order WENO reconstruction in **y**, triangular mesh in **x** (258,476 elements) and 2D Cartesian mesh in **y** (64 elements). The results of the simulation are presented in Fig. 34.

4.3   Parallel Algorithm and Parallel Efficiency of the SFVM

In the previous section we have presented a number of simulations of stochastic flows performed with the SFV method. Clearly, simulations with high-order methods involving complicated geometries and flow phenomena are computationally intensive even in the deterministic case and become much more costly in the presence of uncertainly. Therefore all of the described algorithms have been implemented in parallel using the Message Passing Interface (MPI) library. The basic parallelization principle used is the domain decomposition method which is applied in both physical space (on unstructured grid) and stochastic space (on Cartesian grid). The DG method used to approximate the random solution in the physical space allows to keep the approximation stencil compact. On the unstructured triangular mesh the compact consists of four triangles regardless of the order of the method. Therefore, the number of the mesh elements which need to exchange information between the processors is relatively small compared to the total number of elements in one subdomain.

In order to obtain partition of complicated computational domains we use the METIS library. A typical partition generated by METIS is presented in Fig. 35. Here, different colours indicate different subdomains.

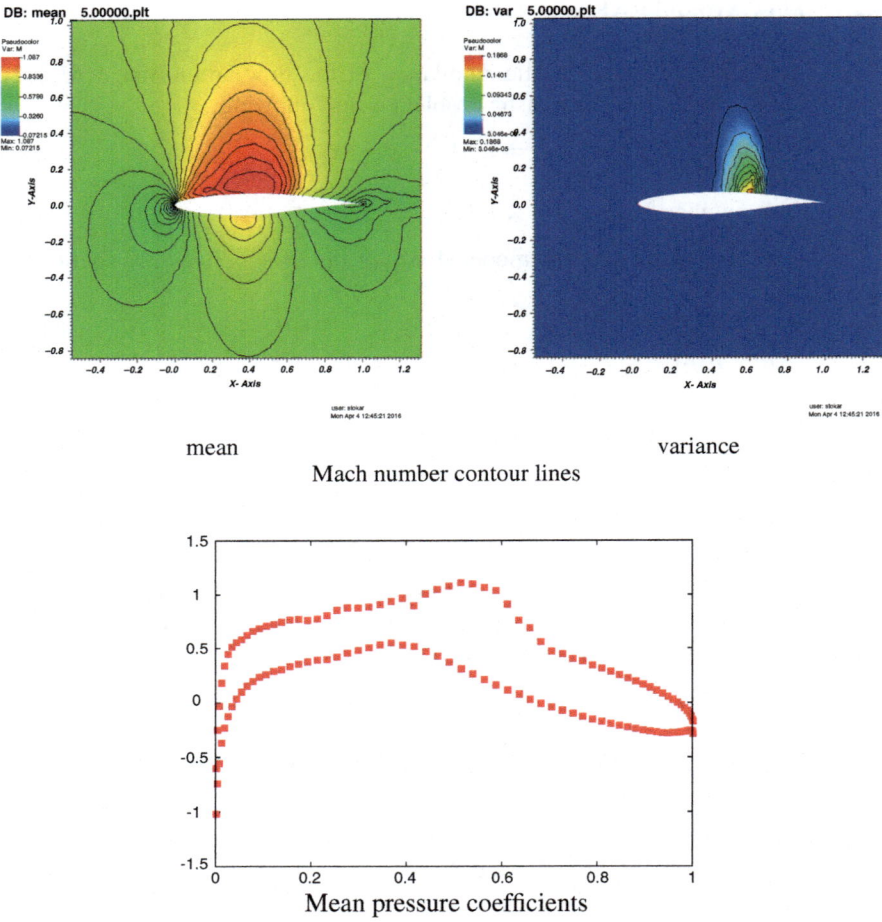

mean   variance

Mach number contour lines

Mean pressure coefficients

Fig. 34 Flow around RAE2822 airfoil

Table 5 lists the computational time per two timesteps, the speedup of the algorithm and efficiency with respect to the number of cores, for the simulation of the transonic flow around NACA0012 airfoil described in the previous section. The physical mesh consists of approximately 90,000 triangles and the stochastic mesh has 8×8 elements. The corresponding plot of the algorithm speedup is presented in Fig. 36.

Therefore, the SFV method can be efficiently parallelized and used for uncertainty quantification in multidimensional systems of conservation laws on complicated physical domains with unstructured meshes.

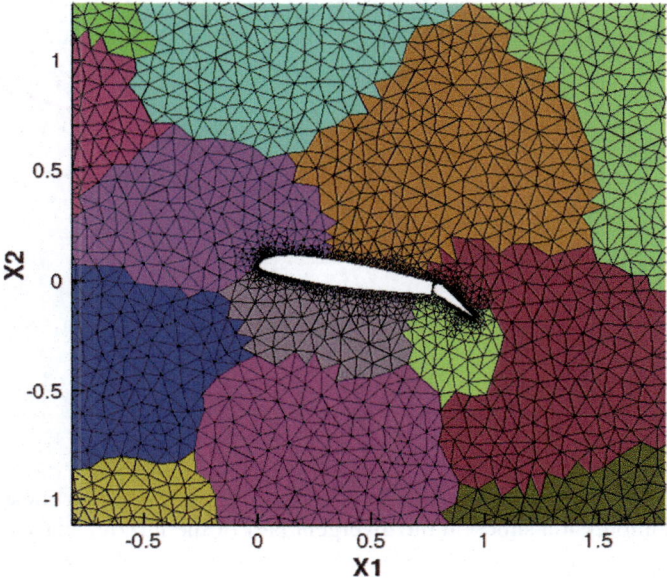

Fig. 35 Partition generated by METIS

Table 5 Parallel efficiency
of SFVM

N	Time (s)	Speedup	Efficiency (%)
1	2663.99	1	100
2	1353.34	1.97	98.42
4	632.71	4.21	105.26
8	334.44	7.97	99.57
16	166.84	15.97	99.8
32	84.88	31.39	98.08
64	48.02	55.47	86.68
128	26.31	101.26	79.11

5 Other Applications

Thanks to its flexibility, this method has several other applications.

5.1 Nozzle Flow with Shock

The steady shocked flow in a convergent-divergent nozzle is taken into account with
a fixed (deterministic) geometry:

$$A(x) = \begin{cases} 1 + 6(x - \frac{1}{2})^2 & \text{for } 0 < x \le \frac{1}{2} \\ 1 + 2(x - \frac{1}{2})^2 & \text{for } \frac{1}{2} < x \le 1 \end{cases} \tag{86}$$

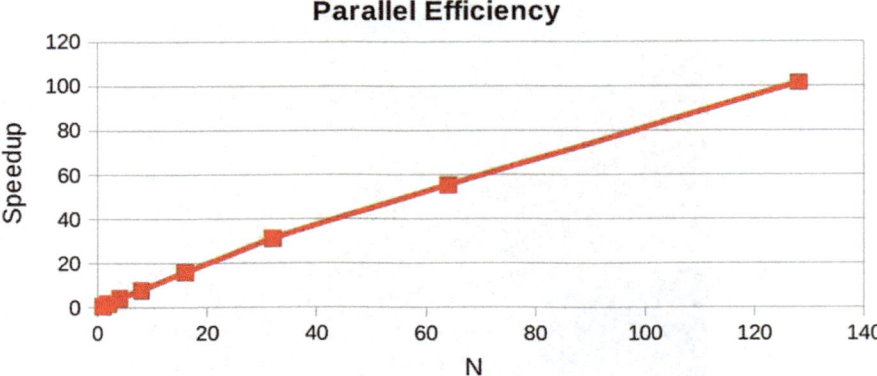

Fig. 36 Algorithm speedup

The outlet pressure (subsonic outlet flow with $p_e = 1.6529$ bar) is chosen in order to have a compression shock in the divergent part of the nozzle, exactly located at $x = 0.75$. For the other boundary conditions a subsonic inlet flow with a stagnation pressure $p_0 = 2$ bar and a stagnation temperature $T_0 = 300$ K are considered. The mean of γ is 1.4. Two test cases are considered. First, an uncertain heat coefficient ratio γ is assumed. The random parameter $\omega = \gamma$ varies within the range $[1.33, 1.47]$, following various choices of pdf (uniform and discontinuous) described below. In the second test-case, two-uncertainties stochastic problem is solved where γ follows a discontinuous pdf and the subsonic outlet flow varies uniformly within the range $[1.6529 \pm 0.98, 1.6529 \pm 1.02]$.

The random parameter ω (defining either the heat ratio or the subsonic outlet flow) ranges between ω_{min} and ω_{max}; the interval $[\omega_{min}, \omega_{max}]$ is mapped onto $[a, b]$ by a linear transformation and the pdf on $[a, b]$ is either:

* uniform with $\omega \in [a, b] = [0, 1]$,
* discontinuous on $[a, b] = [0, 1]$ with a density defined by:

$$f(\gamma) = \frac{1}{M} \times \begin{cases} \dfrac{1 + \cos(\pi x)}{2} & \text{if } x \in [0.5, 1] \\ 10 + \dfrac{1 + \cos(\pi x)}{2} & \text{if } x \in [0, 0.5] \\ 0 & \text{else} \end{cases} \tag{87}$$

and $M = \frac{11}{2}$ to ensure normalization.

Different stochastic methods are used to compute statistic solutions of the supersonic nozzle. Different pdf are used for γ, i.e. uniform in order to compare MC-SOBOL, PC and SI, and the discontinuous pdf (87) in order to compare MC-SOBOL and SI and to demonstrate the flexibility offered by the SI method. After a study on the grid convergence, the 1D physical space is divided in 201 points

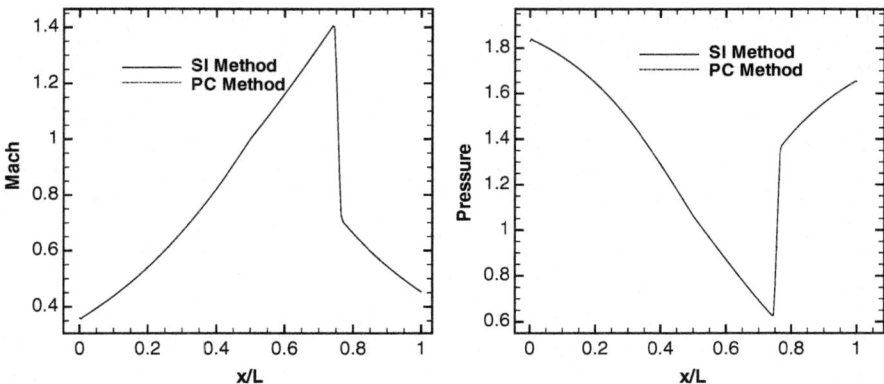

Fig. 37 Nozzle flow with uncertain γ (uniform pdf). Computed mean distribution for the Mach number (left) and the static pressure (right) using the semi-intrusive method with 10 points in the probabilistic space and the PC method with a 10th order polynomial

(with the normalized geometric domain that varies from 0 to 1). A preliminary convergence study with respect to the stochastic estimation has been realized, by using an increasing refinement of the probabilistic space discretization in the case of the SI method, and an increasing polynomial order in the case of PC method. The probabilistic space discretization varies from 5 to 160 points (5, 10, 20, 40, 80, 160), while the polynomial order varies from 2 to 100. Next, the stochastic solutions are compared by computing the mean and the variance of the Mach number and pressure distributions along the nozzle using various choices of pdf for γ. Finally, a comparison in terms of computational cost is performed by computing error ϵ_{L^2} with respect to x.

In Fig. 37, the mean solutions of Mach number and the pressure along the 1D nozzle are reported, where the mean stochastic solutions are computed with the SI method using 10 points in the probabilistic space and the PC method using a 10th order polynomial, with γ described by a uniform pdf (γ varying between 1.33 and 1.47). As it can be observed in Fig. 37, the mean flow is characterized by an isentropic region of increasing speed or Mach number between $x = 0$ and the mean shock location in the divergent (the flow becoming supersonic at the nozzle throat located at $x = 0.5$), followed by a subsonic flow behind the shock with decreasing speed. The mean solutions computed by the two UQ methods are coincident. Next, the standard deviation of the Mach number is computed along the nozzle by using different refinement levels for the probabilistic space in the case of the SI method and different polynomial orders in the case of the PC method, always keeping a uniform pdf for γ. In Table 6, the number of samples required to reach a prescribed error ε_{L^2} is reported for each strategy. We observe that SI method demands fewer points in the stochastic space for a given level of error.

Next, a discontinuous pdf is considered for the stochastic γ. It is interesting to note the innovative contribution the SI method can bring with respect to the PC

Table 6 Number of samples
required for the 1-uncertainty
nozzle problem, uniform pdf

Error level ε_{L^2}	MC-SOBOL	PC	SI
10^{-1}	5	6	5
10^{-2}	24	19	10
10^{-3}	70	59	40

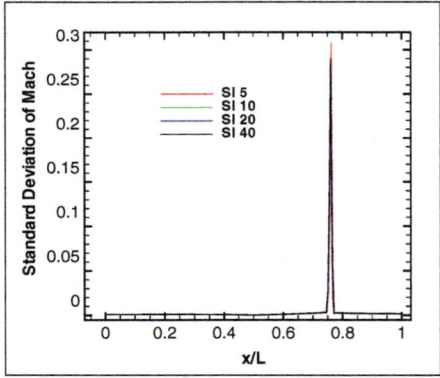 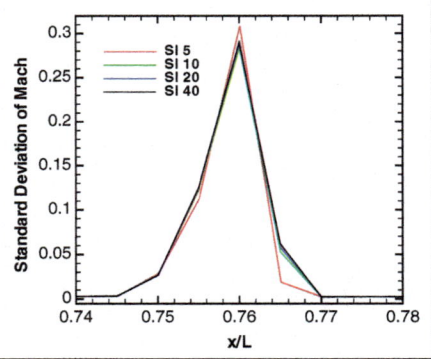

Fig. 38 Nozzle flow with uncertain γ (discontinuous pdf). Convergence study for the standard deviation on the Mach number distribution computed using the SI method

method (in its classical version). To this end, in Fig. 38 the standard deviation of Mach number is reported along the nozzle when the discontinuous pdf (87) is considered. Note that choosing (87) to describe the random variable γ introduces no change whatsoever in the application of the SI method (while the PC method can no longer be used). The standard deviation of the Mach number distribution computed for this discontinuous pdf is plotted in Fig. 38 for several levels of discretization refinement in the probabilistic space: here again the result can be considered as almost converged with no more than a 40-point discretization and fully converged with a 80-point discretization. In Fig. 39, the standard deviation of the Mach number is reported along the nozzle for the discontinuous pdf by using SI and MC-SOBOL methods. The standard deviation distributions computed by means of SI and MC-SOBOL are coincident, even for the maximal standard deviation. The stochastic estimation remains globally very similar for the newly proposed SI approach and the well-established MC-SOBOL method, which allows to validate the SI method results for the case of a discontinuous pdf on γ. Let us estimate the respective computational cost of SI, MC-SOBOL for this case. In Table 7, the number of samples required to reach a prescribed error for ϵ_{L^2} is reported for SI and MC-SOBOL methods. A drastic reduction of the computational cost is obtained by using SI methods with respect to MC-SOBOL solutions.

Next, a two-uncertainties stochastic problem is considered by assuming a discontinuous pdf for γ and a uniform pdf for p_e. In Fig. 40, the standard deviation of the Mach is reported along the nozzle for SI and MC-SOBOL. The standard deviation distributions computed by means of SI and MC-SOBOL are coincident.

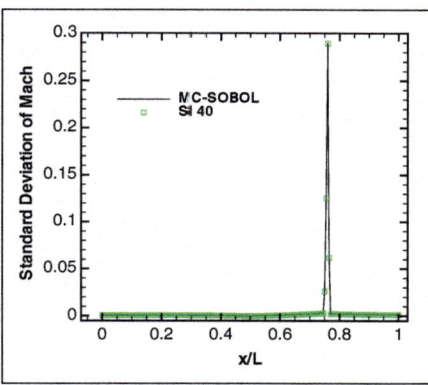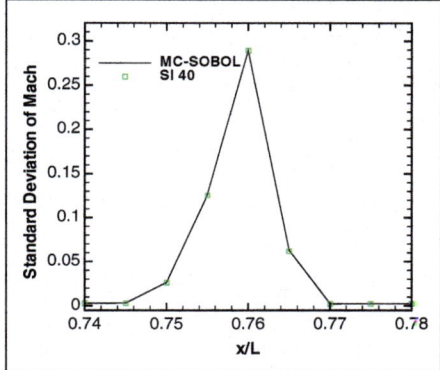

Fig. 39 Nozzle flow with uncertain γ (discontinuous pdf). Standard deviation for the Mach number distribution for MC-SOBOL and SI methods. Left: global view; right: close-up on the shock region

Table 7 Number of samples required for the 1-uncertainty nozzle problem, discontinuous pdf

Error level ε_{L^2}	MC-SOBOL	SI
10^{-1}	4	5
10^{-2}	42	20
10^{-3}	250	40

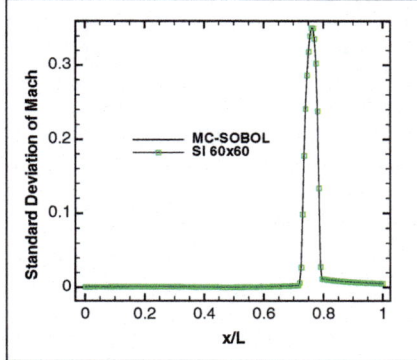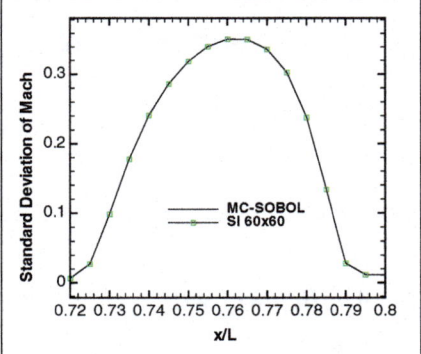

Fig. 40 Nozzle flow with uncertain γ (discontinuous pdf) and p_e (uniform pdf). Standard deviation for the Mach number distribution for MC-SOBOL and SI methods. Left: global view; right: close-up on the shock region

As shown in Table 8, SI method allows strongly reducing the computational cost until six times with respect to MC-SOBOL method.

Table 8 Number of samples
required for the
2-uncertainties nozzle
problem, discontinuous pdf

Error level ε_{L^2}	MC-SOBOL	SI
10^{-1}	35	25
10^{-2}	1000	400
10^{-3}	20,000	3600

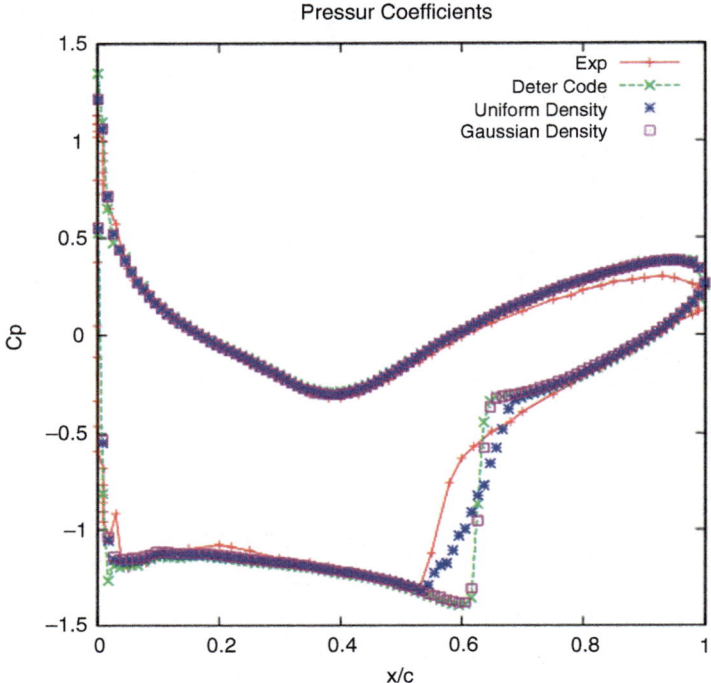

Fig. 41 RAE 2822 airfoil computed with a stochastic residual distribution method

5.2 Application with Other Schemes

The use of the SI/SFV method is not restricted to finite volume or discontinuous
Galerkin schemes, but it is indeed very general. To show this we give the example
of the residual distribution schemes of [1, 4] which can be seen as continuous finite
element methods with non linear stabilisation. Starting from a deterministic method,
one can again write the scheme for any random event and take a conditional average
on any of the stochastic finite volume, in exactly the same spirit as it is sketched in
Sect. 2. Consider for example the flow around RAE 2886 airfoil, with free-stream
Mach number $M_\infty = 0.8$ and velocity U_∞ which is given with 2% of fluctuation
with uniform or centered Gaussian law. The results are presented in Fig. 41. Again,
we see that using approximately five cells in the stochastic direction is enough.

5.3 Overcoming the Curse of Dimensionality

In Sect. 3, the cost of the method has been analysed, and we see exponential growth with respect to the number of random variables. In order to (partially) tackle this problem, a technique issued from the Multi Resolution Analysis of A. Harten [18] has been proposed. The idea is to use multi-resolution analysis in the stochastic dimensions. This technique allows to store only the needed information to reconstruct the random variables, leading potentially to important saving in memory, to the price of added complexity. One can consult [6, 12] for more details.

5.4 Applications for Multiphase Flows

Because of the flexibility of the method, the same technique has been used in multiphase simulation. In [30], a discrete equation method [5] for the simulation of compressible multiphase flows including real-gas effects is coupled to the SI algorithm, using a complex equation of state for both phases. This method is applied to the computational study on the occurrence of rarefaction shock waves (RSW) in a two-phase shock tube with dense vapors of complex organic fluids. Previous studies have shown that a RSW is relatively weak in a single-phase (vapor) configuration, its occurrence and intensity are investigated considering the influence of the initial volume fraction, initial conditions and the thermodynamic model.

6 Conclusions

We have presented the scheme of the Stochastic Finite Volume method (SFVM) and demonstrated the efficiency of the Karhunen–Loève-based adaptation algorithm to construct the anisotropic mesh in the stochastic space. Several application of this generic method has been proposed, from simple ODEs to the fluid mechanics equations. The error estimates for SFVM have been derived. The extension of the SFVM for the DG approximation in the physical space has been proposed. Various numerical examples demonstrating the efficiency and robustness of the implemented algorithms have been presented.

The SFV method studied in this paper appears to be a flexible and effective approach to the solution of stochastic conservation laws. We have shown that the SFV method it is applicable for the uncertainty quantification in a variety of complex problems including systems of conservation laws with random flux coefficients and initial data. The proper adaptation of the stochastic grid significantly reduces the computational cost of the method and improves its convergence.

Acknowledgements The authors acknowledge the contributions of Siddhartha Mishra and Christoph Schwab (Seminar for Applied Mathematics, ETH Zurich, Switzerland) to this work. Pietro Congedo (INRIA, France) and Maria Giovanna Rodio (CEA, France) are also warmly acknowledged for their contributions in the development of this method. This work has been funded in part by the SNF grant #200021_153604 "High fidelity simulation for compressible materials" as well as the AdG ERC Grant "ADDECCO" (contract #226316).

References

1. R. Abgrall, Essentially non oscillatory residual distribution schemes for hyperbolic problems. J. Comput. Phys. **214**(2), 773–808 (2006)
2. R. Abgrall, A simple, flexible and generic deterministic approach to uncertainty quantification in non-linear problems. Technical Report 00325315, INRIA (2007)
3. R. Abgrall, P.M. Congedo, A semi-intrusive deterministic approach to uncertainty quantifications in non-linear fluid flow problems. J. Comput. Phys. **235**, 828–845 (2013)
4. R. Abgrall, D. de Santis, Linear and non-linear high order accurate residual distribution schemes for the discretization of the steady compressible Navier-Stokes equations. J. Comput. Phys. **283**, 329–359 (2015)
5. R. Abgrall, R. Saurel, Discrete equations for physical and numerical compressible multiphase mixtures. J. Comput. Phys. **186**(2), 361–396 (2003)
6. R. Abgrall, P.M. Congedo, G. Geraci, A one-time truncate and encode multiresolution stochastic framework. J. Comput. Phys. **257**, 19–56 (2014)
7. N. Agarwal, N.R. Aluru, A domain adaptive stochastic collocation approach for analysis of MEMs under uncertainties. J. Comput. Phys. **228**, 7662–7688 (2009)
8. T.J. Barth, On the propagation of statistical model parameter uncertainty in CFD calculations. Theor. Comput. Fluid Dyn. **26**(435–457) (2012)
9. B. Cockburn, C.W. Shu, The Runge-Kutta discontinuous Galerkin method for conservation laws V: multidimensional systems. J. Comput. Phys. **141**, 199–224 (1998)
10. C.M. Dafermos, *Hyperbolic Conservation Laws in Continuum Physics*. Fundamental Principles of Mathematical Sciences, vol. 325 (Springer, Berlin, 2010)
11. J. Foo, X. Wan, G.E. Karniadakis, The multi-element probabilistic collocation method (ME-PCM): error analysis and applications. J. Comput. Phys. **227**(22), 9572–9595 (2008)
12. G. Geraci, P.M. Congedo, R. Abgrall, G. Iaccarino, A novel weakly-intrusive non-linear multiresolution framework for uncertainty quantification in hyperbolic partial differential equations. J. Sci. Comput. **66**(1), 358–405 (2016)
13. M. Gerritsma, J. van der Steen, P. Vos, G.E. Karniadakis, Time-dependent generalized polynomial chaos. J. Comput. Phys. **229**(22), 8333–8363 (2010)
14. R. Ghanem, P. Spanos (eds.), *Stochastic Finite Elements: A Spectral Approach* (Dover, New York, 2003)
15. E. Godlewski, P. Raviart, *Hyperbolic Systems of Conservation Laws* (Ellipses, Paris, 1995)
16. E. Godlewski, P.-A. Raviart, *Numerical Approximation of Hyperbolic Systems of Conservation Laws*. Applied Mathematical Sciences, vol. 118 (Springer, Berlin, 1996)
17. D. Gottlieb, D. Xiu, Galerkin method for wave equations with uncertain coefficients. Commun. Comput. Phys. **3**, 505–518 (2008)
18. A. Harten, Multiresolution representation of data: a general framework. SIAM J. Numer. Anal. **33**(3), 1205–1256 (1996)
19. R.H. Kraichnan, Direct-interaction approximation for a system of several interacting simple shear waves. Phys. Fluids **6**(11), 1603–1609 (1963)
20. A. Kuznetsov, Accuracy of some approximate methods for computing the weak solutions of a first-order quasilinear equation. USSR Comput. Math. Math. Phys. **16**, 105–119 (1976)
21. R. LeVeque, *Numerical Methods for Conservation Laws* (Birkhäuser, Berlin, 1992)

22. G. Lin, C.-H. Su, G.E. Karniadakis, Predicting shock dynamics in the presence of uncertainties. J. Comput. Phys. **217**, 260–276 (2006)
23. G. Lin, C.-H. Su, G.E. Karniadakis, Stochastic modelling of random roughness in shock scattering problems: theory and simulations. Comput. Methods Appl. Mech. Eng. **197**, 3420–3434 (2008)
24. X. Ma, N. Zabaras, An adaptive hierarchical sparse grid collocation algorithm for the solution of stochastic differential equations. J. Comput. Phys. **228**, 3084–3113 (2009)
25. S. Mishra, Ch. Schwab, Sparse tensor multi-level monte carlo finite volume methods for hyperbolic conservation laws with random initial data. Math. Comput. **81**, 1979–2018 (2012)
26. S. Mishra, Ch. Schwab, J. Šukys, Multi-level monte carlo finite volume methods for nonlinear systems of conservation laws in multi-dimensions. J. Comput. Phys. **231**, 3365–3388 (2012)
27. S. Mishra, N.H. Risebro, C. Schwab, S. Tokareva, Numerical solution of scalar conservation laws with random flux functions. SIAM/ASA J. Uncertain. Quantif. **4**, 552–591 (2016)
28. S.A. Orszag, L.R. Bissonnette, Dynamical properties of truncated Wiener-Hermite expansions. Phys. Fluids **10**(12), 2603–2613 (1967)
29. G. Poëtte, B. Després, D. Lucor, Uncertainty quantification for systems of conservation laws. J. Comput. Phys. **228**, 2443–2467 (2009)
30. M.G. Rodio, P.M. Congedo, R. Abgrall, Two-phase flow numerical simulation with real-gas effects and occurrence of rarefaction shock waves. Eur. J. Mech. B. Fluids **45**, 20–35 (2014)
31. Ch. Schwab, S. Tokareva, High order approximation of probabilistic shock profiles in hyperbolic conservation laws with uncertain initial data. ESAIM: Math. Model. Numer. Anal. **47**, 807–835 (2013)
32. J. Smoller, *Shock Waves and Reaction-Diffusion Equations*. Fundamental Principles of Mathematical Sciences, vol. 258 (Springer, Berlin, 1994)
33. H.A. Sturges, The choice of a class interval. J. Am. Stat. Assoc. **21**, 65–66 (1926)
34. S. Tokareva, Ch. Schwab, S. Mishra, High order SFV and mixed SDG/FV methods for the uncertainty quantification in multidimensional conservation laws, in *High Order Nonlinear Numerical Schemes for Evolutionary PDEs*, ed. by R. Abgrall, H. Beaugendre, P. Congedo, C. Dobrzynski, V. Perrier, M. Ricchiuto. Lecture Notes in Computational Sciences and Engineering, vol. 99 (Springer, Berlin, 2014)
35. J. Troyen, O. Le Maître, M. Ndjinga, A. Ern, Intrusive Galerkin methods with upwinding for uncertain nonlinear hyperbolic systems. J. Comput. Phys. **229**, 6485–6511 (2010)
36. J. Troyen, O. Le Maître, M. Ndjinga, A. Ern, Roe solver with entropy corrector for uncertain hyperbolic systems. J. Comput. Phys. **235**, 491–506 (2010)
37. X. Wan, G.E. Karniadakis, An adaptive multi-element generalized polynomial chaos method for stochastic differential equations. J. Comput. Phys. **209**(2), 617–642 (2005)
38. X. Wan, G.E. Karniadakis, Multi-element generalized polynomial chaos for arbitrary probability measure. SIAM J. Sci. Comput. **28**(3), 901–928 (2006)

Uncertainty Modeling and Propagation in Linear Kinetic Equations

Guillaume Bal, Wenjia Jing, and Olivier Pinaud

Abstract This paper reviews recent work in two complementary aspects of uncertainty quantification of linear kinetic models. First, we review the modeling of uncertainties in linear kinetic equations as high frequency limits of models for wave propagations in heterogeneous media. Second, we analyze the propagation of stochasticity from the constitutive coefficients to the solutions of linear kinetic equations. Such uncertainty quantifications find many important applications, e.g. in physics-based modeling of errors in inverse problems measurement.

1 Introduction

Uncertainties in measurements of interest come from a very broad spectrum of reasons. One such component arises from the propagation of uncertainty in the constitutive coefficients of a differential equation to the solutions of said equation. Some recent results obtained in the context of elliptic equations were summarized in the review [4]. Here, we consider such a propagation in the context of (phase space linear) transport equations and summarize results obtained primarily by the authors.

G. Bal (✉)
Department of Applied Physics and Applied Mathematics, Columbia University, New York, NY, USA
e-mail: gb2030@columbia.edu

W. Jing
Yau Mathematical Sciences Center, Tsinghua University, Haidian District, Beijing, People's Republic of China
e-mail: wjjing@mail.tsinghua.edu.cn

O. Pinaud
Department of Mathematics, Colorado State University, Fort Collins, CO, USA
e-mail: pinaud@math.colostate.edu

© Springer International Publishing AG, part of Springer Nature 2017
S. Jin, L. Pareschi (eds.), *Uncertainty Quantification for Hyperbolic and Kinetic Equations*, SEMA SIMAI Springer Series 14,
https://doi.org/10.1007/978-3-319-67110-9_2

Two different propagations, occurring at physically different scales, are presented. The first one concerns the derivation of the transport equation from models of high frequency waves propagating in heterogeneous (scattering) media. Transport equations are deterministic models for wave field correlations (or wave field energies) of waves propagating in heterogeneous media modeled as random media. Their derivation may be seen as a homogenization (law of large number) result for phase space descriptions of field-field correlations; see the review [19]. As such, however, the field-field correlations are modeled by a deterministic equation even though the underlying wave fields are inherently random. Characterizing the random fluctuations (random corrections) in the field-field correlations remains a relatively little studied subject. Section 2 present several results obtained in this direction.

Once a kinetic model has been derived, either as an approximation for the energy of wave fields as described above or by any other means, it typically involves constitutive coefficients, such as scattering and absorbing coefficients, that depend on phase space variables and are typically not perfectly known. Such uncertainties have an effect on the transport solution. Recent results obtained in this direction are summarized in Sect. 4 after relevant material and notation on the transport equation are presented in Sect. 3. Several results in Sect. 4 are based on moment estimates proved in [5] for specific random models. These moment estimates are presented in detail and generalized to a large class of sufficiently mixing coefficients in Sect. 4.4.2.

The characterization of the random fluctuations in a transport solution is a problem of independent interest, and allows us to quantify the uncertainty in various functionals of the transport solution of interest. As for the elliptic case considered in [4], we mention two additional applications. The first one pertains to the calibration of upscaling numerical codes. We refer the reader to [8, 9] for such applications in the context of elliptic equations. The second one concerns inverse problems. In typical inverse problems, the reconstruction of the high frequency of the constitutive coefficients is unaccessible from inevitably noisy measurements. Yet, such not-reconstructed components, which we may as well model as random, have an influence on the solutions and hence the available measurements. Uncertainty propagation provides quantitative, physics-based, models for such an influence and allow for more accurate reconstructions of the low frequency components of the coefficients. For an application in the reconstruction of potentials from spectral information, see [15].

2 Uncertainties in the Derivation of Kinetic Equations

2.1 Setting of the Problem

In the context of wave propagation in heterogeneous media, kinetic equations generally describe quadratic quantities in the wavefield, for instance the wave energy. They are derived in the high frequency limit, and offer therefore an approximate description of the propagation up to some errors due to the finiteness of the frequency. Our goal in this section is to quantify these errors, in particular to obtain optimal convergence estimates, and when possible to characterize the first order corrector. We first review in Sect. 2.3 the derivation of transport models and address the corrector analysis in Sect. 2.4.

Kinetic models can be derived for several types of waves, e.g. acoustic, electromagnetic, quantum, and elastic waves. We will focus here on acoustic waves described by the scalar wave equation; see [47] for more general models. For $p(t, x)$ the wavefield, $\rho(x)$ the medium density, and $\kappa(x)$ its compressibility, our starting point is the wave equation

$$\frac{\partial^2 p}{\partial t^2} = \kappa(x)^{-1} \nabla \cdot \rho(x)^{-1} \nabla p + f(t, x), \qquad x \in \mathbb{R}^d, \ t > 0, \tag{1}$$

supplemented with initial conditions $p(t = 0, x)$ and $\frac{\partial p}{\partial t}(t = 0, x)$. Above, f is a source term and d is spatial dimension. While the large scale features of the underlying heterogeneous medium are often known, or at least can be reconstructed, the fine details might not be accessible and are therefore modeled by a random medium with a given statistics. We then assume the following form for ρ and κ: $\rho(x) = \rho_0 = 1$ for simplicity (generalizations are possible), and

$$\kappa(x)^{-1} = \kappa_0(x)^{-1} \left(1 + \sigma_0 V\left(\frac{x}{\ell_c}\right)\right).$$

In the latter equation, κ_0 is the background compressibility modeling the large scale structure of the medium (that we recall is supposed to be known), and V accounts for random fluctuations of strength σ_0 and correlation length ℓ_c, which model the fine details. The term V is a mean zero, stationary random field with correlation function

$$\mathbb{E}\{V(x)V(y)\} = R(x - y).$$

Above, $\mathbb{E}\{\cdot\}$ denotes ensemble average over the different realizations of the random medium.

We will present the kinetic equations derived from (1) in Sect. 2.3 further. The wave equation is often reduced to a simpler model of propagation in order to make the mathematical analysis more amenable. This is done in the paraxial approximation that we introduce below.

2.2 The Paraxial Regime

The main assumption in this regime is that the waves propagate along a privileged direction, say z, and that backscattering is negligible. We then write $x = (z, x_\perp)$ accordingly, for $x_\perp \in \mathbb{R}^{d-1}$. We suppose moreover that κ_0 is constant, and introduce

$c(x) = (\rho(x)\kappa(x))^{-1/2}$ as well as $c_0 = (\rho_0\kappa_0)^{-1/2}$. We also assume that the source f is supported in the region $z < 0$, and that the initial conditions vanish, that is $p(t = 0, x) = 0$ and $\frac{\partial p}{\partial t}(t = 0, x) = 0$. For $\hat{p}_\omega(z, x_\perp)$ the Fourier transform of p with respect to the variable t (after appropriate extension to $t < 0$) and with dual variable ω, we obtain the following Helmholtz equation:

$$\partial_z^2 \hat{p}_\omega + \Delta_{x_\perp} \hat{p}_\omega + \frac{\omega^2}{c(x)^2} \hat{p}_\omega = 0, \qquad z > 0. \tag{2}$$

For $k_\omega = \omega/c_0$, plugging the ansatz $\hat{p}_\omega(z, x_\perp) = e^{ik_\omega z} \psi_\omega(z, x_\perp)$ in (2), where the function ψ_ω is assumed to vary slowly in the z variable, and neglecting as a consequence the term $\partial_z^2 \psi_\omega$, one obtains the Schrödinger equation

$$ik_\omega \partial_z \psi_\omega + \Delta_{x_\perp} \psi_\omega + \sigma_0 k_\omega^2 V \psi_\omega = 0, \qquad z > 0. \tag{3}$$

The equation is augmented with an initial condition $\psi_\omega(z = 0, x_\perp) = \psi_\omega^0(x_\perp)$, that depends on the source term f. See [22, 50] for more details about the paraxial approximation in heterogeneous media.

In the next section, we present the kinetic models obtained from asymptotics of (1) and (3). We start with the wave equation, and continue with the Schrödinger equation.

2.3 High Frequency Limit

2.3.1 The Wave Equation

We give here a formal derivation following the lines of [47]. Comments and references about rigorous results are given at the end of the section. We begin with the scalings.

We suppose here that the source term vanishes, i.e. $f = 0$, with non zero initial conditions. The kinetic limit is done in the regime of weak coupling [26, 48], where it is assumed that the strength of the fluctuations σ_0 is weak and that the correlation length of the medium ℓ_c and the wavelength of the initial condition λ are same order. The stochastic homogenization case $\lambda \gg \ell_c$ leads to waves propagating in an effective medium, see [2], while the case $\lambda \ll \ell_c$ leads to random Liouville equations [17]. If L is a typical distance of propagation, we then set

$$\frac{\ell_c}{L} = \frac{\lambda}{L} = \sigma_0^2 = \varepsilon \ll 1.$$

The fact that $\sigma_0 = \sqrt{\varepsilon}$ ensures the random medium has a non negligible effect at the macroscopic level. We rewrite (1) to obtain the following high frequency wave equation:

$$\frac{\partial^2 p^\varepsilon}{\partial t^2} = c_\varepsilon^2(x)\Delta p^\varepsilon, \qquad p^\varepsilon(t=0,x) = p_0\left(\frac{x}{\varepsilon}\right), \qquad \frac{\partial p^\varepsilon}{\partial t}(t=0,x) = p_1\left(\frac{x}{\varepsilon}\right),$$
(4)

where $c_\varepsilon(x) = (\rho_0 \kappa_\varepsilon(x))^{-1/2}$ with

$$\kappa_\varepsilon(x)^{-1} = \kappa_0(x)^{-1}\left(1 + \sqrt{\varepsilon}V\left(\frac{x}{\varepsilon}\right)\right).$$

The asymptotic analysis of (4) as $\varepsilon \to 0$ is done by means of Wigner transforms. We recast first the scalar wave equation as a first-order hyperbolic system on the acoustic field $\mathbf{u}^\varepsilon = (\mathbf{v}^\varepsilon, p^\varepsilon)$, where $\mathbf{v}^\varepsilon$ is velocity, and obtain the following system:

$$\rho_0 \frac{\partial \mathbf{v}^\varepsilon}{\partial t} + \nabla p^\varepsilon = 0, \qquad \kappa_\varepsilon \frac{\partial p^\varepsilon}{\partial t} + \nabla \cdot \mathbf{v}^\varepsilon = 0.$$

The system is augmented with initial conditions $p^\varepsilon(t=0,x)$ and $\mathbf{v}^\varepsilon(t=0,x) = \nabla\varphi^\varepsilon(x)$ where the pressure potential φ^ε is obtained by solving

$$\Delta\varphi^\varepsilon = -\kappa_\varepsilon \frac{\partial p^\varepsilon}{\partial t}(t=0,\cdot).$$

Kinetic Equations

Wigner transforms provide a phase space description of the propagation of the wave energy, see [36, 44] for a detailed mathematical analysis of their properties. In the high frequency limit, the wave energy satisfies transport equations whose constitutive parameters are deduced from the sound speed c_ε. The Wigner transform of the field $\mathbf{u}^\varepsilon$ is defined as the following matrix-valued function,

$$W^\varepsilon(t,x,k) = \frac{1}{(2\pi)^d}\int_{\mathbb{R}^d} e^{ik\cdot y}\, \mathbf{u}^\varepsilon\left(t, x - \frac{\varepsilon y}{2}\right) \otimes \mathbf{u}^\varepsilon\left(t, x + \frac{\varepsilon y}{2}\right) dy.$$

It is shown in [47] that in the limit $\varepsilon \to 0$, the expectation of the Wigner transform $\mathbb{E}\{W^\varepsilon\}$ converges to a measure W admitting the following decomposition (there are no vortical modes because of the form of the initial condition):

$$W(t,x,k) = a^+(t,x,k)\,\mathbf{b}^+(k) \otimes \mathbf{b}^+(k) + a^-(t,x,k)\,\mathbf{b}^-(k) \otimes \mathbf{b}^-(k),$$

where we have defined

$$\mathbf{b}^\pm(k) = \frac{1}{\sqrt{2\rho_0}}\begin{pmatrix} \hat{k} \\ \pm\rho_0 c_0^{-1} \end{pmatrix}, \qquad \hat{k} = \frac{k}{|k|}.$$

The amplitude a^+ solves the radiative transfer equation below,

$$\frac{\partial a^+}{\partial t} + c_0 \hat{k} \cdot \nabla_x a^+ + \Sigma(k) a^+ = Q(a^+), \tag{5}$$

where Q and Σ^{-1} are the collision operator and the mean free time, respectively. For δ the Dirac measure, they are given by,

$$Q(a)(k) = \int_{\mathbb{R}^d} a(p)\sigma(k,p)\delta(c_0|p| - c_0|k|)dp,$$

and

$$\Sigma(k) = \int_{\mathbb{R}^d} \sigma(k,p)\delta(c_0|p| - c_0|k|)dp.$$

The cross section $\sigma(k,p)$ appearing in these expressions is

$$\sigma(k,p) = \frac{\pi c_0^2 |k|^2}{2(2\pi)^d}\hat{R}(k-p),$$

where $\hat{R}$ is the Fourier transform of the correlation function R with the convention

$$\hat{R}(p) = \int_{\mathbb{R}^d} e^{-ip\cdot x} R(x)dx.$$

A similar equation is obtained for $a^-(t,x,k) = a^+(t,x,-k)$. It is interesting to notice that the transport equation depends on the fluctuations of the random medium only through its power spectrum $\hat{R}$. The amplitude a^+ is related to the wave energy as follows. Defining the latter by

$$\mathcal{E}^\varepsilon(t,x) = \frac{1}{2}\left(\kappa_\varepsilon(x)(p^\varepsilon(t,x))^2 + \rho_0|\mathbf{v}^\varepsilon(t,x)|^2\right),$$

we have

$$\lim_{\varepsilon \to 0} \mathbb{E}\{\mathcal{E}^\varepsilon\}(t,x) = \int_{\mathbb{R}^d} a^+(t,x,k)dk.$$

The initial condition for (5) is the limit of the Wigner transform of the initial condition after appropriate projection, see [47].

Rigorous Results

The main question is to justify of the convergence of W^ε to W, and to define in which sense this takes place. This is a difficult matter, and up to our knowledge there are only two references in the literature. The article [45] deals with lattice waves, described by an equation of the form (4) with Laplacian replaced by its finite differences approximation. Under various assumptions on the random potential V, it is shown that the average $\mathbb{E}\{W^\varepsilon\}$ converges in the distribution sense to W. The thesis [25] concerns precisely (4). The obtained results are much stronger: it is proved that the random process W^ε converges almost surely and in the distribution sense to the deterministic W. The quantity W^ε is then referred to as statistically stable, in the sense that it is weakly random for ε small. Such a property is at the core of transport-based imaging techniques in random media, see [12, 14]. Note that the proofs presented in [45] and [25] are technically involved and based on diagrammatic expansions.

2.3.2 The Schrödinger Equation

The analysis is somewhat simpler than in the previous case. There are many possible scalings for (3), we present here only one and point the reader to [19] for other regimes. We consider a semi-classical Schrödinger equation of the form

$$i\varepsilon\partial_z\psi_\varepsilon + \frac{\varepsilon^2}{2}\Delta_{x_\perp}\psi_\varepsilon + \sqrt{\varepsilon}V\left(\frac{z}{\varepsilon}, \frac{x_\perp}{\varepsilon}\right)\psi_\varepsilon = 0, \qquad z > 0, \tag{6}$$

augmented with $\psi_\varepsilon(z = 0, x_\perp) = \psi_\varepsilon^0(x_\perp)$. The Wigner transform is now scalar and given by

$$W^\varepsilon(z, x_\perp, k) = \frac{1}{(2\pi)^{d-1}}\int_{\mathbb{R}^{d-1}} e^{ik\cdot y}\,\psi_\varepsilon(z, x_\perp - \frac{\varepsilon y}{2})\psi_\varepsilon^*(z, x_\perp + \frac{\varepsilon y}{2})\, dy.$$

It is then shown that $\mathbb{E}\{W^\varepsilon\}$ converges to W, solution to

$$\frac{\partial W}{\partial z} + k\cdot\nabla_{x_\perp}W + \Sigma(k)W = Q(W), \tag{7}$$

where Q and Σ now read

$$Q(W)(k) = \frac{1}{(2\pi)^{d-1}}\int_{\mathbb{R}^{d-1}} \hat{R}\left(\frac{|p|^2}{2} - \frac{|k|^2}{2}, p - k\right) W(p)dp$$

$$\Sigma(k) = \frac{1}{(2\pi)^{d-1}}\int_{\mathbb{R}^{d-1}} \hat{R}\left(\frac{|p|^2}{2} - \frac{|k|^2}{2}, p - k\right) dp.$$

In these definitions, the power spectrum $\hat{R}(\omega, p)$ is still the Fourier transform of $R(x)$, written as $R(x) = R(z, x_\perp)$. The initial condition for (7) is the limit of the Wigner transform of the initial condition ψ_ε^0.

Rigorous Results

More results are available than in the case of the wave equation, see [19] for a review. It is generally assumed that the random potential V is either a Markov process in the z variable, or a gaussian field. With these hypotheses, the proof of convergence follows from martingale and perturbed test functions methods, see e.g. [16, 32, 37]. The convergence holds in probability and in distribution. When V is independent of z, the analysis is more delicate, and a proof of convergence of $\mathbb{E}\{W^\varepsilon\}$ can be found in [31].

2.4 Corrector Analysis

Now that the convergence of W^ε to W is established, we turn to the main topic of this section of the chapter, that is the error analysis. We are not aware of any results about the wave equation, and concentrate therefore on the paraxial regime. The first set of results concerns an even simpler description of the wave propagation offered by the Itô-Schrödinger equation. In a second step, we present the results that are available for more general models.

2.4.1 The Itô-Schrödinger Regime

In this regime, the fluctuations of the random medium in the direction of propagation are supposed to be much faster than in the transverse plane. After an appropriate rescaling, and based on central limit type arguments, the random potential can be approximated (in a statistical sense) by a Brownian field. This leads to a Schrödinger equation of the form

$$id\psi_\varepsilon + \frac{\varepsilon}{2}\Delta_{x_\perp}\psi_\varepsilon + \psi_\varepsilon \circ dB\left(z, \frac{x_\perp}{\varepsilon}\right) = 0, \qquad z > 0, \tag{8}$$

where $\circ$ stands for the Stratonovich product and B is a Brownian field with autocorrelation

$$\mathbb{E}\{B(z, x'_\perp)B(z', x_\perp + x'_\perp)\} = \min(z, z')R_0(x_\perp).$$

This is the most amenable regime for an error analysis since Itô calculus can be used and yields closed-form equations. See [1, 33, 38] for a rigorous derivation of the Itô-Schrödinger equation. With W^ε the Wigner transform as before and W its limit, the goal is to quantify the error $W^\varepsilon - W$ as $\varepsilon \to 0$.

Full Characterization of the Corrector for a Particular Initial Condition

The following results are taken from [43]. From (8), direct calculations show that the Wigner transform satisfies the equation

$$dW^\varepsilon(z, x_\perp, k) + k \cdot \nabla_{x_\perp} W^\varepsilon(z, x_\perp, k)$$
$$= \frac{1}{(2\pi)^{d-1}} \int_{\mathbb{R}^{d-1}} e^{ip \cdot x_\perp / \varepsilon} \left(W^\varepsilon(z, x_\perp, k - \frac{p}{2}) - W^\varepsilon(z, x_\perp, k + \frac{p}{2}) \right) \circ d\hat{B}(z, p),$$

where $\hat{B}(z, p)$ is the Fourier transform of $B(z, x_\perp)$ in the variable $x_\perp$. Suppose that initial Wigner transform satisfies

$$W^\varepsilon(z = 0, x_\perp, k) = \delta(x_\perp)\varphi(k), \tag{9}$$

where δ is the Dirac measure and φ is a smooth function, and define the corrector

$$Z^\varepsilon(z, x_\perp, k) = \frac{W^\varepsilon(z, x_\perp, k) - W(z, x_\perp, k)}{\sqrt{\varepsilon}}.$$

The main result of [43] characterizes the limit of Z^ε as follows, see therein for the mathematical details and technical assumptions: the process Z^ε converges weakly in law to a process Z solution to the radiative transfer equation

$$\frac{\partial Z(z, x_\perp, k)}{\partial z} + k \cdot \nabla_{x_\perp} Z(z, x_\perp, k)$$
$$= \frac{1}{(2\pi)^{d-1}} \int_{\mathbb{R}^{d-1}} \hat{R}_0(k - p) \left(Z(z, x_\perp, p) - Z(z, x_\perp, k) \right) dp, \tag{10}$$

with initial condition

$$Z(0, x_\perp, k) = \delta(x_\perp)X(k),$$

where $X(k)$ is a distribution valued Gaussian random variable. Its (somewhat complex) expression can be found in [43], the main information that it yields is that X is linear in the random potential B. The result can then be interpreted as follows: the leading instabilities are created in a boundary layer around the initial position $z = 0$, and then propagate according to the kinetic equation (10). These instabilities are generated by the single scattering of the ballistic part of the limit W (i.e. the part of W that propagates freely in the medium and is exponentially decreasing) by the potential B.

This behavior is actually not universal as we will see in the next section. Indeed, other types of initial conditions lead to instabilities of different amplitudes and nature.

Characterization of the Covariance of the Corrector for More General Initial Conditions

The results presented here are taken from [11]. The error $W^\varepsilon - W$ is recast as $W^\varepsilon - \mathbb{E}\{W^\varepsilon\} + \mathbb{E}\{W^\varepsilon\} - W$. It is direct to see that the second piece is simply the propagation via the radiative transfer equation of the difference between the initial Wigner transform and its limit. We therefore focus on the first contribution $W^\varepsilon - \mathbb{E}\{W^\varepsilon\}$ to the error that is the most interesting. We analyze the covariance of the corrector $W^\varepsilon - \mathbb{E}\{W^\varepsilon\}$ and not the process itself, which allows us to consider more general settings. More specifically, we introduce the scintillation function J^ε defined by

$$J^\varepsilon(z, x_\perp, k, x'_\perp, k') = \mathbb{E}\{(W^\varepsilon - \mathbb{E}\{W^\varepsilon\})(z, x_\perp, k)(W^\varepsilon - \mathbb{E}\{W^\varepsilon\})(z, x'_\perp, k')\}.$$

In the Itô-Schrödinger regime, J^ε satisfies a closed-form equation (it does not for other classes of potentials), that reads

$$\left(\frac{\partial}{\partial z} + \mathscr{T}_2 + 2R_0(0) - Q_2 - \mathscr{K}_\varepsilon\right)J^\varepsilon = \mathscr{K}_\varepsilon a_\varepsilon \otimes a_\varepsilon, \tag{11}$$

equipped with vanishing initial conditions $J^\varepsilon(0, x_\perp, k, x'_\perp, k') = 0$ when the initial condition of the Schrödinger equation is deterministic. Here, we have defined

$$a_\varepsilon = \mathbb{E}\{W^\varepsilon\}$$
$$\mathscr{T}_2 = k \cdot \nabla_{x_\perp} + p \cdot \nabla_{x'_\perp}$$
$$Q_2 h = \int_{\mathbb{R}^{2(d-1)}} \left(\hat{R}_0(k - p)\delta(k' - p') + \hat{R}_0(k' - p')\delta(k - p)\right) \frac{h(x_\perp, p, x'_\perp, p')dpdp'}{(2\pi)^{2(d-1)}}$$
$$\mathscr{K}_\varepsilon h = \sum_{\epsilon_i, \epsilon_j = \pm 1} \epsilon_i \epsilon_j \int_{\mathbb{R}^{2(d-1)}} \hat{R}_0(u)e^{i\frac{(x_\perp - x'_\perp) \cdot u}{\varepsilon}} h\left(x_\perp, k + \epsilon_i \frac{u}{2}, x'_\perp, k' + \epsilon_j \frac{u}{2}\right) \frac{du}{(2\pi)^{d-1}}.$$

Equation (11) is obtained by computing the fourth moment of the wave function. Consider an initial condition of the Schrödinger equation that has the form of a coherent state,

$$\psi_\varepsilon^0(x_\perp) = \frac{1}{\varepsilon^{\frac{(d-1)\alpha}{2}}} \chi\left(\frac{x_\perp}{\varepsilon^\alpha}\right)e^{i\frac{x_\perp \cdot k_0}{\varepsilon}},$$

where $\alpha \in [0, 1]$, χ is a smooth function with compact support, and $k_0 \in \mathbb{R}^{d-1}$ is the direction of propagation in the transverse plane. The associated Wigner transform reads

$$W_0^\varepsilon(x_\perp, k) = \frac{1}{\varepsilon^{d-1}} W_0\left(\frac{x_\perp}{\varepsilon^\alpha}, \frac{k - k_0}{\varepsilon^{1-\alpha}}\right), \tag{12}$$

where W_0 is the Wigner transform of the rescaled initial condition $\psi_{\varepsilon=1}^0$ with $k_0 = 0$. The parameter α measures the concentration of the initial conditions in the spatial variables which, according the uncertainty principle, quantifies as well for coherent states the concentration in the momentum variables. Note that the initial condition (9) of the previous section corresponds essentially to the case $\alpha = 1$.

The limit of J^ε as $\varepsilon \to 0$ for initial conditions of the form (12) was characterized in [11]. We summarize here the most important points and refer the reader to [11] for complete results and formulas. We focus on the physical case $d = 3$.

- The most stable case corresponds to $\alpha = 0$, with a scintillation of order $\varepsilon^{d-1} = \varepsilon^2$, while the least stable case corresponds to $\alpha = 1$, with a scintillation of order ε. This is a somewhat intuitive result as when the support of the initial condition ψ_ε^0 grows, we can expect the instabilities to be averaged over a larger domain.
- When $\alpha > 1/2$, the (appropriately rescaled) limit of J^ε satisfies a kinetic equation of the form (11) with $\mathcal{K}_\varepsilon = 0$ and a non vanishing initial condition. As in the previous section, instabilities are created in a boundary layer around the initial z position and then propagates according to a transport equation. When $\alpha \le 1/2$, the limit of J^ε still satisfies a kinetic equation of the form (11) with $\mathcal{K}_\varepsilon = 0$, with now a vanishing initial condition and a non zero right-hand side in (11). In the latter configuration, instabilities are created as the wave propagates and not just around the initial position.
- There is a transition in the nature of the corrector defined by a critical value $\alpha^\star(\varepsilon)$ solution to the equation $\varepsilon^{2-3\alpha^\star(\varepsilon)} = \log \varepsilon^{2\alpha^\star(\varepsilon)-1}$ ($\alpha^\star$ is close to $2/3$): when $\alpha < \alpha^\star(\varepsilon)$, the source term (when $\alpha \le 1/2$) and the initial condition (for $\alpha > 1/2$) for the transport equation satisfied by the limit of J^ε are quadratic in the power spectrum $\hat{R}_0$; when $\alpha > \alpha^\star(\varepsilon)$, the initial condition is linear in $\hat{R}_0$. This is an interesting result as it shows that the leading instabilities are generated by the fraction of the wave that was scattered at most twice by the random medium. An interpretation of this fact is that instabilities are created by the most singular components of the wave, and since higher order scattering terms are more regular, they lead to negligible contributions.

As a conclusion, the main factor that influences the size of the corrector and its structure is the regularity of the initial condition, which was measured here in terms of concentration in phase space. The instabilities satisfy transport equations driven by either a source term or initial conditions. See [34, 35] for additional references on scintillation analysis. In the next section, we present a few results available in situations other than the Itô-Schrödinger regime we just considered.

2.4.2 Other Regimes

Schrödinger Equations with an Ornstein-Uhlenbeck Potential

It is proved in [42], that as in the Itô-Schrödinger case addressed earlier, the corrector Z^ε converges weakly in law to a process Z_1 that has a similar structure as the limit Z solution to (10). Namely, Z_1 verifies a radiative transfer equation with a random initial condition. The proof is more involved than [43], as now one cannot rely on Itô calculus and one has to resort to diagrammatic expansions.

Schrödinger Equations with z-Independent Random Potentials

This case is the most difficult to study as there is not direct averaging induced by the z component of the potential. The Wigner transform of a solution to (6) satisfies the equation

$$\frac{\partial W^\varepsilon(z, x_\perp, k)}{\partial z} + k \cdot \nabla_{x_\perp} W^\varepsilon(z, x_\perp, k)$$

$$= \frac{i}{(2\pi)^{d-1}\sqrt{\varepsilon}} \int_{\mathbb{R}^{d-1}} e^{ip\cdot x_\perp/\varepsilon} \left(W(z, x_\perp, k - \frac{p}{2}) - W(z, x_\perp, k + \frac{p}{2}) \right) \hat{V}(p) dp,$$

where $\hat{V}$ is the Fourier transform of the potential $V(x_\perp)$. The analysis of the scintillation J^ε is considerably more difficult in this situation since it does satisfy a closed-form equation. Motivated by the results of the previous section where it was shown that only the single and double scattering of the wave contribute to the instabilities in the Itô-Schrödinger regime, the terms in J^ε linear and quadratic in the power spectrum $\hat{R}$ were characterized in [13, 20]. The results are similar to those of the Itô-Schrödinger case as the largest corrector is of order ε and the smallest of order ε^{d-1}. There is also a transition for a critical α in the structure of the instabilities.

The case of long-range correlations in the underlying random medium is addressed in [13, 20]. This situation corresponds to a non-integrable correlation function and is modeled by a power spectrum of the form

$$\hat{R}(k) = \frac{S(k)}{|k|^\gamma}, \qquad 0 < \gamma < d.$$

The central observation made in [13] is that, contrary to the single scattering case of [20] where the scintillation is approximately of order $\sqrt{\varepsilon}$ when $\gamma \sim d$, the scintillation of the double scattering is of order $\varepsilon^{d-\delta}$ and therefore close to one when $\gamma \sim d$. This indicates that correlations in the medium have a strong influence on the size of the corrector. Determining whether or not high order scattering is statistically stable in media with long-range correlations requires the analysis of the whole series of W^ε in terms of $\hat{R}$ and remains an open problem.

3 Transport Equation

For the rest of the paper, we consider the following boundary value problem for the stationary linear transport equation:

$$\begin{cases} v \cdot \nabla_x u(x, v) + a(x)u(x, v) = \int_V k(x, v, v')u(x, v')dv', & (x, v) \in X \times V, \\ u(x, v) = g(x, v), & (x, v) \in \Gamma_-. \end{cases}$$
$$(13)$$

Besides the applications in wave propagation considered in the preceding section, such equations model the transport and scattering of particles in the spatial domain X with velocities in V in the steady state regime. Here, $X \subset \mathbb{R}^d$, $d = 2, 3$, is a bounded convex open set with smooth boundary ∂X. The velocity space V is in general a subset of $\mathbb{R}^d$ (typically excluding an open vicinity of 0) although only the case $V = S^{d-1}$ (with velocities constrained to the unit sphere) is considered here for simplicity. The particle density is prescribed on the incoming part of the boundary which, together with the outgoing boundary, is defined as

$$\Gamma_\pm := \{(x, v) : x \in \partial X, \ v \in V, \ \text{and} \ \pm n_x \cdot v > 0\}. \tag{14}$$

Here, for a point x on the spatial boundary ∂X, n_x is the outer normal vector at the point. In particular, the projection of Γ_- onto its first component is precisely ∂X, and given $x \in \partial X$, the projection of Γ_- at x onto its second component consists of velocity vectors that point into the interior of X.

3.1 Decomposition of the Transport Solution

We review some basic theoretical results regarding the stationary linear transport equation with vanishing data on the incoming boundary Γ_- and with a source term f in X:

$$v \cdot \nabla_x u(x, v) + a(x)u(x, v) - \int_V k(x, v, v')u(x, v)dv' = f(x, v). \tag{15}$$

Such a situation arises naturally when differences of two solutions with the same incoming data are concerned; see [30, Chap. XXI] and [10, 23, 27, 49] for the detailed mathematical theory. In particular, the solution of the transport equation can be decomposed into a ballistic part, a single scattering part and the remaining multiple scattering part. The mathematical foundation for this decomposition can be explained as follows. The components of the integro-differential operator involved in (15) can be identified as

$$T_0\phi = v \cdot \nabla_x\phi, \quad A_1\phi = a\phi, \quad A_2\phi = -\int_V k(x, v, v')\phi(x, v')dv'.$$

$$T_1 = T_0 + A_1, \quad T = T_1 + A_2.$$

An appropriate functional setting, for $1 \le p \le \infty$, involves the spaces

$$\mathscr{W}^p := \{\phi \in L^p(X \times V), \ T_0\phi \in L^p(X \times V)\},$$

and the following differential or integro-differential operators:

$$\mathbf{T}_1\phi = T_1\phi \ \mathbf{T}\phi = T\phi, \ D(\mathbf{T}_1) = D(\mathbf{T}) = \{\phi \in \mathscr{W}^p, \ \phi|_{\Gamma_-} = 0\}.$$

The fact that a function in $\mathscr{W}^p$ has a trace on $\Gamma_\pm$ is proved in [27, 30].
    Equation (15) is then understood as

$$\mathbf{T}u = f.$$

For simplicity, we restrict the problem to the so-called subcritical setting where the constitutive coefficients (a_r, k) are assumed to satisfy:

(S1) $a, k \ge 0$ a.e. on X, $a, k \in L^\infty(X)$ and k is isotropic.
(S2)   There exists $\beta > 0$, such that $a - \pi_d k \ge \beta$ a.e. on X.

Here and below, π_d is the volume of the unit sphere S^{d-1}. When there is no scattering, the equation reduces to $\mathbf{T}_1 u = f$, and its solution is explicitly given by

$$u(x, v) = \mathbf{T}_1^{-1}f = \int_0^{\tau_-(x,v)} E(x, x - tv)f(x - tv, v)dt,$$

where E is a function defined by $E(x, y) := \exp\{-\int_0^{|x-y|} a(x - s\frac{x-y}{|x-y|})ds\}$. Under condition (S2), it is easy to verify (see e.g. [10]) that $\mathbf{T}_1^{-1}$ is a bounded linear transform on $L^p(X \times V)$, with an operator norm bounded by $\mathrm{diam}(X) \exp(-\beta \mathrm{diam}(X))$, where β is the positive constant in (S2) and $\mathrm{diam}(X)$ denotes the maximal distance between points in X.
    The problem (15) can be viewed as a "perturbation" of the non-scattering transport. By using semi-group techniques, it is proved in [30] that $\mathbf{T}$ is invertible for all $1 \le p \le \infty$. The operator norm of $\mathbf{T}^{-1}$ has an upper bound that only depends on the parameter β in (S2). Moreover, when the constitutive coefficients are random, this bound can be made independent of ω as long as the parameter $\beta > 0$ in (S2)

can be made uniform for almost all realizations; Let

$$\mathcal{K}u := A_2 \mathbf{T}_1^{-1} u = -\int_V \int_0^{\tau_-(x,v')} E(x, x - tv')k(x, v, v')u(x - tv', v')dtdv'$$

$$= -\int_X \frac{E(x, y)k(x, v, v')}{|x - y|^{d-1}} u(y, v')dy,$$

with $v' = (x - y)/|x - y|$. Then one checks that

$$\mathbf{T}^{-1} = \mathbf{T}_1^{-1}(I + \mathcal{K})^{-1}, \tag{16}$$

holds in the L^1 settings. This relation can be expanded to a truncated Neumann series with controlled remainder term. More precisely, one has

$$\mathbf{T}^{-1}f = \mathbf{T}_1^{-1}(f - \mathcal{K}f + \widetilde{\mathcal{K}}\mathcal{K}f). \tag{17}$$

The term $\mathbf{T}_1^{-1}f$ corresponds to the non-scattering transport and is referred to as the ballistic part, the term $-\mathbf{T}_1^{-1}\mathcal{K}f$ is the first order scattering part which takes considerations of particle trajectories that are scattered once, and $\mathbf{T}_1^{-1}\widetilde{\mathcal{K}}\mathcal{K}f$ corresponds to the multiple scattering part. Moreover, $\mathcal{K}$ is a weakly singular integral operator on $L^1(X)$ with a kernel bounded by $C|x - y|^{-d+1}$ and, hence, the multiple scattering part is smoothing. The proof of this decomposition of $\mathbf{T}^{-1}$ can be found, for example, in [5, 10].

Remark 1 A parallel theory can be developed for the adjoint transport equation. Denote

$$\mathbf{T}_1^* u = -T_0 u + A_1 u, \quad \mathbf{T}^* u = \mathbf{T}_1^* u - A_2' u,$$

and $D(\mathbf{T}^*_1) = D(\mathbf{T}^*) = \{u \in \mathcal{W}^p, u|_{\Gamma_+} = 0\}$, and A_2' is of the same form as A_2 but with the variables v and v' of k swapped. When k is assumed to be isotropic, then $A_2' = A_2$. In particular, $\mathbf{T}^{*-1}$ is a bounded linear transform on $L^p(X \times V)$ for all $p \in [1, \infty]$. The bound on the operator norm of $\mathbf{T}^{*-1}$ and the expansion formula still hold, provided that $\mathcal{K}$ is replaced by its formal adjoint. For any Hölder pair (p, q), $u \in L^p$ and $w \in L^q$, it holds that $\langle u, \mathbf{T}^{-1}w \rangle = \langle \mathbf{T}^{*-1}u, w \rangle$.

4 Uncertainty Propagation in Transport Equations

In this section, we review the propagation of uncertainty from the constitutive coefficients to the solutions of kinetic equations. We start by modeling the uncertainty in the coefficients decomposed as a smoothly varying deterministic part plus a highly oscillating (in space) random part. We assume that the random parts are sufficiently mixing, or roughly speaking, very short-range correlated. We then review some homogenization and corrector theory that show how the uncertainty of the coefficients propagates to the random fluctuations in the kinetic solutions.

4.1 Uncertainty Modeling for the Constitutive Coefficients

The constitutive coefficients consist of two parts: a deterministic part that we assume to be known and an uncertain part, which we model as random fields. In many situations and in particular with the applications to inverse problems [15] in mind, we assume the uncertainty part is noise-like, in the sense that it models fluctuations around some average occurring on a very small scale $0 < \varepsilon \ll 1$. Hence, we set:

$$a_{r\varepsilon}\left(x, \frac{x}{\varepsilon}, \omega\right) = a_{r0}(x) + \mu\left(\frac{x}{\varepsilon}, \omega\right), \qquad k_\varepsilon\left(x, \frac{x}{\varepsilon}, \omega\right) = k_0(x) + \nu\left(\frac{x}{\varepsilon}, \omega\right), \qquad (18)$$

for the intrinsic absorption coefficient and the isotropic scattering cross section, respectively. The deterministic part of the coefficients are $a_{r0}(x)$ and $k_0(x)$, and the uncertainty parts are modeled as random fields scaled from $\mu(y, \omega)$ and $\nu(y, \omega)$, which are mean-zero random fields defined on some probability space $(\Omega, \mathscr{F}, \mathbb{P})$. The apparent total attenuation coefficient is then

$$a_\varepsilon\left(x, \frac{x}{\varepsilon}, \omega\right) = a_{r\varepsilon} + \int_{S^{d-1}} k_\varepsilon(x) d\theta = a_0(x) + \mu\left(\frac{x}{\varepsilon}, \omega\right) + \pi_d \nu\left(\frac{x}{\varepsilon}, \omega\right), \qquad (19)$$

where $a_0(x) = a_{r0}(x) + \pi_d k_0(x)$. We henceforth refer to (13), with $a(x)$ replaced by $a_\varepsilon(x, x/\varepsilon, \omega)$ and with $k(x)$ replaced by $k_\varepsilon(x, x/\varepsilon, \omega)$, as the random linear transport equation. For notational convenience, the dependence of a_ε and k_ε on x/ε and ω is usually suppressed.

4.1.1 Stationarity, Ergodicity, and Mixing Properties

At the macroscopic length scale (much larger than ε), the random fluctuations of the constitutive coefficients have a homogenized effect. This means that the solution of the random transport equation can be well approximated by that of an effective equation with homogenized coefficients. The (stochastic) error of this approximation should then be quantified. To rigorously establish such results, we assume that the random fields $\mu(x, \omega)$ and $\nu(x, \omega)$ are stationary, ergodic, and sufficiently mixing as defined below.

Stationarity and Ergodicity A random process $\mu(x, \omega)$ on $(\Omega, \mathscr{F}, \mathbb{P})$ is called $\mathbb{R}^d$-stationary if for any positive integer k, for any k-tuple $(x_1, x_2, \cdots, x_k) \in (\mathbb{R}^d)^k$, and for any $z \in \mathbb{R}^d$,

$$(\mu(x_1, \cdot), \mu(x_2, \cdot), \cdots, \mu(x_k, \cdot)) \text{ and } (\mu(x_1 + z, \cdot), \mu(x_2 + z, \cdot), \cdots, \mu(x_k + z, \cdot),$$

have the same probability distribution. This means the statistics of the random field is homogeneous with respect to spatial translations.

An equivalent formulation can be done on the canonical probability space of μ, which is still denoted by $(\Omega, \mathscr{F}, \mathbb{P})$. Then $\mathbb{R}^d$-stationarity means: there exist a random variable $\hat{\mu} : \Omega \to \mathbb{R}$ and a group action $\{\tau_z\}_{z \in \mathbb{R}^d}$ that is $\mathbb{P}$-preserving, and $\mu(x, \omega) = \hat{\mu}(\tau_x \omega)$. The set $\{\tau_z\}_{z \in \mathbb{R}^d}$ being a $\mathbb{P}$-preserving group action means: for every $z \in \mathbb{R}^d$, $\tau_z : \Omega \to \Omega$ preserves the measure $\mathbb{P}$, i.e. $\mathbb{P}(A) = \mathbb{P}(\tau^{-1}A)$ for every $A \in \mathscr{F}$, and $\tau_{z+y} = \tau_z \tau_y$. We say μ is ergodic if the underlying group action $\{\tau_z\}_{z \in \mathbb{R}^d}$ is ergodic. That is, if $A \in \mathscr{F}$ and $\tau_z A = A$ for all $z \in \mathbb{R}^d$, then $\mathbb{P}(A) \in \{0, 1\}$.

Stationary and ergodicity are the essential properties of random fields that yield qualitative homogenization result which, in some sense, only captures the mean effect of the uncertainty. To quantify the convergence and to further study the uncertainties in the solutions, however, stronger assumption on the decorrelation structure of the random field, such as mixing properties, is often needed.

Mixing Properties We quantify the decorrelation structure of μ by the so-called "maximal correlation coefficient" ϱ, which is a decreasing function $\varrho : [0, \infty) \to [0, 1]$, $\rho(r) \to 0$ as $r \to \infty$, and for each $r > 0$, $\varrho(r)$ is the smallest value such that the bound

$$\mathbb{E}\left(V_1(\mu)V_2(\mu)\right) \leq \varrho(r)\sqrt{\mathbb{E}\left(V_1^2(\mu)\right)\mathbb{E}\left(V_2^2(\mu)\right)} \tag{20}$$

holds for any two compact sets $K_1, K_2 \in \mathscr{C}$ satisfying $d(K_1, K_2) \geq r$ and for any two random variables of the form $V_i(\mu)$, $i = 1, 2$, such that $V_i(q)$ is $\mathscr{F}_{K_i}$-measurable and $\mathbb{E}V_i(q) = 0$.

Here, $\mathscr{C}$ denotes the set of compact sets in $\mathbb{R}^d$. Given $K \subset \mathscr{C}$, $\mathscr{F}_K$ is the σ-algebra generated by the random variables $\{\mu(x) : x \in K\}$. For K_1, K_2 in $\mathscr{C}$, the distance $d(K_1, K_2)$ is defined to be

$$d(K_1, K_2) = \min_{x \in K_1, y \in K_2} |x - y|.$$

Remark 2 It is an important fact that ϱ-mixing fields are ergodic. For a stationary random field μ, the autocorrelation function is defined by

$$R(x) := \mathbb{E}\mu(x + y, \cdot)\mu(y, \cdot).$$

Note that by stationarity, the y variable in this definition does not play any role. $R_\mu(x)$ can be bounded by ϱ. Indeed, for any $x \in \mathbb{R}^d$,

$$|R_\mu(x)| = |\mathbb{E}(\mu(x)\mu(0))| \leq \varrho(|x|)\|\mu\|_{L^2(\Omega)}^2. \tag{21}$$

In view of this relation, the decay rate of mixing coefficient ϱ yields decay rate of second order moments of μ. Such relations for higher order moments will be derived in Sect. 4.4.2.

Main Assumptions We impose the following assumptions on the intrinsic absorption and scattering random fields $a_{r\varepsilon}$ and k_ε.

(A) Let $d = 2, 3$. We assume that the random fields $\mu(x, \omega)$ and $\nu(x, \omega)$ in (18) are $\mathbb{R}^d$-stationary and they admit a maximal correlation function ϱ satisfying $\varrho^{\frac{1}{8}} \in L^1(\mathbb{R}_+, r^{d-1}dr)$, that is

$$\int_0^\infty \varrho^{\frac{1}{8}}(r)r^{d-1}dr < \infty.$$

(B) Let the deterministic part (a, k) satisfies condition (S). Let μ and ν be uniformly bounded so that $a_r + \mu \geq \beta > 0$ a.e. in Ω.

Hypothesis (B), by the discussion in Sect. 3, guarantees that a.e. in Ω, the random transport equations are well posed. Let $(\mathbf{T}_\varepsilon)^{-1}$ be the inverse transport operator, it has bounded operator norm on $L^p(X \times V)$ essentially uniformly in Ω.

Hypothesis (A), in view of (21), implies that R_μ is integrable. Random fields with correlation functions satisfying such decay properties are referred to as short-range correlated. Since μ is stationary, the autocorrelation function $R_\mu(x, y)$ is a non-negative definite function in the sense that, for any $N \in \mathbb{N}$ and for any $x_i \in \mathbb{R}^n$, $y_j \in \mathbb{R}^n$, $i, j = 1, 2, \cdots, N$, the matrix $(M_{ij}) \in \mathbb{R}^{N \times N}$ given by $M_{ij} := R_\mu(x_i, y_j)$ is non-negative definite. Let

$$\sigma_\mu^2 = \int_{\mathbb{R}^n} R_\mu(x, 0)dx, \quad \sigma_\nu^2 = \int_{\mathbb{R}^n} R_\nu(x, 0)dx.$$

Then by (A) and (21), σ_μ^2 and σ_ν^2 are finite real numbers and, thanks to Bôchner's theorem, they are nonnegative. Throughout the paper, we assume those numbers are positive.

When the decay rate of ϱ is much weaker, so that (A) is violated, the random variations of the coefficients are in a different setting, and the quantitative results for the random fluctuations in the transport solutions will be changed; see Sect. 4.4.1.

Remark 3 (Poisson Bumps Model) Assumptions on the mixing coefficient ϱ of random media have been used in [3, 7, 39]; we refer to these papers for explicit examples of random fields satisfying the assumptions.

A widely used mixing random field model is the so-called Poisson bumps model; see e.g. [5]. The model is constructed as follows. We start with a *spatial Poisson point process* with intensity $\rho > 0$, which is a countable random subset $Y_\rho(\omega) := \{y_j(\omega) : j \in \mathbb{N}\} \subseteq \mathbb{R}^d$ defined on an abstract probability space $(\Omega, \mathscr{F}, \mathbb{P})$ satisfying, for any bounded Borel set $A \subseteq \mathbb{R}^d$, that the random variable $N(A)$, which is the cardinality of $A \cap Y_\rho$, follows the Poisson distribution with intensity $\rho|A|$, i.e.,

$$\mathbb{P}\{N(A) = m\} = \frac{e^{-\rho|A|}(\rho|A|)^m}{m!}. \tag{22}$$

See [28] for details. For any disjoint Borel sets $A_1, \cdots, A_n$, $n \geq 2$, the random variables $N(A_i)$, $1 \leq i \leq n$ are independent. The Poisson bumps model for the constitutive coefficients are then defined by

$$a_r(x; \omega) = \sum_j \psi(x - y_j(\omega)), \qquad k(x; \omega) = \sum_j \phi(x - y_j(\omega)). \tag{23}$$

Here, ψ and ϕ are smooth functions satisfying

$$0 \leq \psi, \phi \leq 1, \quad \phi(0) = \psi(0) = 1, \quad \text{and } \psi, \phi \text{ have compact supports.}$$

Using the properties of Poisson point process [28], one can show that a_r and k so defined are stationary and have finite range correlations and, hence, are mixing and ergodic. The mean values $\mathbb{E}(a_r(x, \cdot))$ and $\mathbb{E}k(x, \cdot)$ are constant; in fact, they are given by $\rho\|\psi\|_{L^1}$ and $\rho\|\phi\|_{L^1}$, respectively. Let

$$\mu(x, \omega) = a_r(x, \omega) - \mathbb{E}(a_r(x, \cdot)) \quad \text{and} \quad \nu(x, \omega) = k(x, \omega) - \mathbb{E}k(x, \cdot).$$

They are the mean-zero random parts of a_r and k.

Finally, by scaling the spatial variables, the formulas in (18) provide random field models for the constitutive coefficients which satisfies (A) and (B).

4.2 Homogenization Result and the Convergence Rate

It is well known that the random transport equation homogenizes if the random fluctuations of the constitutive coefficients are periodic or stationary ergodic random fields, see e.g. [23, 29], and the effective coefficients are then given by the statistical average. The assumption (A) gives further quantitative information of the random fluctuations, such as the convergence rates. Throughout this section, u_ε denotes the solution to the random transport equation, and u_0 denotes the solution for the transport equation with the effective (mean) coefficients.

Theorem 1 *Assume (A) and (B). Assume further that the boundary data $g \in L^\infty(\Gamma_-)$ so $u_0 \in L^\infty(X \times V)$. Then there exists some constant $C > 0$ depending only on the diameter of X, $\|g\|_{L^\infty}$ and β such that, as $\varepsilon \to 0$,*

$$\left(\mathbb{E}\|u_\varepsilon - u_0\|_{L^2(X \times V)}^2\right)^{\frac{1}{2}} \leq C\varepsilon^{\frac{1}{2}}. \tag{24}$$

This theorem implies that the convergence rate of the homogenization is of order $\sqrt{\varepsilon}$ in the energy norm $L^2(\Omega, L^2(X \times V))$. The proof of this theorem can be found in [5]. In particular, the main contribution comes from the energy of the ballistic parts which amounts to weighted average of the random fluctuations in the coefficients over lines. Since the central limit scaling of the average of random fields is $\varepsilon^{\partial/2}$, where ∂ is the dimension over which the average is taken, the scaling $\sqrt{\varepsilon}$ above is reasonable.

The leading terms in the homogenization error $u_\varepsilon - u$, i.e. the so-called correctors, contain two parts that play dominating roles at different scales. The mean-zero random part χ_ε is defined by

$$v \cdot \nabla_x \chi_\varepsilon + a(x)\chi_\varepsilon - k(x) \int_V \chi_\varepsilon(x, v')dv' = -\mu_\varepsilon(x)u_0(x, v) + v_\varepsilon(x) \int_V u_0(x, v')dv'. \tag{25}$$

We simply refer to it as $\mathbf{T}^{-1}A_\varepsilon u_0$, where the definition of A_ε can be read from above. χ_ε is the response of the transport equation to a source term given by $A_\varepsilon u_0$. The other part is deterministic and solves

$$v \cdot \nabla_x U + a(x)U - k(x) \int_V U(x, v')dv' = q(x, v), \tag{26}$$

where the source term $q(x, v)$ is given by:

$$\int_{\mathbb{R}} \left(R_\mu(tv)u_0(x, v) - R_{\mu v}(tv)\bar{u}_0(x) - \int_V \left(R_{\mu v}(tw)u_0(x, w) - R_v(tw)\bar{u}_0(x) \right)dw \right)dt.$$

Here, R_μ, R_v and $R_{\mu v}$ correspond to, respectively, the auto-correlation functions of μ, v, and the cross-correlation function of the two. Note also U is essentially the average of $\mathbf{T}^{-1}A_\varepsilon \chi_\varepsilon$. We have the following result.

Theorem 2 *Under the assumptions of Theorem 1, for any test function φ, we have*

$$|\langle \mathbb{E}(u_\varepsilon - u_0) - \varepsilon U, \varphi \rangle| \lesssim \varepsilon^2. \tag{27}$$

This theorem concerns the mean value of the homogenization error which lives on the large scale. Clearly, the term χ_ε is mean zero and does not contribute to $\mathbb{E}u_\varepsilon$, so the deterministic corrector U is indeed the dominating term. In the subsection below, we check that the variance of the homogenization error, integrated with a test function over space and velocity, is of order ε^d, and χ_ε is responsible for the main contribution.

4.3 Central Limit Theory for the Random Fluctuations

In this section, we study the random fluctuation $u_\varepsilon - \mathbb{E}u_\varepsilon$ and characterize their probability distribution at various observation scales.

Observations of the Solutions We consider three different scales of observation of the transport solutions:

- the pointwise data $u_\varepsilon(x, v)$ for fixed (x, v),
- the angularly averaged data $J_\varepsilon(x) := \int_V u_\varepsilon(x, v)\varphi(v)dv$ for fixed $x \in \overline{X}$ and some averaging kernel φ over V,
- the fully averaged data $\int_{X \times V} u_\varepsilon(x, v)\varphi(x, v)$ for some averaging kernel φ over $X \times V$.

When the application to inverse problems is considered, the angularly averaged or the fully averaged data are not uncommon. Below, we focus on the setting of fully averaged data since it is simpler to present. We first briefly comment on other settings of interest.

An interesting fact is that the size of the random fluctuations in u_ε depend on the observation scales. In [6], it is shown that, for pointwise data, the variance of $u_\varepsilon(x, v)$ is of order ε for all dimensions $d \geq 2$. This property arises from integrating random fields along (one-dimensional) lines. For the angularly averaged data $J_\varepsilon(x, \omega)$ defined above, its variance is of order $\varepsilon^2 |\log \varepsilon|$ in dimension two and ε^2 in dimension $d \geq 3$. Angular averaging introduces additional mixing along the angular direction and therefore significantly reduces the variance of the corrector. Finally, as we show in detail below, the variance of the fully averaged data $\langle u_\varepsilon, \varphi \rangle$ is of order ε^d in dimension $d \geq 2$. The random corrector is therefore of smallest size when averaged over the whole phase space. This is consistent with the central limit theorem.

Fluctuation Theory for the Fully Averaged Data We take a sufficiently smooth functions φ and consider the fully averaged data $\langle \varphi, u_\varepsilon - \mathbb{E}u_\varepsilon \rangle$. To describe the limiting probability distribution of this random variable, we introduce additional notation. Let $\psi := \mathbf{T}^{*-1}\varphi$; in other words, ψ is the solution of the following adjoint transport equation:

$$\begin{cases} -v \cdot \nabla_x \psi + a\psi - \int_V k(x, v, v')\psi(x, v')dv' = \varphi, & (x, v) \in X \times V, \\ \psi(x, v) = 0, & (x, v) \in \Gamma_+. \end{cases} \quad (28)$$

We denote by $\rho_{\mu,\nu}$ the correlation factor between the random fields μ and ν; it is defined by $\sigma_\mu^{-1}\sigma_\nu^{-1}\int_{R^n} \mathbb{E}\mu(x, \cdot)\nu(0, \cdot)dx$. Let Σ denote the non-negative matrix

$$\Sigma := \begin{pmatrix} \sigma_\mu^2 & \rho_{\mu,\nu}\sigma_\mu\sigma_\nu \\ \rho_{\mu,\nu}\sigma_\mu\sigma_\nu & \sigma_\nu^2 \end{pmatrix}.$$

We have the following result.

Theorem 3 *Let the dimension $d = 2, 3$. Under the same condition of Theorem 1, we have*

$$\left\langle \varphi, \frac{u_\varepsilon - \mathbb{E}u_\varepsilon}{\varepsilon^{d/2}} \right\rangle \xrightarrow[\varepsilon \to 0]{\text{distr.}} \int_X \left[\int_V \begin{pmatrix} -\psi(x, v)u_0(x, v) \\ \psi(x, v)[-\pi_d u_0(x, v) + \langle u_0 \rangle_V(x)] \end{pmatrix} dv \right] \sqrt{\Sigma} dW(y).$$

$$(29)$$

Here, $W(y) = (W_a(y), W_k(y))'$ is a two dimensional 2d-parameter Wiener process and $\langle u_0 \rangle_V(x)$ denotes the angular average $\int_V u_0(x, v')dv'$.

This theorem says that the random fluctuations of the fully averaged data are of order $\varepsilon^{\frac{d}{2}}$, satisfy a functional central limit theory with limiting Gaussian distribution. The right-hand side of (29) is a Gaussian distribution $\mathcal{N}(0, \sigma^2)$ with the variance σ^2 given by

$$\int_X \left[\int_V \begin{pmatrix} -\psi(x, v)u_0(x, v) \\ \psi(x, v)[-\pi_d u_0(x, v) + \langle u_0 \rangle_V(x)] \end{pmatrix} dv \right]^T \Sigma$$

$$\times \left[\int_V \begin{pmatrix} -\psi(x, v)u_0(x, v) \\ \psi(x, v)[-\pi_d u_0(x, v) + \langle u_0 \rangle_V(x)] \end{pmatrix} dv \right] dx.$$

We refer the reader to [41] for more details on multi-parameter Wiener process. An $\mathbb{R}^2$ valued Wiener process is needed in the limit since two random fields are involved in the random transport equation. For the Poisson bumps model of Remark 3, the correlation factor $\rho_{\mu,\nu}$ above is 1, but more general situations still based on Poisson point process may be considered; see [5].

Finally, it is worth mentioning that for $d = 2, 3$, the term $\mathbb{E}u_\varepsilon$ in the statement of the theorem above can be replaced by εU of Theorem 2. In other words, εU is the only term in the mean error $\mathbb{E}(u_\varepsilon - u_0)$ that is larger than the random fluctuations. The competition between the deterministic and the random parts of the homogenization error is, hence, clearly characterized in dimensions $d \leq 3$.

Outline of the Proof We now outline the proofs of the main theorems. It clearly appears from these proofs that the key ingredients about the random coefficients that allow one to quantify the homogenization are moments estimates.

The required moments estimates and their decay rates were obtained in [5] for the Poisson bumps model and strongly depended on the statistical properties of Poisson point processes. At the end of this section, we extend such moments estimates to general random fields whose mixing coefficient ϱ, defined in (20), decays sufficiently fast. The results of [5] thus generalize to the versions stated in this review.

To get the convergence rate in Theorem 1, we observe that χ_ε defined in (25) satisfies

$$u_\varepsilon - u = \chi_\varepsilon + \mathbf{T}_\varepsilon^{-1} A_\varepsilon \chi_\varepsilon.$$

Since $\mathbf{T}_\varepsilon^{-1}$ and A_ε are uniformly bounded (in ε and $\omega \in \Omega$) linear transformations on $L^2(X \times V)$, due to assumption (B), it suffices to show that $\mathbb{E}\|\chi_\varepsilon\|_{L^2}^2 \lesssim \varepsilon$. Using the representation formula (17) of $\mathbf{T}^{-1}$, we have rather explicit formula for χ_ε and quantifying its mean square norm is straightforward, given the second order moments of the random fields.

For Theorems 2 and 3, the starting point is the following expansion formula:

$$
\begin{aligned}
u_\varepsilon - u_0 = {} & \mathbf{T}^{-1}A_\varepsilon u_0 + \mathbf{T}^{-1}A_\varepsilon \mathbf{T}^{-1}A_\varepsilon u_0 + \mathbf{T}^{-1}A_\varepsilon \mathbf{T}^{-1}A_\varepsilon \mathbf{T}^{-1}A_\varepsilon u_0 \\
& + \mathbf{T}^{-1}A_\varepsilon \mathbf{T}^{-1}A_\varepsilon \mathbf{T}^{-1}A_\varepsilon(u_\varepsilon - u_0).
\end{aligned}
\tag{30}
$$

This formula is obtained by iterating the relation $u_\varepsilon - u_0 = \mathbf{T}^{-1}A_\varepsilon u_\varepsilon$ three times, and the iteration process could be continued further. We focus on the case $d = 2$ below. For the case $d = 3$, another iteration is needed for the argument below to hold.

Consider the inner product of (30). The first term on the right-hand side is mean-zero and has no contribution. The second term is $\langle A_\varepsilon \mathbf{T}^{-1}A_\varepsilon u_0, \psi \rangle$ where $\psi = \mathbf{T}^{*-1}\varphi$. Using second order moments again, the main contribution to its statistical mean comes from $\mathbb{E}\langle A_\varepsilon \mathbf{T}_1^{-1}A_\varepsilon u_0, \psi \rangle$, and it converges to $\langle U, \varphi \rangle$. Using the third and higher order moments of mixing random fields obeying (A), all the remainder terms are of smaller orders than ε^2. Theorem 2 is then proved. Note that the remainder terms in the mean is even smaller than the central limit scaling $\varepsilon^{d/2}$.

For the fluctuation theory, we consider the variances of the terms in (30) after taking inner products with the test function and subtracting the mean values. The first term has variance of order ε^d, agreeing with the central limit setting. To prove Theorem 3, it suffices to establish

$$
\mathrm{Var}\left(\langle \mathbf{T}^{-1}A_\varepsilon \mathbf{T}^{-1}A_\varepsilon u_0, \varphi \rangle\right) \ll O(\varepsilon^d),
\tag{31}
$$

$$
\mathrm{Var}\left(\langle \mathbf{T}^{-1}A_\varepsilon \mathbf{T}^{-1}A_\varepsilon \mathbf{T}^{-1}A_\varepsilon u_0, \varphi \rangle\right) \ll O(\varepsilon^d),
\tag{32}
$$

$$
\mathbb{E}|\langle \mathbf{T}^{-1}A_\varepsilon \mathbf{T}^{-1}A_\varepsilon \mathbf{T}^{-1}A_\varepsilon(u_\varepsilon - u_0), \varphi \rangle| \ll O(\varepsilon^{\frac{d}{2}}),
\tag{33}
$$

$$
\varepsilon^{-d/2}\langle \mathbf{T}^{-1}A_\varepsilon u_0, \varphi \rangle \text{ converges in distribution to the right hand side of (29)}.
\tag{34}
$$

Let us start with the estimate (33) for the remainder. This is an $L^1(\Omega)$ estimate, which is sufficient to show that this term, divided by $\varepsilon^{d/2}$, converges to zero in distribution. We apply Hölder inequality and get

$$
\mathbb{E}|\langle \mathbf{T}^{-1}A_\varepsilon \mathbf{T}^{-1}A_\varepsilon \mathbf{T}^{-1}A_\varepsilon(u_\varepsilon - u_0), \varphi \rangle| \le C\mathbb{E}\{\|u_\varepsilon - u_0\|_{L^2}\|\mathbf{T}^{*-1}A_\varepsilon \mathbf{T}^{*-1}A_\varepsilon \psi\|_{L^2}\}
$$

$$
\le C\left(\mathbb{E}\|u_\varepsilon - u_0\|_{L^2}^2\right)^{\frac{1}{2}}\left(\mathbb{E}\|\mathbf{T}^{*-1}A_\varepsilon \mathbf{T}^{*-1}A_\varepsilon \psi\|_{L^2}^2\right)^{\frac{1}{2}}.
$$

This reduce (33) to $\mathbb{E}\|\mathbf{T}^{*-1}A_\varepsilon\mathbf{T}^{*-1}A_\varepsilon\psi\|_{L^2}^2 \ll O(\varepsilon^{d-1})$. Using fourth-order moments, this term is of order ε^2, and hence the desired result follows for $d = 2$. This estimate also shows that, for $d = 3$, another iteration is needed in (30).

For the second and the third terms in the expansion (30), by similar estimates as above, we see that their $L^1(\Omega)$ norms are not small enough. Hence we need the variance estimates, which are enough for the convergence in distribution. For (31), we appeal to the fourth-order variance estimate in Proposition 3, and for (32), we appeal to the sixth-order variance estimate in Proposition 6. From a technical point of view, the variances are of much smaller order because, in the corresponding variance estimates (37) and (44) for the products of the random fields, the terms which are responsible for the largeness of the $L^1(\Omega)$ controls are eliminated.

Finally, the limiting distribution of $\varepsilon^{-d/2}\langle u_\varepsilon - u_0, \varphi\rangle$ is given by the limiting distribution of $\frac{1}{\sqrt{\varepsilon^d}}\langle \mathbf{T}^{-1}A_\varepsilon u_0, \varphi\rangle$, which has the expression

$$\frac{1}{\sqrt{\varepsilon^d}} \int_X \mu\left(\frac{x}{\varepsilon}, \omega\right) \langle\psi u_0\rangle_V(x) + \nu\left(\frac{x}{\varepsilon}, \omega\right) [-\pi_d\langle\psi\rangle_V(x)\langle u_0\rangle_V(x) + \langle\psi u_0\rangle_V(x)]\, dx.$$

Again, $\langle\psi u_0\rangle_V$ denotes the angular average of the pointwise product ψu_0. This is an oscillatory integral of random fields with short range correlations, and the central limit theorem of such integrals was proved in [3, 24]. The above integral hence has the desired limit as in (29). This proves Theorem 3.

In three dimensions, $d = 3$, the analysis is more involved. A further iteration should be added to (30). The resulted remainder term, and the first three terms are controlled as above. An additional term appears, which, after taking inner product with the test function, becomes $\langle\mathbf{T}^{-1}A_\varepsilon\mathbf{T}^{-1}A_\varepsilon\mathbf{T}^{-1}A_\varepsilon\mathbf{T}^{-1}A_\varepsilon u_0, \varphi\rangle$. We consider its $L^2(\Omega)$ norm

$$\mathbb{E}\left|\langle\mathbf{T}^{-1}A_\varepsilon\mathbf{T}^{-1}A_\varepsilon\mathbf{T}^{-1}A_\varepsilon\mathbf{T}^{-1}A_\varepsilon u_0, \varphi\rangle\right|^2 = \mathbb{E}\left|\langle A_\varepsilon\mathbf{T}^{-1}A_\varepsilon\mathbf{T}^{-1}A_\varepsilon\mathbf{T}^{-1}A_\varepsilon u_0, \psi\rangle\right|^2.$$

By appealing to the eighth-order moment estimates, one can check that this term is of order $o(\varepsilon^3)$ and hence does not contribute to the limit.

4.4　Further Remarks

4.4.1　Long Range Correlated Random Media

When the random fluctuations in the constitutive coefficients have long range correlations and the assumptions (A) is violated, it is still possible to quantify the convergence rate and to analyze the random part of the homogenization error, if sufficient quantitative information about the random fields is given.

For instance, in [6], random fields μ and ν of the form $\Phi(g(x, \omega))$ are studied, where $g(x, \omega)$ is some underlying Gaussian random field with heavy tail $R_g(x)$, and $R_g(r) \sim r^{-\alpha}$ for some $0 < \alpha < 1$ asymptotically at infinity. Here, Φ is a bounded real function with sufficient regularity and of Hermite rank one; see [8, 46]. It is then shown that the pointwise data have random fluctuations of order $\sqrt{\varepsilon^{\alpha}}$, which is much larger than $\sqrt{\varepsilon}$ for the short range setting. We refer the reader to [18, 21, 40] for more discussions on random fields with long range correlations.

4.4.2 Moments Estimates for Mixing Random Fields

In [5], Theorems 1–3 were proved only for the Poisson bumps model. We now show that such results hold for more general random fields satisfying (A) as the moment and variance estimates enjoyed by the Poisson points model also hold for generalized mixing random fields.

Below, we first recall the crucial moments formulae (hence moments estimates) of the Poisson bumps model, and then show that those estimates hold for sufficiently mixing random fields. Even though much more general results can be produced, only the first several (up to eighth-order) moments estimates are provided.

Moments Formulas for the Poisson Bumps Model Let n be a positive integer and let I_n denote the index set $\{1, 2, \cdots, n\}$. Given a set $\{x_1, x_2, \cdots, x_n\} \subset \mathbb{R}^d$ and $J \subseteq I_n$, x_J denotes the subset $\{x_j : j \in J\}$. For a random field $\mu(x, \omega)$, we are interested in getting estimates for

$$\Phi_{\mu}^{(n)}(x_{I_n}) := \mathbb{E}[\mu(x_1)\mu(x_2) \cdots \mu(x_n)],$$

which is the nth order moments of the random field μ evaluated at x_{I_n}. We note that Φ is viewed as a function of set-valued arguments, since the order of the elements in the set plays no role. In the sequel, the dependence on μ is omitted when the random field under study is clear.

For the set I_n, we say $(n_1, n_2, \cdots, n_k)$ is a k-partition of I_n if $1 \leq n_1 \leq n_2 \leq \cdots \leq n_k$ and $\sum_{i=1}^{k} n_i = n$. A partition is called *non-single* if $n_1 \geq 2$. We denote by $\mathscr{G}_n$ the set of all non-single partitions of I_n. Given $(n_1, n_2, \cdots, n_k)$, there are finitely many possible ways to divide I_n (hence x_{I_n}) into k disjoint subsets $J_1, \cdots, J_k$ of cardinalities $n_1, n_2, \cdots, n_k$, respectively. We denote this finite number by $C_n^{n_1, n_2, \cdots, n_k}$, and we order those possibilities following the dictionary order of the array formed by $(\max J_1, \cdots, \max J_k)$. The ℓ-th choice is hence denoted by $(J_1^{\ell}, \cdots, J_k^{\ell})$.

Proposition 1 ([5]) *Let $\mu(x, \omega)$ be the mean-zero part of the Poisson bumps potential. Fix a positive integer n. For any integer $k \leq n$ and any subset $J \subseteq I_n$ with cardinality k, define*

$$T^{(k)}(x_J) := \nu \int \prod_{j \in J} \psi(x_j - z)dz. \tag{35}$$

Then we have the following formula for $\Phi^{(n)}(x_{I_n})$

$$\Phi^{(n)}(x_{I_n}) = \sum_{(n_1,\cdots,n_k)\in\mathscr{G}_n} \sum_{\ell=1}^{C_n^{n_1,\cdots,n_k}} \prod_{j=1}^{k} T^{(n_j)}(x_{J_j^\ell}). \tag{36}$$

In [5], we also need to have estimates on variances of products of μ evaluated at several points. More precisely, if p is a positive integer such that $p \leq n/2$, we are interested in

$$\Psi^{(p,n-p)}(x_{I_p}, x_{I_n\setminus I_p}) := \mathbb{E}\left[\left(\prod_{j\in I_p}\mu(x_j) - \Phi^p(x_{I_p})\right)\left(\prod_{k\in I_n\setminus I_p}\mu(x_k) - \Phi^{n-p}(x_{I_n\setminus I_p})\right)\right],$$

which is the covariance of the random variables $\prod_{j\in I_p}\mu(x_j)$ and $\prod_{k\in I_n\setminus I_p}\mu(x_k)$. It is clear that formulae for those covariance functions can be read from the formulae for moments of μ.

Moments Estimates for Mixing Random Fields In the rest of this section, we fix a random field $\mu(x,\omega)$ satisfying the assumption (A). All of the moment and variance functions Φ and Ψ are understood as those of μ. The first theorem deals with the third order moment.

Proposition 2 *Let $C = \|\mu\|_{L^2(\Omega)}\|\mu^2\|_{L^2(\Omega)}$, and let $\eta(r) = \sqrt{\rho(r/2)}$. Then*

$$\left|\Phi^{(3)}(\{x_1, x_2, x_3\})\right| \leq C\eta(|x_2 - x_1|)\eta(|x_3 - x_1|).$$

From the proof below, it is clear that after a permutation of (x_1, x_2, x_3) on the right hand side, the resulted estimate still holds. Hence, the right hand side can be thought as depending only on the set $\{x_1, x_2, x_3\}$.

Before proceeding to the proof of this result, we introduce some more notation. Recall that $I_n := \{1, 2, \cdots, n\}$. We consider dividing the set I_n into two subsets of positive cardinality p and $n - p$. For each $0 < p \leq n/2$, let

$$\mathscr{I}_n^p = \{J : J \subseteq I_n, \text{card}(J) = p\}.$$

Given a $J \in \mathscr{I}_n^p$, let J^c denote the complement of J in I_n. Then (J, J^c) corresponds to a division of I_n, and (x_J, x_{J^c}) corresponds to a division of the set x_{I_n} into two subsets of cardinality p and $n - p$. Define

$$L_p = \max_{J\in\mathscr{I}_n^p} \text{dist}(x_J, x_{J^c}).$$

Then L_p is the maximum separation distance among divisions of x_{I_n} into two subsets of cardinality p and $n - p$. Finally, let

$$L = \max_{0<p\leq n/2} L_p.$$

Then L is the maximum separation distance between possible divisions of x_{I_n} into two subsets of positive cardinalities.

Proof (Proof of Proposition 2) In view of the discussion above, since $n = 3$, the largest integer smaller or equal to $n/2$ is 1. So, $L = L_1$. Let $J \in \mathscr{I}_3^1$ be $\arg\max L_1$. We consider two scenarios.

If $x_1 \in x_J$, without loss of generality, we assume $L = |x_1 - x_2|$. It is then clear that $|x_3 - x_1| \leq 2L$ because if otherwise, $|x_3 - x_2| \geq |x_3 - x_1| - |x_2 - x_1| > L$ and hence $\{x_3\} \cup \{x_1, x_2\}$ would be a division yielding larger L_1. Applying the mixing condition, we get

$$
\begin{aligned}
\Phi^{(3)}(x_1, x_2, x_3) &= \mathbb{E}\{\mu(x_1)[\mu(x_2)\mu(x_3) - \mathbb{E}(\mu(x_2)\mu(x_3))]\} \\
&\leq \varrho(L)\|\mu\|_{L^\infty(\Omega)}^3 \\
&\leq \varrho^{1/2}(|x_1 - x_2|)\varrho^{1/2}(|x_1 - x_3|/2)\|\mu\|_{L^\infty(\Omega)}^3 \\
&\leq \eta(|x_1 - x_2|)\eta(|x_1 - x_3|).
\end{aligned}
$$

If $x_1 \in x_{J^c}$, without loss of generality, assume the division is given by $\{x_2\} \cup \{x_1, x_3\}$. We consider two further sub-cases. If the separating distance is $L = |x_2 - x_1|$, then $|x_1 - x_3| \leq L$. If, instead, the separating distance is $L = |x_2 - x_3|$, then $|x_1 - x_3| \leq L$ and $L \leq |x_1 - x_2| \leq 2L$. In both cases, we have

$$
\begin{aligned}
\Phi^{(3)}(x_1, x_2, x_3) &= \mathbb{E}\{V(x_2)[V(x_1)V(x_3) - \mathbb{E}(V(x_1)V(x_3))]\} \\
&\leq \varrho(L)\|\mu\|_{L^\infty(\Omega)}^3 \leq \eta(|x_1 - x_2|)\eta(|x_1 - x_3|).
\end{aligned}
$$

The conclusion of the theorem, hence, is established.

The fourth-order moment was considered in [39] and in [3]; see also [8]. For the convenience of the reader, we recall the following estimate of [39].

Proposition 3 *Let* $C = 4(\|\mu\|_{L^2(\Omega)}\|\mu^3\|_{L^2(\Omega)} + \|\mu^2\|_{L^2(\Omega)}^2)$ *and let* $\eta(r) = \sqrt{\rho(r/3)}$. *Then*

$$
\begin{aligned}
&\left|\Phi^{(2,2)}(x_1, x_2, x_3, x_4) - R(x_1 - x_2)R(x_3 - x_4)\right| \\
&\leq C\eta(|x_1 - x_3|)\eta(x_2 - x_4) + C\eta(|x_1 - x_4|)\eta(|x_2 - x_3|).
\end{aligned}
\tag{37}
$$

Note that the estimate (21) implies $|R(x - y)| \leq \eta(|x - y|)$. The variance estimate above also yields the following moment estimate:

$$
\left|\Phi^{(4)}(x_{I_4})\right| \leq \frac{C}{2} \sum_{J \in \mathscr{I}_4^2} \eta(x_J)\eta(x_{J^c}).
\tag{38}
$$

For more general results, the following fact will be helpful.

Proposition 4 *Let $n \in \mathbb{N}$ and $n \geq 2$, and let I_n and L be defined as above, then*

$$\mathrm{dist}(x_j, x_k) \leq 2(n-1)L, \qquad \forall j, k \in I_n. \tag{39}$$

Proof Let L be achieved by a division (x_J, x_{J^c}). Let $p = \mathrm{card}(J)$ and, without loss of generality, $L = \mathrm{dist}(x_1, x_2)$ for some $x_1 \in J$, $x_2 \in J^c$. We show that

$$\mathrm{dist}(x_j, \{x_1, x_2\}) \leq (n-1)L, \qquad \forall j \in I_n. \tag{40}$$

If this fails, say $\mathrm{dist}(x_3, x_2) > (n-1)L$, then since $L \geq L_1$, there must be a point, say x_4, such that $x_4 \in \overline{B}_L(x_3)$. Indeed, if otherwise, then $L_1 \geq \mathrm{dist}(x_3, x_{I_n \setminus \{3\}}) > L$, which is impossible. Similarly, since $L \geq L_2$, there must be another point, say x_5, such that $x_5 \in \overline{B}_L(x_3) \cup \overline{B}_L(x_4)$. Note that $x_5 \in \overline{B}_{2L}(x_3)$. By repeating this argument, we hence find p points

$$\{x_3, x_4, \cdots, x_{3+p-1}\} \subseteq \overline{B}_{(p-1)L}(x_3). \tag{41}$$

Let $q = n - p$. Applying the same argument above with $\{x_3\}$ replaced by $\{x_1, x_2\}$, we find that the remaining set $x_{I_n \setminus \{3, \cdots, 3+p-1\}}$, which contains $\{x_1, x_2\}$, must satisfy

$$\{x_{3+p}, \cdots, x_n\} \subseteq \overline{B}_{(q-2)L}(x_1) \cup \overline{B}_{(q-2)L}(x_1) \subseteq \overline{B}_{(q-1)L}(x_2). \tag{42}$$

Let $K = \{3, 4, \cdots, 3+p-1\}$ and $K^c = \{1, 2, 3+p, \cdots, n\}$. Then (41), (42) and the assumption $\mathrm{dist}(x_2, x_3) > (n-1)L$ imply that $\mathrm{dist}(x_K, x_{K^c}) > L$, which is impossible. Therefore, (40) holds and, by an application of triangle inequality, (39) is established.

We move to the fifth-order moments, and obtain the following estimate.

Proposition 5 *Let C denote the constant $\|\mu\|_{L^2(\Omega)}\|\mu^4\|_{L^2(\Omega)} + \|\mu^2\|_{L^2(\Omega)}\|\mu^3\|_{L^2(\Omega)}$ $+ \|\mu\|_{L^2(\Omega)}^3\|\mu^2\|_{L^2(\Omega)}$. Let η be defined as in Proposition 2 and define $\psi^{(3)}(\{x, y, z\})$ as $\varrho^{\frac{1}{4}}(|x-z|)\varrho^{\frac{1}{4}}(|y-z|)$. Then*

$$\left|\Phi^{(5)}(x_{I_5})\right| \leq C \sum_{J \in \mathscr{I}_5^2} \eta(\tfrac{1}{4}|x_{j_1} - x_{j_2}|)\psi^{(3)}(\tfrac{1}{8}\{x_{J^c}\}). \tag{43}$$

Proof Recall the definition of L, L_1 and L_2 after the statement of Proposition 2. We only need to consider two cases.

Case 1 $L = L_1 \geq L_2$. Without loss of generality, assume that $L = \mathrm{dist}(x_1, x_2) = \mathrm{dist}(\{x_1\}, \{x_2, \cdots, x_5\})$. Then we have, due to the mixing property of the random field μ,

$$\left|\Phi^{(5)}(x_{I_5})\right| = \left|\mathbb{E}\left(\mu(x_1)\left[\prod_{j=2}^5 \mu(x_j) - \Phi^{(4)}(\{x_2, \cdots, x_5\})\right]\right)\right|$$
$$\leq \varrho(L)\|\mu\|_{L^2(\Omega)}\|\mu^4\|_{L^2(\Omega)}.$$

In view of (39), the above is bounded by $C\varrho(|x_1 - x_2|)\psi^{(3)}(\{x_3, x_4, x_5\})$. Since ϱ and hence η are decreasing functions, this bound is smaller than some of the terms of the right-hand side of (43).

Case 2 $L = L_2 > L_1$. Let L be maximized by the division given by $x_J = \{y_1, y_2\}$ and $x_{J^c} = \{y_3, y_4, y_5\}$. Then, by the mixing property,

$$\left|\Phi^{(5)}(x_{I_5}) - R(y_1 - y_2)\Phi^{(3)}(\{y_3, y_4, y_5\})\right| \le \varrho(L)\|\mu^2\|_{L^2(\Omega)}\|\mu^3\|_{L^2(\Omega)}.$$

We can find some $J' = (j'_1, j'_2) \in \mathscr{I}_5^2$ such that $J' \ne J$. Then in view of (39), the right-hand side above can be bounded by $C\varrho^{1/2}(\frac{1}{8}|x_{j'_1} - x_{j'_2}|)\psi^{(3)}(\frac{1}{8}x_{(J')^c})$.

Moreover, in view of (39), we have

$$\left|R(y_1 - y_2)\Phi^{(3)}(\{y_3, y_4, y_5\})\right| \le C\varrho(|y_1 - y_2|)\psi^{(3)}(\{y_3, y_4, y_5\}).$$

The bounds we have for the preceding two quantities correspond to two different terms in the right-hand side of (43). The desired result is hence established.

Next, we study the sixth-order moments of μ. We first derive a variance type estimate, from which the moment estimate follows easily. Let $\mathscr{I}_6^{2,2,2}$ denote the set

$$\{(J_1, J_2, J_3) : \mathrm{card}(J_1) = \mathrm{card}(J_2) = \mathrm{card}(J_3) = 2, \cup_{i=1}^3 J_i = I_6\},$$

which is the collection of partitions of I_6 into three disjoint subsets of cardinality 2.

Proposition 6 *Let C denote the constant* $\|\mu\|_{L^2(\Omega)}\|\mu^5\|_{L^2(\Omega)} + \|\mu^2\|_{L^2(\Omega)}(\|\mu^4\|_{L^2(\Omega)}$ $+ \|\mu\|_{L^3(\Omega)}^3\|\mu\|_{L^2(\Omega)}) + \|\mu^3\|_{L^2(\Omega)}^2$. *Let η be defined as in Proposition 2 and let $\psi^{(3)}$ be defined as in Proposition 5. Define $\xi^{(2)} = \eta^{\frac{2}{3}}$. Then*

$$\left|\Psi^{(3,3)}(\{x_1, x_2, x_3\}, \{x_4, x_5, x_6\})\right| \le \sum_{(J_1, J_2, J_3) \in \mathscr{I}_6^{2,2,2}} C\xi^{(2)}\left(\frac{x_{J_1}}{20}\right)\xi^{(2)}\left(\frac{x_{J_2}}{20}\right)\xi^{(2)}\left(\frac{1}{20}x_{J_3}\right)$$

$$+ \frac{C}{2} \sum_{K \in \mathscr{I}_6^3 \setminus \{\{1,2,3\},\{4,5,6\}\}} \psi^{(3)}\left(\frac{1}{20}x_K\right)\psi^3\left(\frac{1}{20}x_{K^c}\right).$$

$$(44)$$

We note that, as in the variance estimate (37), the partition $\{1, 2, 3\} \cup \{4, 5, 6\}$ does not appear on the right-hand side.

Proof Recall the definition of L, L_1, L_2 and L_3 after the statement of Proposition 2. We study several cases.

Case 1 $L = L_1 \geq \max(L_2, L_3)$. Without loss of generality, let $\arg\max L = \{1\}$. Then we have

$$\left|\Phi^{(6)}(x_{I_6})\right| = \left|\mathbb{E}(\mu(x_1)[\prod_{j\in\{1\}^c}\mu(x_j) - \Phi^{(5)}(x_{\{1\}^c})])\right| \leq \varrho(L)\|\mu\|_{L^2(\Omega)}\|\mu^5\|_{L^2}.$$

Meanwhile, $\text{dist}(x_1, x_j) \geq L$ for $j = 2, 3$. It follows that $\text{dist}(\{x_1\}, \{x_2, x_3\}) \geq L$ and

$$\left|\Phi^{(3)}(\{x_1, x_2, x_3\})\Phi^{(3)}(\{x_4, x_5, x_6\})\right| \leq \|\mu\|^3_{L^3(\Omega)}\|\mu\|_{L^2(\Omega)}\|\mu^2\|_{L^2(\Omega)}\varrho(L). \tag{45}$$

Finally, in view of (39), we may choose any $J \in \mathscr{I}_6^3$ such that $J \neq \{1, 2, 3\}$, and we verify that

$$\varrho(L) \leq \left(\varrho^{\frac{1}{4}}(\tfrac{1}{10}|x_{j_1} - x_{j_2}|)\varrho^{\frac{1}{4}}(\tfrac{1}{10}|x_{j_1} - x_{j_3}|)\right)\left(\varrho^{\frac{1}{4}}(\tfrac{1}{10}|x_{j_4} - x_{j_5}|)\varrho^{\frac{1}{4}}(\tfrac{1}{10}|x_{j_4} - x_{j_6}|)\right),$$

where $J = \{j_1, j_2, j_3\}$ and $J^c = \{j_4, j_5, j_6\}$. It follows that in this case,

$$\left|\Psi^{(3,3)}(\{x_1, x_2, x_3\}, \{x_4, x_5, x_6\})\right| \leq C\psi^{(3)}(\tfrac{1}{10}x_J)\psi^{(3)}(\tfrac{1}{10}x_{J^c}),$$

which is a term on the right hand side of (44).

Case 2 $L = L_2 \geq L_3$ and $L_2 > L_1$. Renaming the points, we assume L is obtained by the division $x_J = \{y_1, y_2\}$, $x_{J^c} = \{y_3, \cdots, y_6\}$ and $\text{dist}(y_2, y_3) = L$. Then we note that

$$\left|\Phi^{(6)}(x_{I_6}) - \Phi^{(2)}(\{y_1, y_2\})\Phi^{(4)}(\{y_3, \cdots, y_6\})\right| \leq \varrho(L)\|\mu^2\|_{L^2(\Omega)}\|\mu^4\|_{L^2(\Omega)}.$$

Moreover, if $x_J \subseteq \{x_1, x_2, x_3\}$ or $x_J \subseteq \{x_4, x_5, x_6\}$, then (45) holds. If otherwise, then we may assume that $y_2 = x_1$ and $y_3 = x_4$. Since $\{x_2, x_3\}$ must contain at least one point from x_{J^c}, and because $\text{dist}(\{x_1\}, x_{J^c}) = L$, we conclude that $\max_{j=2,3}\text{dist}(x_1, x_j) \geq L$. It follows that

$$\max_{j\in I_3}\text{dist}(\{x_j\}, x_{I_3} \setminus \{x_j\}) \geq L/2. \tag{46}$$

Then we have

$$\left|\Phi^{(3)}(\{x_1, x_2, x_3\})\Phi^{(3)}(\{x_4, x_5, x_6\})\right| \leq \|\mu\|^3_{L^3(\Omega)}\|\mu\|_{L^2(\Omega)}\|\mu^2\|_{L^2(\Omega)}\varrho(L/2). \tag{47}$$

We then repeat the argument in Case 1 to control the $\varrho(L)$ and $\varrho(L/2)$ terms. Finally, we get

$$\left|\Psi^{(3,3)}(\{x_1, x_2, x_3\}, \{x_4, x_5, x_6\})\right| \leq \left|R(y_1 - y_2)\Phi^{(4)}(\{y_3, y_4, y_5, y_6\})\right|$$
$$+ C\psi^{(3)}(\tfrac{1}{20}x_K)\psi^{(3)}(\tfrac{1}{20}x_{K^c})$$

where both K and K^c are different from $\{1, 2, 3\}$. For the $|R\Phi^{(4)}|$ term above, we combine the estimates (21) and (38) to get

$$\left| R(y_1 - y_2)\Phi^{(4)}(\{y_3, y_4, y_5, y_6\}) \right| \leq \frac{C}{2}\varrho^{1/3}(|y_1 - y_2|) \sum_J \eta(x_J)\eta(x_{\{3,4,5,6\}\setminus J})$$

where J runs in the set $\{J \subset \{3, 4, 5, 6\} \mid \text{card}(J) = 2\}$. Since by definition $\eta \leq \xi^{(2)}$, the error bound above is dominated by some term on the right-hand side of (44).

Case 3 $L = L_3 > \max(L_1, L_2)$. If $\text{dist}(\{x_1, x_2, x_3\}, \{x_4, x_5, x_5\}) = L$, then we have

$$\left| \Psi^{(3,3)}(\{x_1, x_2, x_3\}, \{x_4, x_5, x_6\}) \right| \leq \varrho(L)\|\mu^3\|^2_{L^2(\Omega)}.$$

If otherwise, $\text{dist}(\{x_1, x_2, x_3\}, \{x_4, x_5, x_5\}) < L$ and there exists $x_J = \{y_1, y_2, y_3\}$, with $x_J \neq \{1, 2, 3\}$ but $x_J \cap \{x_1, x_2, x_3\} \neq \emptyset$, such that $\text{dist}(x_J, x_{J^c}) = L$. Then we have

$$\left| \Phi^{(6)}(x_{I_6}) - \Phi^{(3)}(x_J)\Phi^{(3)}(x_{J^c}) \right| \leq \varrho(L)\|\mu^3\|^2_{L^2(\Omega)}.$$

Without loss of generality, assume $x_1 \in J$, $x_4 \in J^c$ and $\text{dist}(x_1, x_4) = L$. Then $\{x_2, x_3\} \cap J^c$ is non-empty, and the element in this intersection has distance larger than L from x_1. Hence, (46) holds, and the rest of the analysis can be carried out as in Case 2.

In all three cases, we can bound the left-hand side of (44) by some terms on the right-hand side, and, hence, the desired result is established.

As a corollary, we have the following estimate for the full sixth-order moments:

Corollary 1 *Let C be defined as in Proposition 6. Then*

$$\frac{1}{C}\left| \Phi^{(6)}(x_{I_6}) \right| \leq \sum_{(J_1, J_2, J_3) \in \mathscr{I}_6^{2,2,2}} \xi^{(2)}(\tfrac{1}{10}x_{J_1})\xi^{(2)}(\tfrac{1}{10}x_{J_2})\xi^{(2)}(\tfrac{1}{10}x_{J_3})$$

$$+ \frac{1}{2}\sum_{K \in \mathscr{I}_6^3} \psi^{(3)}(\tfrac{1}{20}x_K)\psi^3(\tfrac{1}{20}x_{K^c}).$$

Finally, we have the following result for the eighth-order moments. Define

$$\mathscr{I}_8^{2,2,2,2} := \{(J_1, J_2, J_3, J_4) : \text{card}(J_i) = 2, j = 1, 2, 3, 4, \text{ and } \cup_{i=1}^4 J_i = I_8\},$$

$$\mathscr{I}_8^{2,3,3} := \{(J_1, J_2, J_3) : \text{card}(J_1) = 2, \text{card}(J_2) = \text{card}(J_3) = 3, \cup_{i=1}^3 J_i = I_8\}.$$

$\mathscr{I}_8^{2,2,2,2}$ is the collection of partitions of I_8 into four mutually disjoint subsets, each of which has cardinality two. $\mathscr{I}_8^{2,3,3}$ is the collection of partitions of I_8 into four mutually disjoint subsets of cardinalities two, three and three, respectively.

Proposition 7 *Define* $\zeta^{(2)} = \sqrt{\eta}$ *and* $\phi^{(3)} = \sqrt{\psi^{(3)}}$. *Then there exists some constant* $C > 0$ *so that*

$$\frac{1}{C}\left|\Phi^{(8)}(x_{I_8})\right| \leq \sum_{(J_1,J_2,J_3,J_4)\in\mathscr{I}_8^{2,2,2,2}} \zeta^{(2)}(\tfrac{1}{14}x_{J_1})\zeta^{(2)}(\tfrac{1}{14}x_{J_2})\zeta^{(2)}(\tfrac{1}{14}x_{J_3})\zeta^{(2)}(\tfrac{1}{14}x_{J_4})$$

$$+ \sum_{J\in\mathscr{I}_8^{2,3,3}} \eta(\tfrac{1}{14}x_{J_1})\phi^{(3)}(\tfrac{1}{28}x_{J_2})\phi^{(3)}(\tfrac{1}{28}x_{J_3}).$$

This result can be proved using the same methods as in the proofs of Propositions 5 and 6. We do not reproduce the details.

The function $\phi^{(3)}$ defined above has the expression $\phi^{(3)}(\{x_1, x_2, x_3\}) = \varrho^{\frac{1}{8}}(x_2 - x_1)\varrho^{\frac{1}{8}}(x_3 - x_1)$. For the fluctuation theory that will be reviewed in Sect. 4, we need $\phi^{(3)}$ to be integrability for each of its variables, and hence $\varrho^{\frac{1}{8}}$ should decay sufficiently fast. This explains the integrability condition of the maximal correlation function ϱ that is required in assumption (A).

Acknowledgements GB's contribution was partially funded by NSF Grant DMS-1408867 and ONR Grant N00014-17-1-2096. OP's work is supported by NSF CAREER Grant DMS-1452349.

References

1. F. Bailly, J.F. Clouet, J.-P. Fouque, Parabolic and gaussian white noise approximation for wave propagation in random media. SIAM J. Appl. Math. **56**(5), 1445–1470 (1996)
2. G. Bal, Homogenization in random media and effective medium theory for high frequency waves. Discrete Contin. Dyn. Sys. Ser. B **8**(2), 473–492 (electronic) (2007)
3. G. Bal, Central limits and homogenization in random media. Multiscale Model. Simul. **7**(2), 677–702 (2008)
4. G. Bal, *Propagation of Stochasticity in Heterogeneous Media and Applications to Uncertainty Quantification* (Springer International Publishing, Cham, 2016), pp. 1–24
5. G. Bal, W. Jing, Homogenization and corrector theory for linear transport in random media. Discrete Contin. Dyn. Sys. **28**(4), 1311–1343 (2010)
6. G. Bal, W. Jing, Fluctuation theory for radiative transfer in random media. J. Quant. Spectrosc. Radiat. Transf. **112**(4), 660–670 (2011)
7. G. Bal, W. Jing, Corrector theory for elliptic equations in random media with singular Green's function. Application to random boundaries. Commun. Math. Sci. **19**(2), 383–411 (2011)
8. G. Bal, W. Jing, Corrector theory for MsFEM and HMM in random media. Multiscale Model. Simul. **9**, 1549–1587 (2011)
9. G. Bal, W. Jing, Corrector analysis of FEM-based multiscale algorithms for PDEs with random coefficients. ESAIM: Math. Model. Numer. Anal. **48**(2), 387–409 (2014)
10. G. Bal, A. Jollivet, Stability estimates in stationary inverse transport. Inverse Probl. Imag. **2**(4), 427–454 (2008)
11. G. Bal, O. Pinaud, Dynamics of wave scintillation in random media. Commun. Partial Differ. Equ. **35**(7), 1176–1235 (2010)
12. G. Bal, O. Pinaud, Imaging using transport models for wave-wave correlations. Math. Models Methods Appl. Sci. **21**(5), 1071–1093 (2011)

13. G. Bal, O. Pinaud, Analysis of the double scattering scintillation of waves in random media. Commun. Partial Differ. Equ. **38**(6), 945–984 (2013)
14. G. Bal, K. Ren, Transport-based imaging in random media. SIAM Applied Math. **68**(6), 1738–1762 (2008)
15. G. Bal, K. Ren, Physics-based models for measurement correlations. Application to an inverse Sturm-Liouville problem. Inverse Probl. **25**, 055006 (2009)
16. G. Bal, G. Papanicolaou, L. Ryzhik, Self-averaging in time reversal for the parabolic wave equation. Stochastics Dyn. **4**, 507–531 (2002)
17. G. Bal, T. Komorowski, L. Ryzhik, Self-averaging of Wigner transforms in random media. Commun. Math. Phys. **242**(1–2), 81–135 (2003)
18. G. Bal, J. Garnier, S. Motsch, V. Perrier, Random integrals and correctors in homogenization. Asymptot. Anal. **59**(1–2), 1–26 (2008)
19. G. Bal, T. Komorowski, L. Ryzhik, Kinetic limits for waves in random media. Kinetic Related Models **3**(4), 529–644 (2010)
20. G. Bal, I. Langmore, O. Pinaud, Single scattering estimates for the scintillation function of waves in random media. J. Math. Phys. **51**(2), 022903, 18 (2010)
21. G. Bal, J. Garnier, Y. Gu, W. Jing, Corrector theory for elliptic equations with long-range correlated random potential. Asymptot. Anal. **77**(3–4), 123–145 (2012)
22. A. Bamberger, E. Engquist, L. Halpern, P. Joly, Parabolic wave equation approximations in heterogeneous media. SIAM J. Appl. Math. **48**, 99–128 (1988)
23. A. Bensoussan, J.-L. Lions, G.C. Papanicolaou, Boundary layers and homogenization of transport processes. Publ. Res. Inst. Math. Sci. **15**(1), 53–157 (1979)
24. E. Bolthausen, On the central limit theorem for stationary mixing random fields. Ann. Probab. **10**(4), 1047–1050 (1982)
25. M. Butz, Kinetic limit for wave propagation in a continuous, weakly random medium. Ph.D. thesis, TU Munich, 2015
26. S. Chandrasekhar, *Radiative Transfer* (Dover, New York, 1960)
27. M. Choulli, P. Stefanov, An inverse boundary value problem for the stationary transport equation. Osaka J. Math. **36**(1), 87–104 (1999)
28. D.R. Cox, V. Isham, *Point Processes*. Monographs on Applied Probability and Statistics (Chapman & Hall, London, 1980)
29. A.-L. Dalibard, Homogenization of linear transport equations in a stationary ergodic setting. Commun. Partial Differ. Equ. **33**(4–6), 881–921 (2008)
30. R. Dautray, J.-L. Lions, *Mathematical Analysis and Numerical Methods for Science and Technology. Volume 6* (Springer, Berlin, 1993). Evolution problems. II, With the collaboration of Claude Bardos, Michel Cessenat, Alain Kavenoky, Patrick Lascaux, Bertrand Mercier, Olivier Pironneau, Bruno Scheurer and Rémi Sentis, Translated from the French by Alan Craig
31. L. Erdös, H.T. Yau, Linear Boltzmann equation as the weak coupling limit of a random Schrödinger Equation. Commun. Pure Appl. Math. **53**(6), 667–735 (2000)
32. A.C. Fannjiang, Self-averaging scaling limits for random parabolic waves. Arch. Ration. Mech. Anal. **175**(3), 343–387 (2005)
33. J. Garnier, K. Sølna, Coupled paraxial wave equations in random media in the white-noise regime. Ann. Appl. Probab. **19**(1), 318–346 (2009)
34. J. Garnier, K. Sølna, Scintillation in the white-noise paraxial regime. Commun. Partial Differ. Equ. **39**, 626–650 (2014)
35. J. Garnier, K. Sølna, Fourth-moment analysis for wave propagation in the white-noise paraxial regime. Arch. Ration. Mech. Anal. **220**, 37–81 (2016)
36. P. Gérard, P.A. Markowich, N.J. Mauser, F. Poupaud, Homogenization limits and Wigner transforms. Commun. Pure Appl. Math. **50**, 323–380 (1997)
37. C. Gomez, Radiative transport limit for the random Schrödinger equation with long-range correlations. J. Math. Pures. Appl. **98**, 295–327 (2012)
38. C. Gomez, O. Pinaud, Fractional white-noise limit and paraxial approximation for waves in random media. Arch. Ration. Mech. Anal. **226**(3), 1061–1138 (2017)

39. M. Hairer, E. Pardoux, A. Piatnitski, Random homogenisation of a highly oscillatory singular potential. Stoch. Partial Differ. Equ. Anal. Comput. **1**(4), 571–605 (2013)
40. W. Jing, Limiting distribution of elliptic homogenization error with periodic diffusion and random potential. Anal. Partial Differ. Equ. **9**(1), 193–228 (2016)
41. D. Khoshnevisan, *Multiparameter Processes. An Introduction to Random Fields*. Springer Monographs in Mathematics (Springer, New York, 2002)
42. T. Komorowski, L. Ryzhik, Fluctuations of solutions to Wigner equation with an Ornstein-Uhlenbeck potential. Discrete Contin. Dyn. Sys. B **17**, 871–914 (2012)
43. T. Komorowski, S. Peszat, L. Ryzhik, Limit of fluctuations of solutions of Wigner equation. Commun. Math. Phys. **292**(2), 479–510 (2009)
44. P.-L. Lions, T. Paul, Sur les mesures de Wigner. Rev. Mat. Iberoam. **9**, 553–618 (1993)
45. J. Lukkarinen, H. Spohn, Kinetic limit for wave propagation in a random medium. Arch. Ration. Mech. Anal. **183**, 93–162 (2007)
46. V. Pipiras, M.S. Taqqu, Integration questions related to fractional Brownian motion. Probab. Theory Related Fields **118**(2), 251–291 (2000)
47. L. Ryzhik, G. Papanicolaou, J.B. Keller, Transport equations for elastic and other waves in random media. Wave Motion **24**, 327–370 (1996)
48. P. Sheng, *Introduction to Wave Scattering, Localization and Mesoscopic Phenomena* (Academic, New York, 1995)
49. P. Stefanov, G. Uhlmann, An inverse source problem in optical molecular imaging. Anal. Partial Differ. Equ. **1**(1), 115–126 (2008)
50. F. Tappert, The parabolic approximation method, in *Wave Propagation and Underwater Acoustics*, ed. by J.B. Keller, J.S. Papadakis. Lecture Notes in Physics, vol. 70 (Springer, Berlin, 1977), pp. 224–287

Numerical Methods for High-Dimensional Kinetic Equations

Heyrim Cho, Daniele Venturi, and George Em Karniadakis

Abstract High-dimensionality is one of the major challenges in kinetic modeling and simulation of realistic physical systems. The most appropriate numerical scheme needs to balance accuracy and computational complexity, and it also needs to address issues such as multiple scales, lack of regularity, and long-term integration. In this chapter, we review state-of-the-art numerical techniques for high-dimensional kinetic equations, including low-rank tensor approximation, sparse grid collocation, and ANOVA decomposition.

1 Introduction

Kinetic equations are partial differential equations involving probability density functions (PDFs). They arise naturally in many different areas of mathematical physics. For example, they play an important role in modeling rarefied gas dynamics [12, 13], semiconductors [68], stochastic dynamical systems [18, 63, 74–76, 103, 114], structural dynamics [9, 60, 100], stochastic partial differential equations (PDEs) [19, 57, 66, 111, 112], turbulence [35, 71, 72, 90], system biology [30, 85, 123], etc. Perhaps, the most well-known kinetic equation is the Fokker-Planck equation [74, 96, 107], which describes the evolution of the probability

H. Cho (✉)
University of Maryland, College Park, MD, USA
e-mail: hcho1237@math.umd.edu

D. Venturi
University of California, Santa Cruz, CA, USA
e-mail: venturi@ucsc.edu

G. E. Karniadakis
Brown University, Providence, RI, USA
e-mail: george_karniadakis@brown.edu

© Springer International Publishing AG, part of Springer Nature 2017
S. Jin, L. Pareschi (eds.), *Uncertainty Quantification for Hyperbolic and Kinetic Equations*, SEMA SIMAI Springer Series 14,
https://doi.org/10.1007/978-3-319-67110-9_3

density function of Langevin-type dynamical systems subject to Gaussian white noise. Another well-known example of kinetic equation is the Boltzmann equation [115] describing a thermodynamic system involving a large number of interacting particles [13]. Other examples that may not be widely known are the Dostupov-Pugachev equations [26, 60, 103, 114], the reduced-order Nakajima-Zwanzig-Mori equations [24, 112, 127], and the Malakhov-Saichev PDF equations [66, 111] (see Table 1). Computing the numerical solution to a kinetic equation is a challenging task that needs to address issues such as:

1. *High-dimensionality:* Kinetic equations describing realistic physical systems usually involve many phase variables. For example, the Fokker-Planck equation of classical statistical mechanics is an evolution equation for a joint probability density function in n phase variables, where n is the dimension of the underlying stochastic dynamical system, plus time.
2. *Multiple scales:* Kinetic equations can involve multiple scales in space and time, which could be hardly accessible by conventional numerical methods. For example, the Liouville equation is a hyperbolic conservation law whose solution is purely advected (with no diffusion) by the underlying system's flow map. This can easily yield mixing, fractal attractors, and all sorts of complex dynamics.
3. *Lack of regularity:* The solution to a kinetic equation is, in general, a distribution [50]. For example, it could be a multivariate Dirac delta function, a function with shock-type discontinuities [19], or even a fractal object (see Figure 1 in [112]). From a numerical viewpoint, resolving such distributions is not trivial although in some cases it can be done by taking integral transformations or projections [120].
4. *Conservation properties:* There are several properties of the solution to a kinetic equation that must be conserved in time. The most obvious one is mass, i.e.,

Table 1 Examples of kinetic equations in different areas of mathematical physics

Fokker-Planck [74, 96]	$\dfrac{\partial p}{\partial t} + \sum\limits_{i=1}^{n} \dfrac{\partial}{\partial z_i}\left(G_i p\right) = \dfrac{1}{2}\sum\limits_{i,j=1}^{n} \dfrac{\partial^2}{\partial z_i \partial z_j}\left(b_{ij}p\right)$
Boltzmann [12, 67]	$\dfrac{\partial p}{\partial t} + \sum\limits_{j=1}^{3} v_j \dfrac{\partial p}{\partial z_j} = H(p,p)$
Liouville [26, 57, 60, 103]	$\dfrac{\partial p}{\partial t} + \sum\limits_{j=1}^{n} \dfrac{\partial}{\partial z_j}\left(G_j p\right) = 0$
Malakhov-Saichev [66, 111]	$\dfrac{\partial p}{\partial t} + \dfrac{\partial}{\partial z}\left(\sum\limits_{j=1}^{3} G_j \int_{-\infty}^{z} \dfrac{\partial p}{\partial x_j} dz'\right) = -\dfrac{\partial(Hp)}{\partial z}$
Mori-Zwanzig [112, 127]	$\dfrac{\partial p_1}{\partial t} = PLp_1 + PLe^{tQL}p_2(0) + PL\int_0^t e^{(t-s)QL}QLp_1\, ds$

the solution to a kinetic equation always integrates to one. Another property that must be preserved is the positivity of the joint PDF, and the fact that a partial marginalization still yields a PDF.

5. *Long-term integration:* The flow map defined by nonlinear dynamical systems can yield large deformations, stretching and folding of the phase space. As a consequence, numerical schemes for kinetic equations associated with such systems will generally loose accuracy in time. This is known as long-term integration problem and it can be eventually mitigated by using adaptive methods.

Over the years, many different techniques have been proposed to address these issues, with the most efficient ones being problem-dependent. For example, a widely used method in statistical fluid mechanics is the particle/mesh method [77, 89–91], which is based directly on stochastic Lagrangian models. Other methods are based on stochastic fields [109] or direct quadrature of moments [33]. In the case of Boltzmann equation, there is a very rich literature. Both probabilistic approaches such as direct simulation Monte Carlo [8, 97], as well as deterministic methods, e.g., discontinuous Galerkin and spectral methods [15, 16, 31], have been proposed to perform simulations. However, classical techniques such as finite-volumes, finite-differences or spectral methods, are often prohibitive in terms of memory requirements and computational cost. Probabilistic methods such as direct Monte Carlo are extensively used instead because of their very low computational cost compared to the classical techniques. However, Monte Carlo usually yields poorly accurate and fluctuating solutions, which need to be post-processed appropriately, for example through variance reduction techniques. We refer to Di Marco and Pareschi [67] for a recent excellent review.

In this chapter, we review the state-of-the-art in numerical techniques to address the high-dimensionality challenge in both the *phase space* and the *space of parameters* of kinetic systems. In particular, we discuss the sparse grid method [84, 102], low-rank tensor approximation [6, 17, 29, 40, 59, 79, 80], and analysis of variance (ANOVA) decomposition [11, 36, 61, 125] including Bogoliubov-Born-Green-Kirkwood-Yvon (BBGKY) [73] closures. These methods have been established as new tools to address high-dimensional problems in scientific computing during the last years, and here we discuss those in the context of kinetic equations, particularly in the deterministic Eulerian approach. As we will see, most of these methods allow us to reduce the problem of computing high-dimensional PDF solutions to sequences of problems involving low-dimensional PDFs. The range of applicability of the numerical methods is sketched in Fig. 1 as a function of the number of phase variables n and the number of parameters m appearing in the kinetic equation.

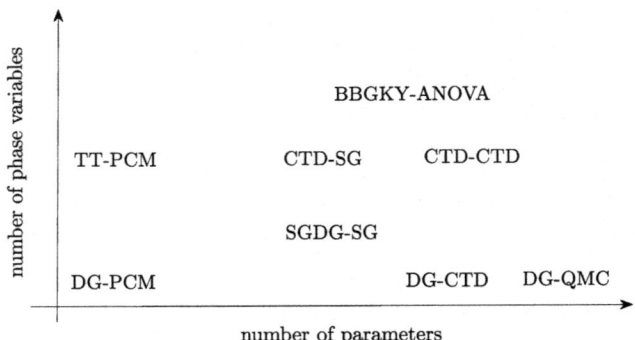

Fig. 1 Range of applicability of different numerical methods for solving kinetic equations as a function of the number of phase variables and the number of parameters appearing in the equation. The first name refers to the numerical method we employ to discretize the phase variables, the second name we employ to discretize the space of parameters. For example, DG-PCM refers to an algorithm in which we discretize the phase variables with discontinuous Galerkin methods (DG), and random parameters with tensor product probabilistic collocation (PCM). Other methods listed are: canonical tensor decomposition (CTD), tensor train (TT), high-dimensional model representation (ANOVA), sparse grids (SG), and quasi Monte Carlo (QMC). Reproduced with permission from [20]

2   Numerical Methods

This chapter discuss three classes of algorithms to compute the numerical solution of high-dimensional kinetic equations. The first class is based on sparse grids, and we discuss its construction in both the phase space and the space of parameters. The second class is based on low-rank tensor approximation and alternating direction methods, such as alternating least squares (ALS). The third class is based on ANOVA decomposition and BBGKY closures.

2.1   Sparse Grids

The sparse grid technique [10, 37] has been developed as a major tool to break the curse of dimensionality of grid-based approaches. The key idea relies on a tensor product hierarchical basis representation, which can reduce the degrees of freedom without losing much accuracy. Early work on sparse grid techniques can be traced back to Smolyak [102], in the context of high-dimensional numerical integration. The scheme is based on a proper balancing between the computational cost and the corresponding accuracy by seeking a proper truncation of the tensor product hierarchical bases, which can be formally derived by solving an optimization problem of cost/benefit ratios [41]. Sparse grid techniques have been incorporated in various numerical methods for high-dimensional PDEs, e.g., in finite element methods [10, 99], finite difference methods [42], spectral methods [38, 101], and collocation methods for stochastic differential equations [64, 78, 117]. More recently, sparse grids have been proposed within the discontinuous Galerkin (DG) framework to simulate elliptic and hyperbolic systems using wavelet bases [43, 116].

The sparse grid formulation is based on a hierarchical set of basis functions in one-dimension. For instance, we can consider basis functions in a space V_k of piecewise polynomials of degree at most q on the k-th level grid that consists of 2^k uniform intervals, i.e.,

$$V_k \doteq \{v \mid v \in P^q(I_k^j), \ I_k^j = [2^{-k}j, \ 2^{-k}(j+1)], j = 0, \cdots, 2^k - 1\},$$

on $\Omega = [0, 1]$. Clearly, we have

$$V_0 \subset V_1 \subset V_2 \subset V_3 \subset \cdots.$$

These basis functions are suitable for the discontinuous Galerkin framework. Then, we define W_k as the orthogonal complement of V_{k-1} on V_k with respect to the L_2 inner product on Ω, that is,

$$V_{k-1} \oplus W_k = V_k, \quad V_{k-1} \perp W_k,$$

with $W_0 = V_0$. This yields the hierarchical representation of

$$V_k = \oplus_{0 \leq j \leq k} W_j.$$

Next, define the multidimensional increment space as defined as $\mathbf{W}_l = W_{l_1, z_1} \otimes W_{l_2, z_2} \otimes \cdots \otimes W_{l_N, z_N}$ with $l = (l_1, \cdots, l_N)$ as the multivariate mesh level. Accordingly, the standard tensor product space $\mathbf{V}_\ell$ can be represented as

$$\mathbf{V}_\ell = \bigcup_{|l|_\infty \leq \ell} \mathbf{W}_l, \tag{1}$$

and the sparse grid approximation space as

$$\widetilde{\mathbf{V}}_\ell = \bigcup_{|l| \leq \ell} \mathbf{W}_l, \tag{2}$$

where $|l|_\infty = \max_i l_i$ and $|l| = \sum_{i=1}^N l_i$. Then, $\widetilde{\mathbf{V}}_\ell \subset \mathbf{V}_\ell$. The number of degrees of freedom of $\widetilde{\mathbf{V}}_\ell$ is significantly smaller than the one of $\mathbf{V}_\ell$. This set of basis functions is also called multi-wavelet basis and it has been employed with the discontinuous Galerkin method to study the Vlasov and the Boltzmann equations [43, 101]. In particular, for sufficiently smooth solutions, it was shown in [101, 116] that a semi-discrete L^2 stability condition and an error estimate of the order $O\left((\log h)^N h^{q+1/2}\right)$ can be obtained. We emphasize that although the computational cost of the sparse grid formulation is significantly smaller than the full tensor product, the curse of dimensionality still remains as the sparse grid level ℓ increases. For this reason, [43, 101, 116] can handle problems with less than ten dimensions in the phase space.

The application of the sparse grid technique in the space of parameters differs from the one we just described only in regard of the choice of the basis functions. In fact, in this case, we are usually interested in computing multi-dimensional integrals in the form

$$p(z) = \int_{\mathbb{R}^m} p(z, \mathbf{b}) d\mathbf{b} \simeq \sum_{k=1}^{q} w^k p(z, \mathbf{b}^k), \tag{3}$$

where $\mathbf{b} = (b_1, \ldots, b_m)$. The collocation points $\mathbf{b}^k = (b_1^k, \ldots, b_m^k)$ and quadrature weights w^k are obtained by suitable cubature rules with high polynomial exactness, e.g., Clenshaw-Curtis or Gauss abscissae [118]. More recent sparse collocation techniques can increase the number of dimensions that can be handled in the space of parameters up to hundreds [119, 122].

2.2   Low-Rank Tensor Approximation

Low-rank tensor approximation has been established as a new tool to overcome the curse of dimensionality in representing high-dimensional functions and the solution to high-dimensional PDEs. The method has been recently applied to stochastic PDEs [25, 29, 56, 69, 79], approximation of high-dimensional Green's functions [44], the Boltzmann equation [48, 55], and Fokker-Planck equation [2, 22, 49, 54]. The key idea of low-rank tensor approximation [17, 40, 81] is to represent a multivariate function in terms of series involving products of low-dimensional functions. This allows us to reduce the problem of computing the solution from high-dimensional PDEs to a sequence of low-dimensional problems that can be solved recursively and in parallel, e.g., by alternating direction algorithms such as alternating least squares [20, 25] and its parallel extension [52]. These algorithms are usually based on low-rank matrix techniques [39], and they have a convergence rate that depends on the type of kinetic equation and on its solution.

The most simplest tensor format is a rank one tensor of an N-dimensional function, $p(z_1, \cdots, z_N) = p_1(z_1) p_2(z_2) \cdots p_N(z_N)$, where $p_j(z_j)$ are one-dimensional functions. Upon discretization we can write p in a tensor notation as

$$\mathbf{p} = \mathbf{p}_1 \otimes \cdots \otimes \mathbf{p}_N, \tag{4}$$

where $\mathbf{p}_j$ is a vector of length q_z corresponding to the discretization of $p_j(z_j)$ with q_z degrees of freedom.[1] More generally, we have a summation of rank-one tensors

$$p(z_1, \cdots, z_N) = \sum_{r=1}^{R} \alpha_r p_1^r(z_1) p_2^r(z_2) \cdots p_N^r(z_N), \tag{5}$$

and

$$\mathbf{p} = \sum_{r=1}^{R} \alpha_r \mathbf{p}_1^r \otimes \mathbf{p}_2^r \otimes \cdots \otimes \mathbf{p}_N^r, \tag{6}$$

where R is the *tensor rank* or *separation rank*. This representation is also known as separated series expansion or canonical tensor decomposition. The main advantage of using a representation in the form (5)–(6) to solve a high-dimensional kinetic PDE relies on the fact that the algorithms to compute $\mathbf{p}_j^r$ and the normalization factors α_r involve operations with one-dimensional functions. In principle, the computational cost of such algorithms scales linearly with respect to the dimension N of the phase space, thus potentially avoiding the curse of dimensionality. The representation can be generalized to any combination of low-dimensional separated functions. Canonical tensor decompositions have been employed to compute the solution to the Malakhov-Saichev kinetic equation [20], the Vlasov-Poisson equation [27], and functional differential equations [110].

More advanced tensor decomposition techniques involve Tucker decomposition, tensor train decomposition (TT), and hierarchical Tucker decomposition (HT). In particular, the tensor train decomposition is in the form of

$$p(z_1, \cdots, z_N) = Q_1(z_1) Q_2(z_2) \cdots Q_N(z_N), \quad Q_j(z_j) \in \mathbb{R}^{R_{j-1} \times R_j}, \tag{7}$$

where the *tensor rank* becomes a tuple of $(R_1, \cdots, R_{N-1})$ with $R_0 = R_N = 1$. In each direction j, the index that runs over $\mathbb{R}^{R_{j-1}}$ and $\mathbb{R}^{R_j}$ takes care of the coupling to the $j-1$-th and the $j+1$-th dimension, respectively. A discretization of (7) with q_z degrees of freedom in each dimensions yields

$$\mathbf{p} = \sum_{r_0=1}^{R_0} \cdots \sum_{r_N=1}^{R_N} \mathbf{Q}_1^{r_0,r_1} \otimes \mathbf{Q}_2^{r_1,r_2} \otimes \cdots \otimes \mathbf{Q}_N^{r_{N-1},r_N}, \tag{8}$$

---

[1]For instance, if we represent $p_j(z_j)$ in terms of an interpolant

$$p_j(z_j) = \sum_{k=1}^{q_z} \mathrm{p}_{j,k} \phi_{j,k}(z_j),$$

then $\mathbf{p}_j = (\mathrm{p}_{j,1}, \cdots, \mathrm{p}_{j,q_z})$.

where $\mathbf{Q}_j^{r_{j-1},r_j}$ is a vector of length q_z. With a payoff of an additional tensor rank dimension, the problem of constructing a tensor train decomposition is closed and it can be solved to any given error tolerance or fixed rank [86]. The algorithm is based on a sequence of SVD applied to the matricizations of the tensor, i.e. the so-called high-order singular value decomposition (HOSVD) [39]. Methods for reducing the computational cost of tensor train are discussed in [82, 87, 126]. Applications to the Vlasov kinetic equation can be found in [23, 46, 58].

2.2.1 Temporal Dynamics

To include temporal dynamics in the low rank tensor representation of a field we can simply add additional time-dependent functions, i.e., represent $p(t, z_1, \ldots, z_N)$ as

$$p(t, z_1, \cdots, z_N) = \sum_{r=1}^{R} \alpha_r p_t^r(t) p_1^r(z_1) p_2^r(z_2) \cdots p_N^r(z_N). \tag{9}$$

This approach has been considered by several authors, e.g., [2, 17], and it was shown to be effective for problems dominated by diffusion. However, for complex transient problems (e.g., hyperbolic dynamics), such approach is not practical as it requires a high resolution in the time domain. To address this issue, a discontinuous Galerkin method in time was proposed by Nouy in [79]. The key idea is to split the integration period into small intervals (finite elements in time) and then consider a space-time separated representation of the solution within each interval.

Alternatively, one can consider an explicit or implicit time-integration schemes [20, 59]. In this case, the separated representation of the solution is computed at each time step. In such representations we look for expansions in the form

$$p(t, z_1, \cdots, z_N) = \sum_{r=1}^{R} \alpha_r(t) p_1^r(z_1, t) p_2^r(z_2, t) \cdots p_N^r(z_N, t). \tag{10}$$

Here, we demonstrate the procedure with reference to the simple Crank-Nicolson scheme. To this end, we consider the linear kinetic equation in the form

$$\frac{\partial p(\mathbf{z}, t)}{\partial t} = L(\mathbf{z}) p(\mathbf{z}, t), \tag{11}$$

where $\mathbf{z} = (z_1, \ldots, z_N)$ is the vector of phase variables and $L(\mathbf{z})$ is a linear operator. For instance, in the case of the Fokker-Planck equation we have

$$L(\mathbf{z}) = -\sum_{j=1}^{N} \left(\frac{\partial G_j}{\partial z_j} + G_j \frac{\partial}{\partial z_j} \right) + \frac{1}{2} \sum_{i,j=1}^{N} \left(\frac{\partial^2 b_{ij}}{\partial z_i \partial z_j} + b_{ij} \frac{\partial^2}{\partial z_i \partial z_j} + 2 \frac{\partial b_{ij}}{\partial z_i} \frac{\partial}{\partial z_j} \right).$$

We discretize (11) in time by using the Crank-Nicolson scheme. This yields

$$\frac{p(\mathbf{z}, t_{k+1}) - p(\mathbf{z}, t_k)}{\Delta t} = \frac{1}{2}\left(L(\mathbf{z})p(\mathbf{z}, t_{k+1}) + L(\mathbf{z})p(\mathbf{z}, t_k)\right) + \tau_k(\mathbf{z}), \qquad \Delta t = t_{k+1} - t_k,$$

i.e.,

$$\left(I - \frac{1}{2}\Delta t L(\mathbf{z})\right)p(\mathbf{z}, t_{k+1}) = \left(I + \frac{1}{2}\Delta t L(\mathbf{z})\right)p(\mathbf{z}, t_k) + \tau_k(\mathbf{z}), \tag{12}$$

where $\tau_k(\mathbf{z})$ is the truncation error arising from the temporal discretization. Assuming that $p(\mathbf{z}, t_k)$ is known, (12) is a linear equation for $p(\mathbf{z}, t_{k+1})$ which can be written concisely (at each time step) as

$$A(\mathbf{z})\, p(\mathbf{z}) = f(\mathbf{z}) + \tau(\mathbf{z}), \tag{13}$$

where

$$A(\mathbf{z}) \doteq \left(I - \frac{1}{2}\Delta t L(\mathbf{z})\right), \qquad f(\mathbf{z}) \doteq \left(I + \frac{1}{2}\Delta t L(\mathbf{z})\right)p(\mathbf{z}, t_k).$$

Note that we dropped the time t_{k+1} in $p(\mathbf{z}, t_{k+1})$ with the understanding that the linear system (13) has to be solved at each time step. We emphasize that other multistep and time-splitting schemes [27, 58]—including geometric integrators [45]—can be used instead of the Crank-Nicolson method.

2.2.2 Alternating Direction Algorithms

The low-rank tensor decomposition is particularly convenient when the system operator $A(\mathbf{z})$ and the right-hand-side $f(\mathbf{z})$ are separable with respect to $\mathbf{z}$, i.e.,

$$A(\mathbf{z}) = \sum_{k=1}^{n_A} A_1^k(z_1) \cdots A_N^k(z_N), \qquad f(\mathbf{z}) = \sum_{k=1}^{n_f} f_1^k(z_1) \cdots f_N^k(z_N). \tag{14}$$

Note that $A(\mathbf{z})$ is separable if $L(\mathbf{z})$ is separable. A simple example of a two-dimensional separable operator $L(\mathbf{z})$ with separation rank $n_L = 3$ is

$$L(z_1, z_2) = z_2 \frac{\partial^2}{\partial z_1 \partial z_2} + \sin(z_1) z_2 \frac{\partial^2}{\partial z_1^2} + e^{-z_1^2} \frac{\partial}{\partial z_2}. \tag{15}$$

Another example is the Liouville operator associated to nonlinear dynamical systems with polynomial nonlinearities. A substitution of the tensor representation

(5) into (12) yields the residual[2]

$$W(\mathbf{z}) = A(\mathbf{z})p(\mathbf{z}) - f(\mathbf{z}), \tag{16}$$

which depends on $\mathbf{z}$ and on all degrees of freedom associated with p_j^r. To determine such degrees of freedom we require that

$$\|W(\mathbf{z})\| = \|A(\mathbf{z})p(\mathbf{z}) - f(\mathbf{z})\| \leq \varepsilon, \tag{17}$$

in an appropriately chosen norm, and for a prescribed target accuracy ε. Ideally, the optimal tensor rank of can be defined as the minimal R such that the solution has an exact tensor decomposition with R terms, i.e., $\epsilon = 0$. However, the storage requirements and the computational cost increase with R, which makes the tensor decomposition attractive for small R. Therefore, we look for a low-rank tensor approximation of the solution to (13), with a reasonable accuracy ϵ. Although there are at present no useful theorems on the size R needed for a general class of functions, there are examples where tensor expansions are exponentially more efficient than one would expect a priori (see [6]).

Many existing algorithms to determine the best low-rank approximation of the solution to (13) are based on alternating direction methods. The key idea is to construct the tensor expansion (5) iteratively by determining $p_j^r(z_j)$ one at a time while freezing the degrees of freedom associated with all other functions. This yields a sequence of low-dimensional problems that can be solved efficiently [5, 6, 59, 79, 80, 83], eventually in parallel [52]. Perhaps, one of the first alternating direction algorithms to compute a low rank representation of the solution of a high-dimensional PDE was the one proposed in [2]. To clarify how the method works in simple terms, suppose we have constructed an approximated solution to the system (12) in the form (5), i.e., suppose we have available $p^R(\mathbf{z})$ with tensor rank R. Then we look for an *enriched* solution in the form

$$p^R(\mathbf{z}) + r_1(z_1) \cdots r_N(z_N),$$

where $\{r_1(z_1), \ldots, r_N(z_N)\}$ are N unknown functions to be determined. In the alternating direction method, such functions are determined iteratively, one at a time. Typical algorithms to perform such iterations are based on alternating least squares (ALS),

$$\min_{r_j} \left\| \sum_{k=1}^{n_A} A_1^k \cdots A_N^k \left(p^R + r_1 \cdots r_N \right) - \sum_{k=1}^{n_f} f_1^k \cdots f_N^k \right\|^2, \tag{18}$$

---

[2]The residual $W(\mathbf{z})$ incorporates both the truncation error arising from the time discretization as well as the error arising from the finite-dimensional expansion (5).

or alternating Galerkin methods,

$$\left\langle q, \sum_{k=1}^{n_A} A_1^k \cdots A_N^k \left(p^R + r_1 \cdots r_N \right) \right\rangle = \left\langle q, \sum_{k=1}^{n_f} f_1^k \cdots f_N^k \right\rangle, \qquad (19)$$

where $\langle \cdot \rangle$ is an inner product (multi-dimensional integral with respect to $\mathbf{z}$), and q is a test function, typically chosen as $q(\mathbf{z}) = r_1(z_1) \cdots \phi_{j,k}(z_j) \cdots r_N(z_N)$ for $k = 1, \ldots, q_z$. In a finite-dimensional setting, the minimization problem (18) reduces to the problem of finding the minimum of a scalar function in as many variables as the number of unknowns we consider in each basis function $r_j(z_j)$, say q_z. Similarly, the alternating direction solution to (19) yields a sequence of low-dimensional linear systems of size $q_z \times q_z$. If $A(\mathbf{z})$ is a nonlinear operator, then we can still solve (18) or (19), e.g., by using Newton iterations. Once the functions $\{r_1(z_1), \ldots, r_N(z_N)\}$ are computed, they are normalized (yielding the normalization factor α_{R+1}) and added to $p^R(\mathbf{z})$ to obtain $p^{R+1}(\mathbf{z})$. The tensor rank is increased until the norm of the residual (16) is smaller than the desired target accuracy ε (see Eq. (17)). We would like to emphasize that it is possible to include additional constraints when solving the linear system (13) with alternating direction algorithms. For example, one can impose that the solution $p(\mathbf{z})$ is positive and it integrates to one [59], i.e., it is a probability density function.

The enrichment procedure just described has been criticized in the literature due to its slow convergence rate, in particular for equations dominated by advection [79]. Depending on the criterion used to construct the tensor decomposition, the enrichment procedure might not even converge. To overcome this problem, Doostan and Iaccarino [25] proposed an alternating least-square (ALS) algorithm with granted convergence properties. The algorithm simultaneously updates the entire rank of the basis set in the j-th direction. In this formulation, the least square approach (18) becomes

$$\min_{\{p_j^1, \ldots, p_j^R\}} \left\| \sum_{k=1}^{n_A} A_1^k \cdots A_N^k \left(\sum_{r=1}^R \alpha_r p_1^r \cdots p_N^r \right) - \sum_{k=1}^{n_f} f_1^k \cdots f_N^k \right\|^2.$$

The computational cost of this method clearly increases compared to (18). In fact, in a finite dimensional setting, the simultaneous determination of $\{p_j^1, \ldots, p_j^R\}$ requires the solution of a $Rq_z \times Rq_z$ linear system. However, this algorithm usually results in a separated solution with a lower tensor rank R than the regular approach, which makes the algorithm more favorable to advection dominated kinetic systems. The basic idea of updating the entire rank of functions depending on a specific variable can be also applied to the alternating Galerkin formulation (19) (see [20]). In Sect. 4 we provide a numerical example of such algorithm—see also Algorithm 1.

Further developments and applications of low-rank tensor approximation methods can be found in the excellent reviews papers [3, 40, 81]. Gradient-based and Newton-like methods modifying and improving the basic ALS algorithm are

---

Algorithm 1 Alternating least squares with canonical tensor decomposition

---
Compute the tensor representation of the initial condition $\mathbf{p}(t_0)$
for $t_1 \leq t_k \leq t_{n_T}$ **do**
    Compute $\mathbf{f}$ by using $\mathbf{p}(t_{k-1})$
    Set $R = 1$
    while $\|A\mathbf{p}^R(t_k) - \mathbf{f}\| > \varepsilon$ **do**
        Initialize $\{\mathbf{p}_1^R(t_k), \ldots, \mathbf{p}_N^R(t_k)\}$ at random
        while $\|A\mathbf{p}^R(t_k) - \mathbf{f}\|$ does not decrease **do**
            Solve $\mathbf{B}_n \mathbf{p}_n^R = \mathbf{g}_n$ (38) in each direction for $1 \leq n \leq N$
        end while
        Normalize the basis set and solve $\mathbf{D}\boldsymbol{\alpha} = \mathbf{d}$ (39) to compute the coefficients $\boldsymbol{\alpha}$
        Set $R = R + 1$
    end while
end for

---

discussed in [1, 14, 28, 34, 53, 88, 93, 105, 106], Convergence of ALS and its parallel implementation has been studied in [21, 52, 70, 108].

2.3 ANOVA Decomposition and BBGKY Hierarchies

Another typical approach to model high-dimensional functions is based on the truncation of interactions. Hereafter we discuss two different methods to perform such approximation, namely, the ANOVA decomposition [11, 36, 61, 125] and the BBGKY (Bogoliubov-Born-Green-Kirkwood-Yvon) technique. Both these methods rely on a representation of multivariate functions in terms of series expansions involving functions with a smaller number of variables. For example, a second-order ANOVA approximation of a multivariate PDF in N variables is a series expansion involving functions of at most two variables. As we will see, both the ANOVA decomposition and the BBGKY technique [73] yield a hierarchy of coupled PDF equations for each given stochastic dynamical system. These methods are especially appropriate for anisotropic problems where dimensional adaptivity can be pursued.

The ANOVA series expansion [11, 41, 121] involves a superimposition of functions with an increasing number of variables. Specifically, the ANOVA decomposition of an N-dimensional PDF takes the from

$$p(z_1, z_2, \ldots, z_N) = p_0 + \sum_{i=1}^{N} p_i(z_i) + \sum_{i<j}^{N} p_{ij}(z_i, z_j) + \sum_{i<j<k}^{N} p_{ijk}(z_i, z_j, z_k) + \cdots.$$

$$(20)$$

The function p_0 is a constant. The functions $p_i(z_i)$, which we shall call first-order interaction terms, give us the overall effects of the variables z_i in p as if they were acting independently of the other variables. The functions $p_{ij}(z_i, z_j)$ represent the interaction effects of the variables z_i and z_j, and therefore they will be called second-order interactions. Similarly, higher-order terms reflect the cooperative effects of an increasing number of variables, and the series is usually truncated at a certain interaction order. These terms can be computed in different ways [92, 124], however, we point out the following procedure,

$$p_K(z_K) = \int p(z)d\mu(z_{K'}) - \sum_{S \subset K} p_T(z_S), \qquad (21)$$

where $S \subset K \subset \{1, \cdots, N\}$, K' is the complement of K in $\{1, \cdots, N\}$, $p_K(z_K) = p_{j_1,\ldots,j_k}(z_{j_1}, \cdots, z_{j_k})$ for $K = \{j_1, \cdots, j_k\}$, and μ is the Lebesgue measure. Due to its construction, this procedure generates ANOVA terms that are orthogonal with respect to μ, that is, $\int p_K(z_K)p_S(z_S)d\mu(z)$, for all $S \neq K$, which provides an effective criterion for dimensional adaptivity [65, 121].

The ANOVA expansion can be readily applied in the space of parameters of kinetic systems since the parameters do not depend on time and each terms computed at the initial time can be updated independently. To pursue a collocation approach similar to the sparse grid collocation method (3), we replace the Lebesgue measure with a Dirac measure $d\mu = \delta(z - c)$ at an appropriate anchor point c, and consider the corresponding collocation scheme [118]. This method is called the anchored-ANOVA method (PCM-ANOVA) [7, 32, 36, 121]. The anchor points are often taken as the mean value of the random variable in each dimension [125]. Then, each PDF equations in Table 1 can be solved at the *PCM-ANOVA collocation* points in the space of parameters.

On the other hand, representing the dependence of the solution PDF on the phase variables through the ANOVA expansion yields a hierarchy of coupled PDF equations that resembles the BBGKY hierarchy of classical statistical mechanics. Let us briefly review the BBGKY technique type with reference to a nonlinear dynamical system in the form

$$\dot{\mathbf{z}}(t) = \mathbf{G}(\mathbf{z}, t), \qquad \mathbf{z}(0) = z_0(\omega), \qquad (22)$$

where $\mathbf{z}(t) \in \mathbb{R}^N$ is a multi-dimensional stochastic process including both phase and parametric variables, $\mathbf{G} : \mathbb{R}^{N+1} \to \mathbb{R}^N$ is a Lipschitz continuous (deterministic) function, and $z_0 \in \mathbb{R}^N$ is a random initial state. The joint PDF of $\mathbf{z}(t)$ evolves according to the Liouville equation

$$\frac{\partial p(\mathbf{z}, t)}{\partial t} + \nabla \cdot [\mathbf{G}(\mathbf{z}, t)p(\mathbf{z}, t)] = 0, \qquad \mathbf{z} \in \mathbb{R}^N, \qquad (23)$$

whose solution can be computed numerically with standard discretization methods only for relatively small N. This leads us to look for PDF equations involving only a reduced number of phase variables, for instance, the PDF of each component $z_i(t)$. Such equations can be formally obtained by marginalizing (23) with respect to different phase variables and discarding terms at infinity. This yields, for example,

$$\frac{\partial p_i(z_i, t)}{\partial t} = -\frac{\partial}{\partial z_i} \int [G_i(\mathbf{y}, t)\delta(z_i - y_i(t))p(\mathbf{y}, t)] \, d\mathbf{y}, \tag{24}$$

$$\frac{\partial p_{ij}(z_i, z_j, t)}{\partial t} = -\frac{\partial}{\partial z_i} \int \left[G_i(\mathbf{y}, t)\delta(z_i - y_i(t))\delta(z_j - y_j(t))p(\mathbf{y}, t) \right] d\mathbf{y}$$

$$- \frac{\partial}{\partial z_j} \int \left[G_j(\mathbf{y}, t)\delta(z_i - y_i(t))\delta(z_j - y_j(t))p(\mathbf{y}, t) \right] d\mathbf{y}. \tag{25}$$

Higher-order PDF equations can be derived similarly. The computation of the integrals in (24) and (25) requires the full joint PDF of $\mathbf{z}(t)$, which is available only if we solve the full Liouville equation (23). Alternatively, we can solve (24) or (25) directly, provided we need to introduce approximations. The most common one is to assume that the joint PDF $p(\mathbf{z}, t)$ can be written in terms of lower-order PDFs, e.g., as $p(\mathbf{z}, t) = p(z_1, t) \cdots p(z_N, t)$ (mean-field approximation). By using integration by parts, this assumption reduces the Liouville equation to a hierarchy of low-dimensional PDF equations (see, e.g., [20, 112]). An example of such approximation will be presented later in this chapter with an application to Lorenz-96 model.

3   Computational Cost

Consider a kinetic partial differential equation with n phase variables and m parameters, i.e., a total number of $N = n + m$ variables. Suppose that we represent the solution by using q_z degrees of freedom in each phase variable and q_b degrees of freedom in each parameter. If we employ a tensor product discretization, the number of degrees of freedom becomes $q_z^n \cdot q_b^m$ and the computational cost grows exponentially as $O(q_z^{2n} \cdot q_b^m)$. Hereafter we compare the computational cost of the methods we discussed in the previous sections. Table 2 summarizes the main results.

Table 2 Number of degrees of freedom and computational cost of solving kinetic equations by using different methods

	Degrees of freedom	Computational cost
Sparse grids	$O(q_z \mid \log(q_z)\mid^{n-1})$	$O\left(q_z^2 \mid \log(q_z)\mid^{2n-2}\right)$
ANOVA or BBGKY	$\displaystyle\sum_{s=0}^{\ell} q_z^s \frac{n!}{(n-s)!s!}$	$O\left(q_z^{2\ell} \frac{n!}{(n-\ell)!\ell!}\right)$
Canonical tensor decomposition	$q_z R n$	$O\left(q_z^2 R^3 n\right)$

Shown are results for sparse grid, ANOVA decomposition, and low-rank canonical tensor decomposition. In the table, n is the phase space dimension in the kinetic equation assuming that the PDF solution is discretized with q_z degrees of freedom in each phase variable. Also, R is the tensor rank and ℓ is the interaction orders of the ANOVA expansion or the BBGKY closure

3.1 Sparse Grids

The computational complexity of sparse grids grows logarithmically with the number of degrees of freedom in each dimension, i.e., $O(q_z|\log_2(q_z)|^{n-1})$. If we employ the multi-wavelet basis we mentioned before in the context of the discontinuous Galerkin framework, then it can be shown that the computational complexity is $O((q_z + 1)^n 2^\ell \ell^{n-1})$, where ℓ is the element level and q_z is the polynomial order in each element (see [43]). In the space of parameters, the sparse grid collocation method yields $2^l(m + l)!/(m!l!)$ points, where l is the sparse grid level and m is the number of parameters. Thus, if we consider sparse grid in both phase and parametric space, the total computational cost can be estimated as $O(q_z^2|\log_2(q_z)|^{2n-2}) \cdot \sum_{l=0}^{\ell} 2^l(m + l)!/(m!l!)$.

3.2 Low-Rank Tensor Approximation

The total number of degrees of freedom in a low-rank tensor decomposition grows *linearly* with both n and m. For instance, we have $R(nq_z + mq_b)$ in the canonical tensor decomposition (6), and $R^2(nq_z + mq_b)$ in the tensor train approach (8). If the tensor rank R turns out to be relatively small, then the tensor approximation is far more efficient than full tensor product, sparse grid, or ANOVA approaches, in terms of memory requirements as well as the computational cost. The classical alternating direction algorithm at the basis of the canonical tensor decomposition can be divided into two steps, i.e., the enrichment and the projection steps (see Algorithm 1). The computational cost of the projection step can be neglected with respect to the one of the enrichment step, as it reduces to solving a linear system of rather small size ($r \times r$). The enrichment step at tensor rank r requires $O((rq_z)^2 + (rq_z)^2)$ operations—provided we employ appropriate iterative linear solvers. If we assume that the average number of iterations is n_{itr}, and sum up the cost for $r = 1, \ldots, R$, the overall computational cost of canonical tensor decomposition can be estimated as $O\left(R^3\left(nq_z^2 + mq_b^2\right)\right) \cdot n_{itr}$. In the tensor train approach, the cost also depends on the matrix rank S that comes from the procedure of HOSVD, and it becomes $O\left(R^2 S^2 nq_z^2 + R^3 S^3 nq_z\right)$ [58].

3.3 ANOVA Decomposition

If we consider the ANOVA expansion or the BBGKY hierarchy, the computational complexity has a factorial dependency on the dimensionality $n + m$ and the interaction orders of the variables [32]. In particular, the total number of degrees of freedom for a fixed interaction order ℓ and assuming $q_b = q_z$ is

$$\sum_{l=0}^{\ell} C(n + m, l, q_z) \quad \text{where} \quad C(N, l, q_z) = q_z^l \frac{N!}{(N - l)!l!}. \tag{26}$$

The computational cost of matrix-vector operations involving discretized variables in each level is $O\left(C(n + m, \ell, q_z^{2\ell})\right)$. It is possible to combine the BBGKY technique with the PCM-ANOVA approach to improve the accuracy, since the interaction order of the phase variables and the parameters, denoted as ℓ and ℓ', can be controlled separately. In this case, the total number of degrees of freedom and the corresponding computational cost become, $(\sum_{l=0}^{\ell} C(n, l, q_z)) \cdot (\sum_{l=0}^{\ell'} C(m, l, q_b))$ and $O\left(C(n, \ell, q_z^{2\ell}) \cdot (\sum_{l=0}^{\ell'} C(m, l, q_b))\right)$, respectively.

4 Applications

In this section, we present numerical examples to illustrate the performance and accuracy of the algorithms we discussed in this chapter. Specifically, we study the alternating Galerkin formulation (canonical tensor decomposition) of a kinetic model describing stochastic advection of a scalar field. We also study the BBGKY hierarchy of the Lorentz-96 model evolving from a random initial state.

4.1 Stochastic Advection of Scalar Fields

Let us consider the following stochastic advection equations

$$\frac{\partial u}{\partial t} + \left(1 + \sum_{k=1}^{m} \frac{1}{2k} \sin(kt)\xi_k(\omega)\right) \frac{\partial u}{\partial x} = 0, \tag{27}$$

$$\frac{\partial u}{\partial t} + \frac{\partial u}{\partial x} = \sin(t) \sum_{k=1}^{m} \frac{1}{5(k+1)} \sin((k+1)x)\xi_k(\omega), \tag{28}$$

where $x \in [0, 2\pi]$ and $\{\xi_1, \ldots, \xi_m\}$ are i.i.d. uniform random variables in $[-1, 1]$. The kinetic equations governing the joint probability density function of $\{\xi_1, \ldots, \xi_m\}$ and the solution to (27) or (28) are, respectively,

$$\frac{\partial p}{\partial t} + \left(1 + \sum_{k=1}^{m} \frac{1}{2k} \sin(kt)b_k\right) \frac{\partial p}{\partial x} = 0, \tag{29}$$

$$\frac{\partial p}{\partial t} + \frac{\partial p}{\partial x} = -\left(\sin(t) \sum_{k=1}^{m} \frac{1}{5(k+1)} \sin((k+1)x)b_k\right) \frac{\partial p}{\partial a}, \tag{30}$$

where $p = p(x, t, a, \mathbf{b})$, $\mathbf{b} = \{b_1, \ldots, b_m\}$ (see [111] for a derivation). Note that this PDF depends on x, t, one phase variable a (corresponding to $u(x, t)$), and m parameters $\mathbf{b}$ (corresponding to $\{\xi_1, \ldots, \xi_m\}$). The analytical solutions to Eqs. (29)

and (30) can be obtained by using the method of characteristics [95]. They are both in the form

$$p(x, t, a, \mathbf{b}) = p_0(x - X(t, \mathbf{b}), a - A(x, t, \mathbf{b}), \mathbf{b}) \tag{31}$$

where

$$X(t, \mathbf{b}) = t - \sum_{k=1}^{m} \frac{(\cos(kt) - 1)b_k}{2k^2}, \qquad A(x, t, \mathbf{b}) = 0 \tag{32}$$

in the case of Eq. (29) and

$$X(t, \mathbf{b}) = t, \quad A(x, t, \mathbf{b}) = \sum_{k=2}^{m+1} \frac{b_{k-1}}{10k} \left(\frac{\sin(kx - t)}{k - 1} - \frac{\sin(kx + t)}{k + 1} - \frac{2\sin(k(x - t))}{(k - 1)(k + 1)} \right) \tag{33}$$

in the case of Eq. (30). Also, $p_0(x, a, \mathbf{b})$ is the joint PDF of $u(x, t_0)$ and $\{\xi_1, \ldots, \xi_m\}$. In our simulations we take

$$p_0(x, a, \mathbf{b}) = \frac{1}{2} \left(\frac{\sin^2(x)}{2\pi\sigma_1} \exp\left[-\frac{(a - \mu_1)^2}{2\sigma_1} \right] + \frac{\cos^2(x)}{2\pi\sigma_2} \exp\left[-\frac{(a - \mu_2)^2}{2\sigma_2} \right] \right),$$

which has tensor rank $R = 2$. Non-separable initial conditions can be approximated in the tensor format (5). Also, we consider different number of parameters in Eqs. (29) and (30), i.e., $m = 3, 13, 24, 54, 84, 114$.

4.1.1 Finite-Dimensional Representations

Let us represent the joint probability density function (5) in terms of polynomial basis functions as

$$p_n^r(z_n) = \sum_{k=1}^{q_z} \mathrm{p}_{n,k}^r \phi_{n,k}(z_n), \tag{34}$$

where q_z is the number of degrees of freedom in each variable. In particular, for (29) and (30), we consider a spectral collocation method in which $\{\phi_{1,j}\}$ and $\{\phi_{2,j}\}$ are trigonometric polynomials, while $\{\phi_{n,j}\}_{n=3}^{N}$ (basis elements for the space of parameters) are Lagrange interpolants at Gauss-Legendre-Lobatto points. The finite-dimensional representation of the joint PDF admits the following canonical tensor form

$$\mathbf{p} = \sum_{r=1}^{R} \alpha_d \mathbf{p}_1^r \otimes \cdots \otimes \mathbf{p}_N^r,$$

where the vector

$$\mathbf{p}_n^r = \left[\mathbf{p}_{n,1}^r, \cdots, \mathbf{p}_{n,q_z}^r\right],$$

collects the (normalized) values of the solution at the collocation points. The fully discrete Galerkin formulation of our kinetic equations takes the form

$$\mathbf{A}\mathbf{p} = \mathbf{f}, \tag{35}$$

where

$$\mathbf{A} = \sum_{k=1}^{n_A} \mathbf{A}_1^k \otimes \cdots \otimes \mathbf{A}_N^k, \qquad \mathbf{f} = \sum_{k=1}^{n_f} \mathbf{f}_1^k \otimes \cdots \otimes \mathbf{f}_N^k, \tag{36}$$

$$\mathbf{A}_n^k[i,j] = \int \phi_{n,i}(z_n) A_n^k(z_n) \phi_{n,j}(z_n) \, dz_n, \qquad \mathbf{f}_n^k[i] = \int f_n^k(z_n) \phi_{n,i}(z_n) \, dz_n. \tag{37}$$

By using a Gauss quadrature rule to evaluate the integrals, we obtain system matrices $\mathbf{A}_n^k$ that are either diagonal or coincide with the classical differentiation matrices of spectral collocation methods [47]. For example, in the case of Eq. (29) we have

$$\mathbf{A}_1^1[i,j] = \mathbf{w}_x[i]\delta_{ij}, \quad \mathbf{A}_1^k[i,j] = \frac{\Delta t}{2}\mathbf{w}_x[i]\mathscr{D}_x[i,j], \ k = 2, \ldots, n_A,$$

$$\mathbf{A}_2^1[i,j] = \mathbf{A}_2^2[i,j] = \mathbf{w}_z[i]\delta_{ij}, \quad \mathbf{A}_2^{k+2}[i,j] = \frac{\sin(kt_{n+1})}{2k}\mathbf{w}_z[i]\delta_{ij}, \ k = 1, \ldots, m,$$

$$\mathbf{A}_3^k[i,j] = \mathbf{w}_b[i]\delta_{ij}, \ k \neq 3, \quad \mathbf{A}_3^3[i,j] = \mathbf{w}_b[i]\mathbf{q}_b[i]\delta_{ij}, \quad \cdots$$

where $\mathbf{q}_b$ denotes the vector of collocation points, $\mathbf{w}_x$, $\mathbf{w}_z$, and $\mathbf{w}_b$ are collocation weights, $\mathscr{D}_x$ is the differentiation matrix, and δ_{ij} is the Kronecker delta function. In an alternating direction setting, we aim at solving the system (35) in a greedy way, by freezing all degrees of freedom except those representing the dimension n. This yields a sequence of linear systems

$$\mathbf{B}_n \mathbf{p}_n^R = \mathbf{g}_n, \tag{38}$$

where $\mathbf{B}_n$ is a block matrix with $R \times R$ blocks of size $q_z \times q_z$, and $\mathbf{g}_n$ is multi-component vector. Specifically, the hv-th block of $\mathbf{B}_n$ and the h-th component of $\mathbf{g}_n$ are obtained as

$$\mathbf{B}_n^{hv} = \sum_{k=1}^{n_A} \left(\prod_{i \neq n}^{N} \left[\mathbf{p}_i^h\right]^T \mathbf{A}_i^k \mathbf{p}_i^v\right) \mathbf{A}_n^k, \qquad \mathbf{g}_n^h = \sum_{k=1}^{n_f} \left(\prod_{i \neq n}^{N} \left[\mathbf{p}_i^h\right]^T \mathbf{f}_i^k\right) \mathbf{f}_n^k.$$

The solution vector

$$\mathbf{p}_n^R = \left[\mathbf{p}_n^1, \dots, \mathbf{p}_n^R\right]^T$$

is normalized as $\mathbf{p}_n^r / \left\|\mathbf{p}_n^r\right\|$ for all $r = 1, .., R$ and $n = 1, \dots, N$. This operation yields the coefficients $\boldsymbol{\alpha} = (\alpha_1, \dots, \alpha_R)$ as a solution to the linear systems

$$\mathbf{D}\boldsymbol{\alpha} = \mathbf{d}, \tag{39}$$

where the entries of the matrix $\mathbf{D}$ and the vector $\mathbf{d}$ are, respectively

$$\mathbf{D}^{hv} = \sum_{k=1}^{n_A} \prod_{i=1}^{N} \left[\mathbf{p}_i^h\right]^T \mathbf{A}_i^k \mathbf{p}_i^v, \qquad \mathbf{d}^h = \sum_{k=1}^{n_f} \prod_{i=1}^{N} \left[\mathbf{p}_i^h\right]^T \mathbf{f}_i^k.$$

The main steps of the computational scheme are summarized in Algorithm 1. We also refer the reader to [21, 70, 108] for a convergence analysis of the alternating direction algorithm.

The iterative procedure at each time step is terminated when the norm of the residual is smaller than a tolerance, i.e., when $\left\|\mathbf{A}\mathbf{p}^R - \mathbf{f}\right\| \le \varepsilon$. This usually involves the computation of an N-dimensional tensor norm, which can be expensive and compromise the computational efficiency of the whole algorithm. To avoid this problem, we replace the condition $\left\|\mathbf{A}\mathbf{p}^R - \mathbf{f}\right\| \le \varepsilon$ with the simpler convergence criterion

$$\max \left\{ \frac{\left\|\widetilde{\mathbf{p}}_1^R - \mathbf{p}_1^R\right\|}{\left\|\mathbf{p}_1^R\right\|}, \dots, \frac{\left\|\widetilde{\mathbf{p}}_N^R - \mathbf{p}_N^R\right\|}{\left\|\mathbf{p}_N^R\right\|} \right\} \le \varepsilon_1, \tag{40}$$

where $\{\widetilde{\mathbf{p}}_1^R, \dots, \widetilde{\mathbf{p}}_N^R\}$ denotes the solution at the previous iteration. This criterion involves the computation of N vector norms instead of one N-dimensional tensor norm.

4.1.2 Numerical Results: Low-Rank Tensor Approximation

We compute the solution to the kinetic equations (29) and (30) by using Algorithm 1. The PDF solution is represented in the canonical tensor format as

$$p(x, t, a, \mathbf{b}) \simeq \sum_{r=1}^{R} \alpha_r(t) p_x^r(x, t) p_a^r(a, t) P_1^r(b_1, t) \cdots P_m^r(b_m, t). \tag{41}$$

We chose the degrees of freedom of the expansion to carefully balance the error between the space and time discretization, as well as the truncation error due to the finite rank R. In particular, x and a are discretized in terms of an interpolant with

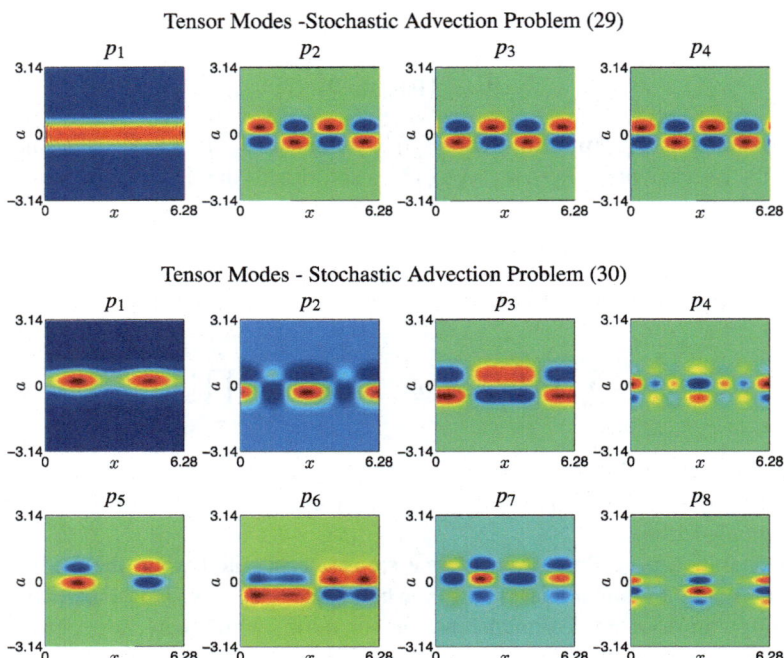

Fig. 2 Tensor modes of the kinetic solution to the stochastic advection equations at $t = 2$. Reproduced with permission from [20]

collocation points $q_z = 50$ in each variable, while the parametric dependence on b_j ($j = 1, .., m$) is represented with Legendre polynomials of order $q_b = 7$.

In Fig. 2 we plot the first few tensor modes $p_r(x, a, t) \doteq p_x^r(x, t)p_a^r(a, t)$ of the solution to Eqs. (29) and (30) at time $t = 2$. Specifically, we considered $m = 54$ in (29) and $m = 3$ in (30). Note that the tensor modes we obtain from Eq. (29), p_r, are very similar to each other for $r \geq 2$, while in the case of Eq. (30) the modes are quite distinct, suggesting the presence of modal interactions and the need of a larger tensor rank to achieve a certain accuracy. This is also observed in Fig. 3, where we plot the normalization coefficients $\{\alpha_1, \dots, \alpha_R\}$, which can be interpreted as the spectrum of the tensor solution. The stochastic advection problem with random forcing yields a stronger coupling between the tensor modes, i.e., a slower spectral decay than the problem of random coefficient.

In Fig. 4 we plot the error of the low-rank tensor approximation of the solution versus the number of parameters m for different tensor rank R. As it is predicted from the spectra shown in Fig. 3, the overall relative error of the solution in the random forcing case is larger than in the random coefficient case (see also Fig. 5 for the convergence with respect to R). This is due to the presence of the time-dependent forcing term in Eq. (28), which injects additional energy in the system and activates new modes. This yields a higher tensor rank for a prescribed

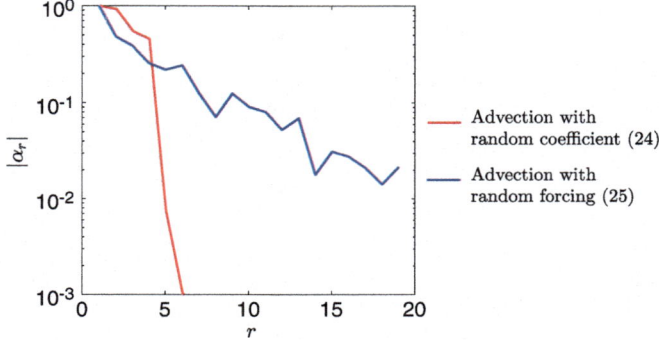

Fig. 3 Spectra of the canonical tensor decomposition of the stochastic advection problem at $t = 2$. Reproduced with permission from [20]

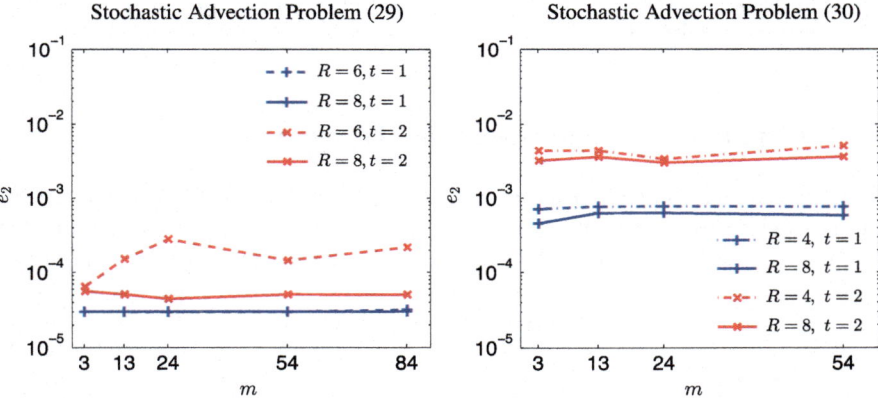

Fig. 4 Relative L_2 errors of the low-rank tensor approximation of the solution with respect to the analytical solution (31). Shown are results for different number of random variables m in (27)–(28) and different tensor ranks R. It is seen that the accuracy of the tensor method mainly depends on the actual tensor rank rather than on the dimensionality. Reproduced with permission from [20]

level of accuracy. In addition, the plots suggest that the accuracy of the low-rank tensor approximation method depends primarily on the tensor rank rather than on the number of parameters of the problem. The choice of the tensor format that yields the smallest possible tensor rank for a specific problem is an open question. Recent studies suggest that the answer is usually problem-dependent. For instance, Kormann [58] has recently shown that a semi-Lagrangian solver for the Vlasov equation in tensor train format achieves best performances if the phase variables are sorted as $(v_1, x_1, x_2, v_2, x_3, v_3)$.

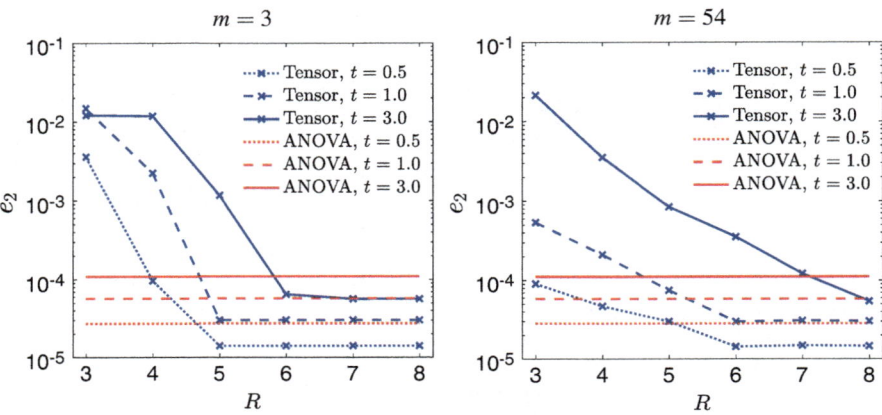

Fig. 5 Relative L_2 errors of the tensor solution and the ANOVA solution of level $\ell = 2$ with respect to the analytical solution (31). Shown are results of the kinetic equation (29) at $t = 0.5$, $t = 1$ and $t = 3$ for different tensor ranks R and dimensionality m. Reproduced with permission from [20]

4.1.3 Comparison Between Tensor Approximation and ANOVA

In this section we compare the accuracy and the computational cost of the low-rank alternating Galerkin method with the ANOVA expansion technique to compute the solution to Eqs. (29) and (30). The PCM-ANOVA representation of the solution is

$$p(x, t, a, \mathbf{b}) \simeq \sum_{|K| \leq \ell} p_K(t, x, a) P_K(b_K). \tag{42}$$

For $\ell = 2$ (level 2) and m parameters, the expansion (42) has $1 + m + m(m-1)/2$ terms.

In Fig. 5 we compare the accuracy of the low-rank tensor approximation and the PCM-ANOVA expansion in computing the solution to the kinetic equation (29). In particular, the convergence of the tensor solution with respect to R is demonstrated. Note that the tensor solution attains the same level of accuracy as the ANOVA decomposition with just five modes for $t \leq 1$. Therefore the low-rank tensor approximation is preferable over ANOVA especially when $m \geq 54$. However, this is not true in the case of Eq. (30) due to its relatively large tensor ranks. To overcome this problem, we developed an adaptive algorithm that sets the separation rank of the solution based on a prescribed target accuracy on the residual of the kinetic equation, or other quantities related to it.

In Fig. 6 (left) we plot the temporal dynamics of the tensor rank $R(t)$ obtained by setting a threshold on the spectral condition number defined as the ratio between the smallest and the largest α_i. Specifically, we increase R by one at $t = t^*$ whenever the following condition is verified $\alpha_R(t^*)/\alpha_1(t^*) > \theta$. For a small threshold θ, we

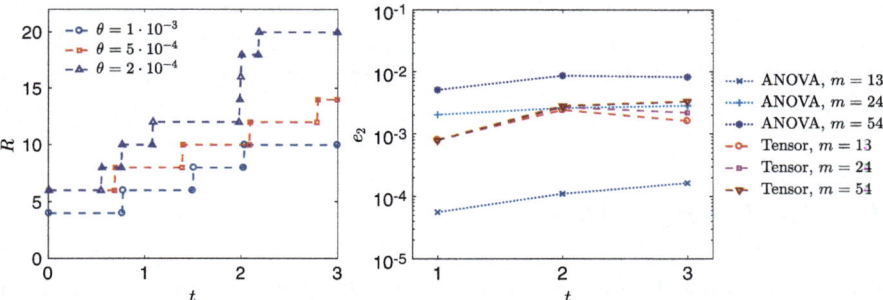

Fig. 6 Comparison between the relative L_2 errors of the adaptive tensor method and the ANOVA method of level $\ell = 2$. Results are for the kinetic equation (30) with threshold $\theta = 5 \times 10^{-4}$. It is seen that the error of the tensor solution is slightly independent of m, while the error of ANOVA level 2 increases as we increase m. Reproduced with permission from [20]

notice that R can increase to 20 and more at later times. This result reveals two key aspects of efficient tensor algorithms in practical applications. It is essential to develop a robust adaptive procedure that can identify the proper tensor rank *on-the-fly* and an effective compression technique that can reduce the tensor rank in time. This is critical especially when computing long term behavior of kinetic systems.

In Fig. 6 (right) we plot the error of the adaptive tensor method and the level 2 ANOVA method versus time. It is seen that error in the tensor method is almost independent of m, while the error of ANOVA increases with m. The accuracy can be improved either by increasing the tensor rank (canonical tensor decomposition) or increasing the interaction order (ANOVA method). Before doing so, however, one should carefully examine the additional computational cost incurred by each method. For example, increasing the interaction order from two to three in the PCM-ANOVA expansion would increase the number of collocation points from 70,498 to 8,578,270 (case $m = 54$). In Fig. 7 we compare the computational cost of canonical

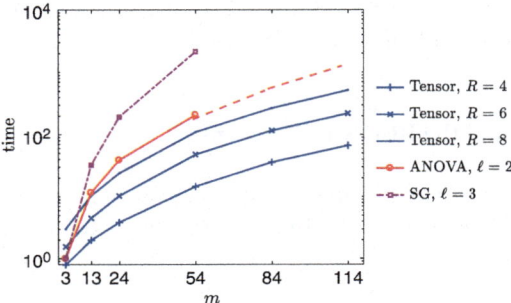

Fig. 7 Computational time of the tensor decomposition, ANOVA level 2, and sparse grid (SG) level 3 with respect to the dimensionality m and the tensor rank R. The results are normalized with respect to the computing time of ANOVA when $m = 3$. Reproduced with permission from [20]

tensor decomposition with different ranks, ANOVA of level two, and sparse grid of level three in computing the solution to Eq. (30). It is seen that the tensor method is the most efficient one, in particular for high dimensions and low tensor rank, e.g., $m \geq 24$ and $R \leq 8$.

4.2   The Lorenz-96 Model

The Lorenz-96 model is a continuous in time and discrete in space model often used in atmospheric sciences to study fundamental issues related to forecasting and data assimilation [51, 62]. The basic equations are

$$\frac{dx_i}{dt} = (x_{i+1} - x_{i-2}) x_{i-1} - x_i + F, \qquad i = 1, \ldots, n. \tag{43}$$

Here we consider $n = 40$, $F = 1$, and assume that the initial state $[x_1(0), \ldots, x_{40}(0)]$ is jointly Gaussian with PDF

$$p_0(z_1, \ldots, z_{40}) = \left(\frac{25}{2\pi}\right)^{20} \prod_{i=1}^{40} \exp\left[-\frac{25}{2}\left(z_i - \frac{i}{40}\right)^2\right]. \tag{44}$$

Without an additional parametric space, the dimensionality of this system is $n = 40$. The kinetic equation governing the joint PDF of the phase variables $[x_1(t), \ldots, x_{40}(t)]$ is

$$\frac{\partial p(\mathbf{z}, t)}{\partial t} = -\sum_{i=1}^{40} \frac{\partial}{\partial z_i} \left[((z_{i+1} - z_{i-2})z_{i-1} - z_i + F) p(\mathbf{z}, t)\right], \quad \mathbf{z} \in \mathbb{R}^{40}. \tag{45}$$

Such hyperbolic conservation law cannot be obviously solved in a classical tensor product representation because of high-dimensionality and possible lack of regularity (for $F > 10$) related to the fractal structure of the attractor [51]. Thus, we are led to look for reduced-order PDF equations.

4.2.1   Truncation of the BBGKY Hierarchy

In this section we illustrate how to compute low order probability density function equations by truncations of the BBGKY hierarchy. To this end, consider the dynamical system

$$\frac{dy_i}{dt} = G_i(\mathbf{y}, t),$$

where

$$G_i(\mathbf{y}, t) = g_{ii}(y_i, t) + \sum_{\substack{k=1 \\ k \neq i}}^{N} g_{ik}(y_i, y_k, t).$$

With such velocity field $G_i(\mathbf{y}, t)$ we can calculate the integrals at the right hand side of the one-point PDF equation (24) exactly as

$$\frac{\partial p_i}{\partial t} = -\frac{\partial}{\partial z_i} \left[g_{ii}(z_i, t)p_i + \sum_{k \neq i}^{N} \int g_{ik}(z_i, z_k, t)p_{ik}dz_k \right], \tag{46}$$

where $p_i = p(z_i, t)$ and $p_{ik} = p(z_i, z_k, t)$. Similarly, the two-point PDF equations (25) can be approximated as

$$\frac{\partial p_{ij}}{\partial t} = -\frac{\partial}{\partial z_i} \left[\left(g_{ii}(z_i, t) + g_{ij}(z_i, z_j, t) \right) p_{ij} + \left(\sum_{k \neq i,j}^{N} \int g_{ik}(z_i, z_k, t)p_{ik}dz_k \right) p_j \right]$$
$$- \frac{\partial}{\partial z_j} \left[\left(g_{jj}(z_j, t) + g_{ji}(z_j, z_i, t) \right) p_{ij} + \left(\sum_{k \neq i,j}^{N} \int g_{jk}(z_j, z_k, t)p_{jk}dz_k \right) p_i \right], \tag{47}$$

where we discarded all contributions from the three-point PDFs and the two-point PDFs except the ones interacting with the i-th variable. A variance-based sensitivity analysis in terms of Sobol indices [98, 104, 113] can be performed to identify the system variables with strong correlations. This allows us to determine whether it is necessary to add the other two-points correlations or the three-points PDF equations for a certain triple $\{x_k(t), x_i(t), x_j(t)\}$, and to further determine the equation for a general form of G_i.

In the specific case of the Lorenz-96 system, we can write Eq. (46) as

$$\frac{\partial p_i}{\partial t} = -\frac{\partial}{\partial z_i} \left[(\langle z_{i+1} \rangle - \langle z_{i-2} \rangle) \langle z_{i-1} \rangle_{i-1|i} - (z_i - F)p_i \right], \tag{48}$$

where $\langle f(\mathbf{z}) \rangle_{i|j} \doteq \int f(\mathbf{z})p_{ij}(z_i, z_j, t)dz_i$. In order to close such a system within the level of one-point PDFs, $\langle z_{i-1} \rangle_{i-1|i}$ could be replaced, e.g., by $\langle z_{i-1} \rangle p_i(z_i, t)$.

Similarly, Eq. (47) can be written for the two adjacent nodes as

$$\frac{\partial p_{ii+1}}{\partial t} = -\frac{\partial}{\partial z_i}\left[z_{i+1}\langle z_{i-1}\rangle_{i-1|i} p_{i+1} - \langle z_{i-2}\rangle\langle z_{i-1}\rangle_{i-1|i} p_{i+1} - (z_i - F)p_{ii+1}\right]$$

$$-\frac{\partial}{\partial z_{i+1}}\left[\langle z_{i+2}\rangle_{i+2|i+1} z_i p_i - \langle z_{i-1}\rangle z_i p_{ii+1} - (z_{i+1} - F)p_{ii+1}\right]. \qquad (49)$$

By adding the two-points closure of one node apart, i.e., $p_{i-1 i+1}(z_{i-1}, z_{i+1}, t)$, the quantity $\langle z_{i-2}\rangle\langle z_{i-1}\rangle_{i-1|i} p_{i+1}$ in the first row and $\langle z_{i-1}\rangle z_i p_{ii+1}$ in the second row can be substituted by $\langle z_{i-2}\rangle_{i-2|i}\langle z_{i-1}\rangle_{i-1|i+1}$ and $\langle z_{i-1}\rangle_{i-1|i+1} z_i p_i$, respectively. In Fig. 8, we compare the mean and the standard deviation of the solution to (43) as computed by the one- and two-points BBGKY closures (Eqs. (48) and (49), respectively) and a Monte Carlo simulation with 50,000 solution samples. It is seen that the mean of both the one-point and the two-points BBGKY closures basically coincide with the Monte Carlo results. On the other hand, the error in standard deviation is slightly different, and it can be improved in the two-points BBGKY closure (Fig. 9).

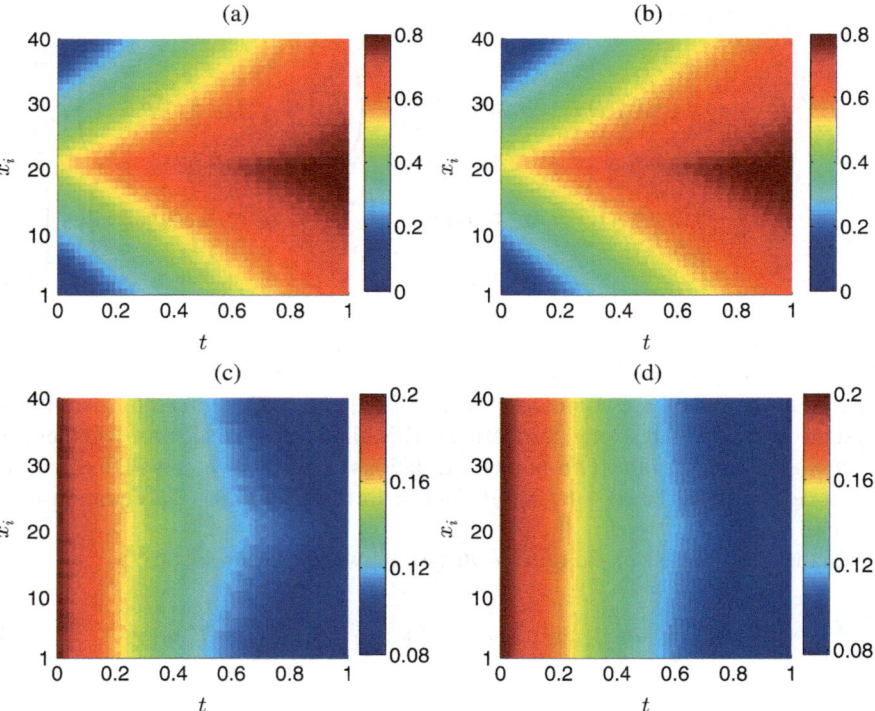

Fig. 8 Mean (**a**, **b**) and standard deviation (**c**, **d**) of the Lorenz-96 system computed by the one-point (**a**) and two-points (**c**) BBGKY closure compared to the Monte-Carlo simulation (**b**, **d**). Reproduced with permission from [20]

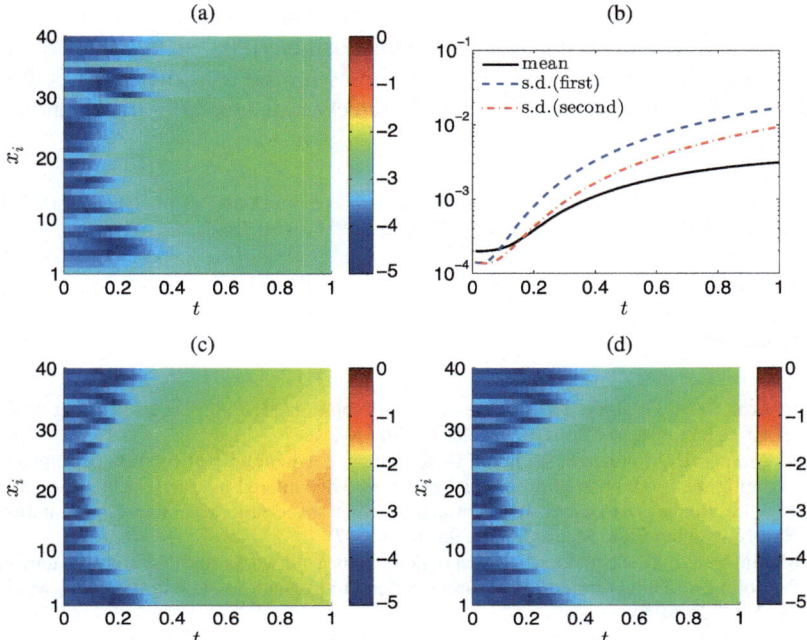

Fig. 9 The absolute error of the mean (**a**) and standard deviation (**c, d**) of the Lorenz-96 system by using the BBGKY closure compared to the Monte-Carlo simulation in log-scale. In (**c**) and (**d**), the results are computed by the one- and two-points BBGKY closure, respectively, and the L_1 error is shown in (**b**). Reproduced with permission from [20]

5 Summary

In this chapter we reviewed state-of-the-art algorithms to compute the numerical solution of high-dimensional kinetic equations. The algorithms are based on low-rank tensor approximation, sparse grids, and ANOVA decomposition. A common feature of these methods is that they allow us to reduce the problem of computing the solution to a high-dimensional PDE to a sequence of low-dimensional problems. The range of applicability of the algorithms is sketched in Fig. 1 as a function of the number of phase variables and the number of parameters appearing in the kinetic equation. The computational complexity ranges from logarithmic (sparse grids) to linear (canonical tensor decomposition) with respect to the dimension of the system. Further extensions of the proposed algorithms can be addressed along different directions. For example, adaptive procedures capable of resolving different phase variables with different accuracy may allow applications to kinetic systems with non-smooth solutions and scaling to extremely high-dimensions. In the context of low-rank tensor approximation methods [20, 27, 58], a fundamental question is the development of effective techniques for rank reduction [4, 94]. This is especially challenging for hyperbolic PDEs, since such equations can yield a slow convergence rate when solved with canonical tensor decompositions [20, 79]. Future work should address the development of adaptive algorithms for the construction of controlled low-rank approximations and an adaptive selection of separation ranks and tensor formats.

Acknowledgements We gratefully acknowledge support from DARPA grant N66001-15-2-4055, ARO grant W991NF-14-1-0425, and AFOSR grant FA9550-16-1-0092.

References

1. E. Acar, D.M. Dunlavy, T.G. Kolda, A scalable optimization approach for fitting canonical tensor decompositions. J. Chemom. **25**, 67–86 (2011)
2. A. Ammar, B. Mokdad, F. Chinesta, R. Keunings, A new family of solvers for some classes of multidimensional partial differential equations encountered in kinetic theory modelling of complex fluids: part II: transient simulation using space-time separated representations. J. Non-Newtonian Fluid Mech. **144**(2), 98–121 (2007)
3. M. Bachmayr, R. Schneider, A. Uschmajew, Tensor networks and hierarchical tensors for the solution of high-dimensional partial differential equations. Found. Comput. Math. **16**, 1423–1472 (2016)
4. C. Battaglino, G. Ballard, T.G. Kolda, A practical randomized CP tensor decomposition (2017). arXiv: 1701.06600
5. G. Beylkin, M.J. Mohlenkamp, Algorithms for numerical analysis in high dimensions. SIAM J. Sci. Comput. **26**(6), 2133–2159 (2005)
6. G. Beylkin, J. Garcke, M.J. Mohlenkamp, Multivariate regression and machine learning with sums of separable functions. SIAM J. Sci. Comput. **31**(3), 1840–1857 (2009)
7. M. Bieri, C. Schwab, Sparse high order FEM for elliptic SPDEs. Comput. Methods Appl. Mech. Eng. **198**, 1149–1170 (2009)
8. G.A. Bird, *Molecular Gas Dynamics and Direct Numerical Simulation of Gas Flows* (Clarendon Press, Oxford, 1994)
9. V.V. Bolotin, *Statistical Methods in Structural Mechanics* (Holden-Day, San Francisco, 1969)
10. H.J. Bungartz, M. Griebel, Sparse grids. Acta Numer. **13**, 147–269 (2004)
11. Y. Cao, Z. Chen, M. Gunzbuger, ANOVA expansions and efficient sampling methods for parameter dependent nonlinear PDEs. Int. J. Numer. Anal. Model. **6**, 256–273 (2009)
12. C. Cercignani, *The Boltzmann Equation and Its Applications* (Springer, New York, 1988)
13. C. Cercignani, U.I. Gerasimenko, D.Y. Petrina, *Many Particle Dynamics and Kinetic Equations*, 1st edn. (Kluwer Academic Publishers, Dordrecht, 1997)
14. Y. Chen, D. Han, L. Qi, New ALS methods with extrapolating search directions and optimal step size for complex-valued tensor decompositions. IEEE Trans. Signal Process. **59**, 5888–5898 (2011)
15. Y. Cheng, I.M. Gamba, A. Majorana, C.W. Shu, A discontinuous Galerkin solver for Boltzmann-Poisson systems in nano devices. Comput. Methods Appl. Mech. Eng. **198**, 3130–3150 (2009)
16. Y. Cheng, I.M. Gamba, A. Majorana, C.W. Shu, A brief survey of the discontinuous Galerkin method for the Boltzmann-Poisson equations. SEMA J. **54**, 47–64 (2011)
17. F. Chinesta, A. Ammar, E. Cueto, Recent advances and new challenges in the use of the proper generalized decomposition for solving multidimensional models. Comput. Methods. Appl. Mech. Eng. **17**(4), 327–350 (2010)

18. H. Cho, D. Venturi, G.E. Karniadakis, Adaptive discontinuous Galerkin method for response-excitation PDF equations. SIAM J. Sci. Comput. **35**(4), B890–B911 (2013)
19. H. Cho, D. Venturi, G.E. Karniadakis, Statistical analysis and simulation of random shocks in Burgers equation. Proc. R. Soc. A **260**, 20140080(1–21) (2014)
20. H. Cho, D. Venturi, G.E. Karniadakis, Numerical methods for high-dimensional probability density function equations. J. Comput. Phys. **305**, 817–837 (2016)
21. P. Comon, X. Luciani, A.L.F. de Almeida, Tensor decompositions, alternating least squares and other tales. J. Chemom. **23**, 393–405 (2009)
22. S.V. Dolgov, B.N. Khoromskij, I.V. Oseledets, Fast solution of parabolic problems in the tensor train/quantized tensor train format with initial application to the Fokker-Planck equation. SIAM J. Sci. Comput. **34**(6), A3016–A3038 (2012)
23. S.V. Dolgov, A.P. Smirnov, E.E. Tyrtyshnikov, Low-rank approximation in the numerical modeling of the Farley-Buneman instability in ionospheric plasma. J. Comput. Phys. **263**, 268–282 (2014)
24. J. Dominy, D. Venturi, Duality and conditional expectations in the Nakajima-Mori-Zwanzig formulation. J. Math. Phys. **58**, 082701(1–26) (2017)
25. A. Doostan, G. Iaccarino, A least-squares approximation of partial differential equations with high-dimensional random inputs. J. Comput. Phys. **228**(12), 4332–4345 (2009)
26. B.G. Dostupov, V.S. Pugachev, The equation for the integral of a system of ordinary differential equations containing random parameters. Automatika i Telemekhanica (in Russian) **18**, 620–630 (1957)
27. V. Ehrlacher, D. Lombardi, A dynamical adaptive tensor method for the Vlasov-Poisson system. J. Comput. Phys. **339**, 285–306 (2017)
28. M. Espig, W. Hackbusch, A regularized Newton method for the efficient approximation of tensors represented in the canonical tensor format. Numer. Math. **122**, 489–525 (2012)
29. M. Espig, W. Hackbusch, A. Litvinenko, H.G. Matthies, P. Wähnert, Efficient low-rank approximation of the stochastic Galerkin matrix in tensor formats. Comput. Math. Appl. **67**(4), 818–829 (2014)
30. A. Fiasconaro, B. Spagnolo, A. Ochab-Marcinek, E. Gudowska-Nowak, Co-occurrence of resonant activation and noise-enhanced stability in a model of cancer growth in the presence of immune response. Phys. Rev. E **74**(4), 041904 (2006)
31. F. Filbet, G. Russo, High-order numerical methods for the space non-homogeneous Boltzmann equations. J. Comput. Phys. **186**, 457–480 (2003)
32. J. Foo, G.E. Karniadakis, Multi-element probabilistic collocation method in high dimensions. J. Comput. Phys. **229**, 1536–1557 (2010)
33. R.O. Fox, *Computational Models for Turbulent Reactive Flows* (Cambridge University Press, Cambridge, 2003)
34. S. Friedland, V. Mehrmann, R. Pajarola, S.K. Suter, On best rank one approximation of tensors. Numer. Linear Algebra Appl. **20**, 942–955 (2013)
35. U. Frisch, Turbulence: the legacy of A. N. Kolmogorov (Cambridge University Press, Cambridge, 1995)
36. Z. Gao, J.S. Hesthaven, On ANOVA expansions and strategies for choosing the anchor point. Appl. Math. Comput. **217**(7), 3274–3285 (2010)
37. J. Garcke, M. Griebel, *Sparse Grids and Applications* (Springer, Berlin, 2013)
38. V. Gradinaru, Fourier transform on sparse grids: code design and the time dependent Schrödinger equation. Computing **80**(1), 1–22 (2007)
39. L. Grasedyck, Hierarchical singular value decomposition of tensors. SIAM J. Matrix Anal. Appl. **31**, 2029–2054 (2010)
40. L. Grasedyck, D. Kressner, C. Tobler, A literature survey of low-rank tensor approximation techniques. GAMM Mitteilungen **36**(1), 53–78 (2013)
41. M. Griebel, Sparse grids and related approximation schemes for higher dimensional problems, in *Foundations of Computational Mathematics Santander 2005*, vol. 331, ed. by L.M. Pardo, A. Pinkus, E. Süli, M.J. Todd (Cambridge University Press, Cambridge, 2006), pp. 106–161

42. M. Griebel, G. Zumbusch, Adaptive sparse grids for hyperbolic conservation laws, in *Hyperbolic Problems: Theory, Numerics, Applications* (Springer, Berlin, 1999), pp. 411–422
43. W. Guo, Y. Cheng, An adaptive multiresolution discontinuous Galerkin method for time-dependent transport equations in multi-dimensions. SIAM J. Sci. Comput. **38**(6), 1–29 (2016)
44. W. Hackbusch, B.N. Khoromskij, Tensor-product approximation to multidimensional integral operators and Green's functions. SIAM J. Matrix Anal. Appl. **30**(3), 1233–1253 (2008)
45. E. Hairer, C. Lubich, G. Wanner, Geometric numerical integration illustrated by the Störmer-Verlet method. Acta Numer. **12**, 399–450 (2003)
46. D.R. Hatch, D. del Castillo-Negrete, P.W. Terry, Analysis and compression of six-dimensional gyrokinetic datasets using higher order singular value decomposition. J. Comput. Phys. **22**, 4234–4256 (2012)
47. J.S. Hesthaven, S. Gottlieb, D. Gottlieb, *Spectral Methods for Time-Dependent Problems* (Cambridge University Press, Cambridge, 2007)
48. I. Ibragimov, S. Rjasanow, Three way decomposition for the Boltzmann equation. J. Comput. Math. **27**, 184–195 (2009)
49. T. Jahnke, W. Huisinga, A dynamical low-rank approach to the chemical master equation. Bull. Math. Biol. **70**, 2283–2302 (2008)
50. R.P. Kanwal, *Generalized Functions: Theory and Technique*, 2nd edn. (Birkhäuser, Boston, 1998)
51. A. Karimi, M.R. Paul, Extensive chaos in the Lorenz-96 model. Chaos **20**(4), 043105(1–11) (2010)
52. L. Karlsson, D. Kressner, A. Uschmajew, Parallel algorithms for tensor completion in the CP format. Parallel Comput. **57**, 222–234 (2016)
53. V.A. Kazeev, E.E. Tyrtyshnikov, Structure of the Hessian matrix and an economical implementation of Newton's method in the problem of canonical approximation of tensors. Comput. Math. Math. Phys. **50**, 927–945 (2010)
54. V. Kazeev, M. Khammash, M. Nip, C. Schwab, Direct solution of the chemical master equation using quantized tensor trains. Semin. Appl. Math. **2013-04**, 2283–2302 (2013)
55. B.N. Khoromskij, Structured data-sparse approximation to high order tensors arising from the deterministic Boltzmann equation. Math. Comput. **76**(259), 1291–1315 (2007)
56. B.N. Khoromskij, I.V. Oseledets, Quantics-TT collocation approximation of parameter-dependent and stochastic elliptic PDEs. Comput. Methods Appl. Math. **10**(4), 376–394 (2010)
57. V.I. Klyatskin, *Dynamics of Stochastic Systems* (Elsevier, Amsterdam, 2005)
58. K. Kormann, A semi-lagrangian Vlasov solver in tensor train format. SIAM J. Sci. Comput. **37**(4), B613–B632 (2015)
59. G. Leonenko, T. Phillips, On the solution of the Fokker-Planck equation using a high-order reduced basis approximation. Comput. Methods Appl. Mech. Eng. **199**(1-4), 158–168 (2009)
60. J. Li, J.B. Chen, *Stochastic Dynamics of Structures* (Wiley, Singapore, 2009)
61. G. Li, S.W. Wang, H. Rabitz, S. Wang, P. Jaffé, Global uncertainty assessments by high dimensional model representations (HDMR). Chem. Eng. Sci. **57**(21), 4445–4460 (2002)
62. E.N. Lorenz, Predictability - a problem partly solved, in *ECMWF Seminar on Predictability*, Reading, vol. 1 (1996), pp. 1–18
63. D. Lucor, C.H. Su, G.E. Karniadakis, Generalized polynomial chaos and random oscillators. Int. J. Numer. Methods Eng. **60**(3), 571–596 (2004)
64. X. Ma, N. Zabaras, An adaptive hierarchical sparse grid collocation method for the solution of stochastic differential equations. J. Comput. Phys. **228**, 3084–3113 (2009)
65. X. Ma, N. Zabaras, An adaptive high-dimensional stochastic model representation technique for the solution of stochastic partial differential equations. J. Comput. Phys. **229**, 3884–3915 (2010)
66. A.N. Malakhov, A.I. Saichev, Kinetic equations in the theory of random waves. Radiophys. Quantum Electron. **17**(5), 526–534 (1974)
67. G.D. Marco, L. Pareschi, Numerical methods for kinetic equations. Acta Numer. **23**, 369–520 (2014)
68. P. Markovich, C. Ringhofer, C. Schmeiser, *Semiconductor Equations* (Springer, Berlin, 1989)

69. H.G. Matthies, E. Zander, Solving stochastic systems with low-rank tensor compression. Linear Algebra Appl. **436**(10), 3819–3838 (2012)
70. M.J. Mohlenkamp, Musings on multilinear fitting. Linear Algebra Appl. **438**, 834–852 (2013)
71. A.S. Monin, A.M. Yaglom, *Statistical Fluid Mechanics*, vol. I (Dover, Mineola, 2007)
72. A.S. Monin, A.M. Yaglom, *Statistical Fluid Mechanics*, vol. II (Dover, Mineola, 2007)
73. D. Montgomery, A BBGKY framework for fluid turbulence. Phys. Fluids **19**(6), 802–810 (1976)
74. F. Moss, P.V.E. McClintock (eds.), *Noise in Nonlinear Dynamical Systems*. Volume 1: Theory of Continuous Fokker-Planck Systems (Cambridge University Press, Cambridge, 1995)
75. F. Moss, P.V.E. McClintock (eds.), *Noise in Nonlinear Dynamical Systems*. Volume 2: Theory of Noise Induced Processes in Special Applications (Cambridge University Press, Cambridge, 1995)
76. F. Moss, P.V.E. McClintock (eds.), *Noise in Nonlinear Dynamical Systems*. Volume 3: Experiments and Simulations (Cambridge University Press, Cambridge, 1995)
77. M. Muradoglu, P. Jenny, S.B. Pope, D.A. Caughey, A consistent hybrid finite-volume/particle method for the PDF equations of turbulent reactive flows. J. Comput. Phys. **154**, 342–371 (1999)
78. F. Nobile, R. Tempone, C. Webster, A sparse grid stochastic collocation method for partial differential equations with random input data. SIAM J. Numer. Anal. **46**(5), 2309–2345 (2008)
79. A. Nouy, A priori model reduction through proper generalized decomposition for solving time-dependent partial differential equations. Comput. Methods Appl. Mech. Eng. **199**(23-24), 1603–1626 (2010)
80. A. Nouy, Proper generalized decompositions and separated representations for the numerical solution of high dimensional stochastic problems. Comput. Methods Appl. Mech. Eng. **17**, 403–434 (2010)
81. A. Nouy, Low-rank tensor methods for model order reduction, in *Handbook of Uncertainty Quantification* (Springer International Publishing, Berlin, 2016), pp. 1–26
82. A. Nouy, Higher-order principal component analysis for the approximation of tensors in tree-based low rank formats. 1–43 (2017). arXiv:1705.00880
83. A. Nouy, O.P.L. Maître, Generalized spectral decomposition for stochastic nonlinear problems. J. Comput. Phys. **228**, 202–235 (2009)
84. E. Novak, K. Ritter, Simple cubature formulas with high polynomial exactness. Constr. Approx. **15**, 499–522 (1999)
85. D. Nozaki, D.J. Mar, P. Grigg, J.J. Collins, Effects of colored noise on stochastic resonance in sensory neurons. Phys. Rev. Lett. **82**(11), 2402–2405 (1999)
86. I.V. Oseledets, Tensor-train decomposition. SIAM J. Sci. Comput. **33**(5), 2295–2317 (2011)
87. A.H. Phan, P. Tichavský, A. Cichocki, CANDECOMP/PARAFAC decomposition of high-order tensors through tensor reshaping. IEEE Trans. Signal Process. **61**, 4847–4860 (2013)
88. A.H. Phan, P. Tichavský, A. Cichocki, Low complexity damped Gauss-Newton algorithms for CANDECOMP/ PARAFAC. SIAM J. Matrix Anal. Appl. **34**, 126–147 (2013)
89. S.B. Pope, A Monte Carlo method for the PDF equations of turbulent reactive flow. Combust. Sci. Technol. **25**, 159–174 (1981)
90. S.B. Pope, Lagrangian PDF methods for turbulent flows. Annu. Rev. Fluid Mech. **26**, 23–63 (1994)
91. S.B. Pope, Simple models of turbulent flows. Phys. Fluids **23**(1), 011301(1–20) (2011)
92. H. Rabitz, Ö.F. Aliş, J. Shorter, K. Shim, Efficient input–output model representations. Comput. Phys. Commun. **117**(1-2), 11–20 (1999)
93. M. Rajih, P. Comon, R.A. Harshman, Enhanced line search: a novel method to accelerate PARAFAC. SIAM J. Matrix Anal. Appl. **30**, 1128–1147 (2008)
94. M. Reynolds, G. Beylkin, A. Doostan, Optimization via separated representations and the canonical tensor decomposition. J. Comput. Phys. **348**(1), 220–230 (2016)
95. H.K. Rhee, R. Aris, N.R. Amundson, *First-Order Partial Differential Equations*. Volume 1: Theory and Applications of Single Equations (Dover, New York, 2001)

96. H. Risken, *The Fokker-Planck Equation: Methods of Solution and Applications* (Springer, Berlin, 1989)
97. S. Rjasanow, W. Wagner, *Stochastic Numerics for the Boltzmann Equation* (Springer, Berlin, 2004)
98. A. Saltelli, K. Chan, M. Scott, *Sensitivity Analysis* (Wiley, New York, 2000)
99. C. Schwab, E. Suli, R.A. Todor, Sparse finite element approximation of high-dimensional transport-dominated diffusion problems. ESAIM: Math. Model. Numer. Anal. **42**, 777–819 (2008)
100. M.F. Shlesinger, T. Swean, *Stochastically Excited Nonlinear Ocean Structures* (World Scientific, Singapore, 1998)
101. R. Shu, J. Hu, S. Jin, A stochastic Galerkin method for the Boltzmann equation with multi-dimensional random inputs using sparse wavelet bases. Numer. Math. Theor. Methods Appl. **10**(2), 465–488 (2017)
102. S. Smolyak, Quadrature and interpolation formulas for tensor products of certain classes of functions. Sov. Math. Dokl. **4**, 240–243 (1963)
103. K. Sobczyk, *Stochastic Differential Equations: With Applications to Physics and Engineering* (Springer, Berlin, 2001)
104. I.M. Sobol, Global sensitivity indices for nonlinear mathematical models and their Monte Carlo estimates. Math. Comput. Simul. **55**, 271–280 (2001)
105. H.D. Sterck, A nonlinear GMRES optimization algorithm for canonical tensor decomposition. SIAM J. Sci. Comput. **34**, A1351–A1379 (2012)
106. H.D. Sterck, K. Miller, An adaptive algebraic multigrid algorithm for low-rank canonical tensor decomposition. SIAM J. Sci. Comput. **35**, B1–B24 (2012)
107. R.L. Stratonovich, Some Markov methods in the theory of stochastic processes in nonlinear dynamical systems, in *Noise in Nonlinear Dynamical Systems*, vol. 1, ed. by F. Moss, P.V.E. McClintock (Cambridge University Press, Cambridge, 1989), pp. 16–68
108. A. Uschmajew, Local convergence of the alternating least squares algorithm for canonical tensor approximation. SIAM J. Matrix Anal. Appl. **33**, 639–652 (2012)
109. L. Valino, A field Monte Carlo formulation for calculating the probability density function of a single scalar in a turbulent flow. Flow Turbul. Combust. **60**(2), 157–172 (1998)
110. D. Venturi, The numerical approximation of functional differential equations. 1–113 (2016). arXiv: 1604.05250
111. D. Venturi, G.E. Karniadakis, New evolution equations for the joint response-excitation probability density function of stochastic solutions to first-order nonlinear PDEs. J. Comput. Phys. **231**(21), 7450–7474 (2012)
112. D. Venturi, G.E. Karniadakis, Convolutionless Nakajima-Zwanzig equations for stochastic analysis in nonlinear dynamical systems. Proc. R. Soc. A **470**(2166), 1–20 (2014)
113. D. Venturi, M. Choi, G.E. Karniadakis, Supercritical quasi-conduction states in stochastic Rayleigh-Bénard convection. Int. J. Heat Mass Transfer **55**(13–14), 3732–3743 (2012)
114. D. Venturi, T.P. Sapsis, H. Cho, G.E. Karniadakis, A computable evolution equation for the probability density function of stochastic dynamical systems. Proc. R. Soc. A **468**, 759–783 (2012)
115. C. Villani, A review of mathematical topics in collisional kinetic theory, in *Handbook of Mathematical Fluid Mechanics*, vol. 1, ed. by S. Friedlander, D. Serre (North-Holland, Amsterdam, 2002), pp. 71–305
116. Z. Wang, Q. Tang, W. Guo, Y. Cheng, Sparse grid discontinuous Galerkin methods for high-dimensional elliptic equations. J. Comput. Phys. **314**, 244–263 (2016)
117. D. Xiu, Efficient collocational approach for parametric uncertainty analysis. Commun. Comput. Phys. **2**(2), 293–309 (2007)
118. D. Xiu, J. Hesthaven, High-order collocation methods for differential equations with random inputs. SIAM J. Sci. Comput. **27**(3), 1118–1139 (2005)
119. L. Yan, L. Guo, D. Xiu, Stochastic collocation algorithms using ℓ_1-minimization. Int. J. Uncertain. Quantif. **2**, 279–293 (2012)

120. Y. Yang, C.W. Shu, Discontinuous Galerkin method for hyperbolic equations involving δ-singularities: negative-order norm error estimate and applications. Numer. Math. **124**, 753–781 (2013)
121. X. Yang, M. Choi, G.E. Karniadakis, Adaptive ANOVA decomposition of stochastic incompressible and compressible fluid flows. J. Comput. Phys. **231**, 1587–1614 (2012)
122. X. Yang, H. Lei, N.A. Baker, G. Lin, Enhancing sparsity of Hermite polynomial expansions by iterative rotations. J. Comput. Phys. **307**, 94–109 (2016)
123. C. Zeng, H. Wang, Colored noise enhanced stability in a tumor cell growth system under immune response. J. Stat. Phys. **141**(5), 889–908 (2010)
124. Z. Zhang, M. Choi, G.E. Karniadakis, Anchor points matter in ANOVA decomposition, in *Proceedings of ICOSAHOM'09*, ed. by E. Ronquist, J. Hesthaven (Springer, Berlin, 2010)
125. Z. Zhang, M. Choi, G.E. Karniadakis, Error estimates for the ANOVA method with polynomial chaos interpolation: tensor product functions. SIAM J. Sci. Comput. **34**(2), 1165–1186 (2012)
126. G. Zhou, A. Cichocki, S. Xie, Accelerated canonical polyadic decomposition by using mode reduction. IEEE Trans. Neural Netw. Learn Syst. **24**, 2051–2062 (2013)
127. R. Zwanzig, Memory effects in irreversible thermodynamics. Phys. Rev. **124**, 983–992 (1961)

From Uncertainty Propagation in Transport Equations to Kinetic Polynomials

Bruno Després

Abstract In view of the modeling of uncertainties which propagate in non linear transport equations and general hyperbolic systems, we review some recent alternatives to the classical moment method. These approaches are obtained by reconsidering the non linear structure with entropy considerations. It is shown that the entropy variable and the kinetic formulation of conservation laws yield new approaches with strong control of the maximum principle. A general minimization principle is proposed for these kinetic polynomials, together with an original reformulation as an optimal control problem. Basic numerical illustrations show the properties of these new techniques. A surprising linked to quaternion algebras is evoked in relation with kinetic polynomials. Natural limitations are discussed in the conclusion.

1 Introduction

It is little less than 80 years since Norbert Wiener's visionary article on "The homogeneous chaos" [39] and some of the questions he addressed are still vividly debated among the community that seeks for a comprehensive framework for uncertainties in fluid mechanics. One question in [39] can be summed up as

Question 1 Is it possible to have a measurement of the dynamics of a flow via polynomial expansions of certain quantities, where the polynomials are optimal with respect to some underlying probability laws?
The engineering and computational community recognized that it is a fundamental issue also for uncertainty calculations in many different fields, see [21] and

B. Després (✉)
Sorbonne Universités, UPMC Univ Paris 06, CNRS UMR 7598, Laboratoire Jacques-Louis Lions, Paris, France

Institut Universitaire de France, Paris, France
e-mail: bruno.despres@upmc.fr

© Springer International Publishing AG, part of Springer Nature 2017
S. Jin, L. Pareschi (eds.), *Uncertainty Quantification for Hyperbolic and Kinetic Equations*, SEMA SIMAI Springer Series 14,
https://doi.org/10.1007/978-3-319-67110-9_4

references therein. Since then, any orthogonal polynomial expansion related to a certain probability law (not only the Hermite expansion [39] well suited for Gaussian processes and turbulence) is called a *chaos polynomial* expansion. There is actually another question asked at the end of Wiener's paper.

Question 2 What is the compatibility of these polynomial expansions with respect to the PDE structure needed for fluid mechanics?
For the Burgers equations which is a paradigm, Wiener quickly realized that loss of regularity may degrade the accuracy of polynomial approximations. This remark looks evident nowadays since the theory [27] of shock waves and discontinuous solutions is well established. It seems at the lecture of Wiener's paper that he wanted to address both questions at the same time, meaning a theory for the development of turbulence—whatever it meant for him—and for the existence of shocks (which degrades the regularity of the solutions so lessens the quality of polynomial approximations).

The purpose of this work is to review some recent progresses which try to answer the second question, and only the second one. It will be presented following a certain chronological order with which the author looked at these problems, so the title of the present contribution. Even if some problems evoked below seem at first sight extremely far from uncertainty propagation (such is the quaternion structure at the end of this paper in Sect. 4), it is hoped the ensemble has a coherent structure and reflects some scientific issues in the modeling of uncertainty propagation in hyperbolic and kinetic equations. In a completely different direction, the reader interested to a modern statistical but PDE based treatment of hyperbolic conservation laws is strongly advised to refer to [19], and therein.

The plan is the following. Section 2 begins with the introduction of standard notations and results about the hyperbolic structure of systems of conservation laws with polynomial modeling of uncertainties. Section 3 takes advantage of the rewriting of scalar conservation laws as the limit of kinetic equations. It will explain nevertheless that another view is possible for polynomial expansions, denoted as kinetic polynomials. Section 3.3 will provided advanced material on kinetic polynomials. Section 3.4 will deal with a first formal extension to isentropic Euler system with $\gamma = 3$. Numerical illustrations are provided in Sect. 4.

Similar notations will sometimes be used for different uses. For example f denotes the flux function in Sect. 2, but refers to the kinetic unknown in Sect. 3: the context makes this abuse non ambiguous. On the contrary the indices are kept the same: d is the space dimension, m the size of the system of conservation laws, p the dimension of the uncertain space and n the polynomial degree.

2   Hyperbolic Structure

The modern mathematical treatment of non viscous compressible fluid mechanics is based of the theory of hyperbolic systems of conservation laws [27]. Let us start with the Euler system of compressible non viscous fluid mechanics written in the domain $\mathbf{x} \in \mathscr{D} \subset \mathbb{R}^d$

$$\begin{cases} \partial_t \rho + \nabla \cdot (\rho \mathbf{u}) = 0, \\ \partial_t(\rho \mathbf{u}) + \nabla \cdot (\rho \mathbf{u} \otimes \mathbf{u}) + \nabla p = 0, \\ \partial_t(\rho e) + \nabla \cdot (\rho u e + p u) = 0, \end{cases} \tag{1}$$

where $\rho(\mathbf{x}, t) > 0$ stands for the density of a gas or a fluid, $\mathbf{u}(\mathbf{x}, t) \in \mathbb{R}^d$ is the velocity and e is the total energy. The total energy is the sum of the internal energy ε and of the kinetic energy, that is $e = \varepsilon + \frac{1}{2}|\mathbf{u}|^2$. Considering that an equation of state (*EOS*) is provided, the pressure law is $p = EOS(\rho, \varepsilon)$. System (1) is rewritten as a system of conservation laws

$$\begin{cases} \partial_t U + \nabla \cdot f(U) = 0, \ \mathbf{x} \in \mathscr{D}, t > 0, \\ U(\mathbf{x}, 0) = U_0(\mathbf{x}), \quad \mathbf{x} \in \mathscr{D}, \end{cases} \tag{2}$$

The unknown $U(\mathbf{x}, t) \in \Omega \subset \mathbb{R}^m$ is assumed to live in the set Ω of admissible states. A minimal requirement for well posedness is to have the hyperbolic structure, which means that the Jacobian matrix

$$A(U) = \nabla_U f(U) \in \mathbb{R}^{m \times m}$$

is diagonalizable in $\mathbb{R}^m$: for all $U \in \Omega$, there is a set of real eigenvectors and eigenvalues. This is guaranteed if one has a smooth entropy-entropy flux pair (S, F) for the system. The entropy function $S : \Omega \to \mathbb{R}$ and the entropy pair function $F : \Omega \to \mathbb{R}^d$ are such that S is strictly convex, that is $\nabla^2 S > 0$, and

$$\nabla S \nabla f = \nabla F, \quad U \in \Omega.$$

The modeling of uncertainty propagation with chaos polynomials techniques is usually performed with another variable, call it $\omega \in \Theta \subset \mathbb{R}^p$. The uncertainty can be in the initial data $U_0(\mathbf{x}, \omega) \in \mathbb{R}^m$ or in the flux function $f_\omega : \Omega \to \mathbb{R}^{m \times d}$ which displays a dependency with respect to ω. One obtains the system of conservation laws

$$\begin{cases} \partial_t U + \nabla \cdot f_\omega(U) = 0, \quad \mathbf{x} \in \mathscr{D}, \omega \in \Theta, t > 0, \\ U(\mathbf{x}, 0; \omega) = U_0(\mathbf{x}, t; \omega), \ \mathbf{x} \in \mathscr{D}, \omega \in \Theta, \end{cases} \tag{3}$$

where the unknown $U(\mathbf{x}, t; \omega)$ is function of the space-time variables and of the uncertain variable. The mathematical structure of (3) is extremely simple since it can be seen as an infinite collection of decoupled systems like (47), but for different ω.

For the simplicity of the exposure the function f is now considered as independent of ω. It is not really a restriction with respect to the main mathematical issues since it is possible to rewrite (3) as an augmented system

$$\partial_t \begin{pmatrix} U \\ \omega \end{pmatrix} + \nabla \cdot \begin{pmatrix} f_\omega(U) \\ 0 \end{pmatrix} = 0, \quad \mathbf{x} \in \mathscr{D}, \quad \omega \in \Theta, \quad t > 0. \tag{4}$$

Up to the definition of an augmented flux function $f^{\text{aug}}(U, \omega) = \begin{pmatrix} f_\omega(U) \\ 0 \end{pmatrix}$, the system (4) is made of $m + p$ conservation laws. An entropy can be defined under natural conditions [17].

Since the number of variables of the generic system (3) is large, indeed the dimension of the space of static variables is

$$\dim (\text{physical space}) + \dim (\text{uncertain space}) = m + p,$$

the idea of model reduction is appealing. This is performed below with what is called chaos expansion or chaos polynomials [8, 13, 21, 24, 28, 34, 39, 41]. For this task, one adds for convenience one extremely important information which is the a **priori knowledge of an underlying probability law** $d\mu(\omega)$: one has that $\int_\Theta d\mu(\omega) = 1$. One can argue that, for a practical problem, no such probability law is a priori known. This is true in general, but there exists situations where the probability law can be characterized by physical experiments. Three different examples are ICF (Inertial Confinement Fusion) modeling [33], discussion of EOS for ICF modeling [7] and in another direction signal processing [9].

The idea behind chaos polynomials is to use this information with optimal accuracy [2, 10]. The procedure is as follows: one determines firstly a family of orthogonal polynomial with increasing degree (partial or total)

$$\int_\Omega p_i(\omega) p_j(\omega) d\mu(\omega) = \delta_{ij}.$$

A basic example is Legendre polynomials

$$p_i(x) = \frac{1}{2^i i!} \frac{d^i}{dx^i} (x^2 - 1)^i$$

which are orthogonal for the uniform law $d\mu(\omega) = \mathbf{I}_{\{-1 < \omega 1\}}$

$$\int_{-1}^1 p_i(x) p_j(x) dx = \frac{2}{2i + 1} \delta_{ij}.$$

For the simplicity of notations, the polynomials ordering is the simplest one, that is $i \in \mathbb{N}$. All this motivates the definition of a truncated unknown

$$U^n(\mathbf{x}, t; \cdot) \in P^n := \text{Span}_{0 \le i \le n} \{p_i\}, \tag{5}$$

that is

$$U^n(\mathbf{x}, t; \omega) = \sum_{i=0}^{n} U_i^n(\mathbf{x}, t) p_i(\omega) \text{ where } U_i^n(\mathbf{x}, t) = \int_{\Omega} U^n(\mathbf{x}, t; \omega) p_i(\omega) d\mu(\omega).$$

(6)

Since U_i^n is the ith moment of U^n with respect to p_i, this modeling is strongly related to two classical theories: the first one is the classical problem of moments [1, 11] and the second one is the closure problem of moments for kinetic equations [12, 23, 26, 30].

If correctly solved, the closure problem yields a closed system for the evolution of $\left(U_n^i\right)_{0 \le i \le n}$. A naive method is to close readily as

$$\partial_t U_i^n(\mathbf{x}, t) + \nabla \cdot \int_{\Omega} f(U^n(\mathbf{x}, t; \omega)) p_i(\omega) d\mu(\omega) = 0, \quad 0 \le i \le n.$$

(7)

When using such structure for calculations on computers, the numerical evaluation of the integrals $\int_{\Omega} f(U^n(\mathbf{x}, t; \omega)) p_i(\omega) d\mu(\omega)$ is needed. These integrals are highly non linear for many problems of interest. Some prescriptions can be found in [34]. Discarding these practical issues, there is a bad news [17].

Lemma 2.1 *Take the uniform law $d\mu = d\omega$ on the interval $\Omega = (0, 1)$. When applied to the Euler system (48) or to the system of shallow water, the system (7) with the closure (6) may be non hyperbolic even for physical correct datas.*

So far, the only possibility to have an hyperbolic closure is to modify the expansion (5) using the entropy variable $V = \nabla S$ which induces a diffeomorphism written as $\varphi(U) = \nabla S = V$. The expansion writes

$$V^n(\mathbf{x}, t; \cdot) \in P^n := \text{Span}_{0 \le i \le n} \{p_i\},$$

(8)

that is

$$V^n(\mathbf{x}, t; \omega) = \sum_{i=0}^{n} V_i^n(\mathbf{x}, t) p_i(\omega) \text{ where } V_i^n(\mathbf{x}, t) = \int_{\Omega} V^n(\mathbf{x}, t; \omega) p_i(\omega) d\mu(\omega)$$

(9)

is the moment of the entropy variable. The closure is now written as

$$U^n(\mathbf{x}, t; \omega) = \varphi^{-1}(V^n(\mathbf{x}, t; \omega)).$$

This method introduces additional non linearity in the model. There is however a good news.

Theorem 2.1 (Proof in [17]) *The system of conservation laws (7) with the closure (9) is hyperbolic unconditionally for $U \in \Omega$. It admits the entropy-entropy pair $(\mathscr{S}, \mathscr{F})$*

$$\mathscr{S} = \int_{\Omega} S(U^n(\boldsymbol{\omega})) d\omega \text{ and } \mathscr{F} = \int_{\Omega} F(U^n(\boldsymbol{\omega})) d\omega.$$

Many different probability measures are available. An issue with such extended systems is the discretization procedure, since the simplicity of the coding is a highly desirable property. Since the situation is not very different from moment models, efficient implementation is possible [17]. A variant adapted to quasi-linear systems is proposed in [40], with a simpler implementation. Convergence to the limit entropy solution with respect to n is challenging to establish [22]: results are only partial. It seems that no theoretical result of convergence is available so far after the appearance of shocks in the solution, see [17] with a weak-strong technique. In the rest of this work, an alternative to chaos polynomials is considered. Following [15], it is called kinetic polynomials.

3   Kinetic Structure

The kinetic formulation of conservation laws [4, 29, 31, 32] is another possibility to model uncertainties. Let $\varepsilon > 0$ be a small parameter which ultimately tends to zero. The kinetic formulation of the conservation law with flux $F : \mathbb{R} \to \mathbb{R}$

$$\partial_t u + \partial_x F(u) = 0 \tag{10}$$

writes as a Boltzmann equation for $t \geq 0$, $x \in \mathbb{R}^d$ and $\xi \geq 0$, in a BGK (relaxation) form,

$$\begin{cases} \partial_t f_\varepsilon + a(\xi).\nabla f_\varepsilon + \frac{1}{\varepsilon} f_\varepsilon = \frac{1}{\varepsilon} M(u_\varepsilon; \xi), \qquad a = \nabla F, \\[2mm] u_\varepsilon(x, t) = \displaystyle\int f_\varepsilon(x, \xi, t) d\xi, \\[2mm] f_\varepsilon(t = 0) = M\big(u^{\text{init}}; \xi\big). \end{cases} \tag{11}$$

The right hand side

$$M(u; \xi) = \mathbf{I}_{\{0 < \xi < u\}} \tag{12}$$

is called a Maxwellian. The initial condition satisfies

$$0 \leq u(x, \omega, 0) = u^{\text{init}}(x, \omega) \leq u_{\max}, \qquad \int u^{\text{init}}(x, \omega) dx d\mu(\omega) < \infty. \tag{13}$$

The non negativity $u \geq 0$ is needed for (12) to make sense in our context. That is why we assume the initial data is non negative $u^{\text{init}} \geq 0$. This assumption simplifies some non essential technicalities and allows to disregard the negative part of M; the reader can find in [29, 31, 32] the adaptation to general signs as well as convergence proofs when $\varepsilon \to 0^+$: typically $u_\varepsilon \to u$ and $f_\varepsilon \to M(u)$ in natural functional spaces.

The Maxwellian $M(u; \xi)$ is a universal minimizer for a family of entropy based functionals [4, 29, 31, 32]. For all convex functionals $S(\xi)$, one has that

$$M(u; \cdot) = \operatorname*{argmin}_{u = \int g d\xi, \, 0 \le g \le 1} \int S'(\xi) g d\xi. \tag{14}$$

To model uncertainties, the idea is now to write (11) for all ω, and then to modify it in a polynomial manner so as to consider the intrusive kinetic formulation

$$\begin{cases} \partial_t f_\varepsilon^n + a(\xi).\nabla f_\varepsilon^n + \frac{1}{\varepsilon} f_\varepsilon^n = \frac{1}{\varepsilon} M^n(u_\varepsilon^n; \xi, \omega), \\[2mm] u_\varepsilon^n(x, \omega, t) = \int f_\varepsilon^n(x, \xi, \omega, t) d\xi, \\[2mm] f_\varepsilon^n(t = 0) = M^n(u^{\text{init}}; \xi, \omega), \end{cases} \tag{15}$$

where $0 \le M^n(u_\varepsilon^n; \xi, \omega) \le 1$ is a suitable polynomial modification of the Maxwellian M. Notice that $\int f_\varepsilon^n(t = 0) d\xi d\omega = \int u^{\text{init}} d\omega$ but the initial data needs not be at equilibrium since u^{init} usually does not belong to P_ω^n. The solutions of (15) depend now on two parameters ε and n. Depending on the way M^n is defined, it is possible to get various estimates which explain the theoretical interest of the method.

3.1 Convolution Techniques

A first series of polynomial Maxwellian is obtained with suitable convolution techniques. One seeks M^n under the form

$$M^n(u_\varepsilon^n; \xi, \omega) = G^n *_\omega M(u_\varepsilon^n; \xi) := \int G^n(\omega, \omega') M(u_\varepsilon^n(\omega'); \xi) d\mu(\omega')$$

where the convolution kernel G^n is decomposed as

$$G^n(\omega, \omega') = \sum_{i=0}^{n} c_i p_i(\omega) p_i(\omega'), \tag{16}$$

where c_i are appropriate coefficients and G^n satisfies

$$G^n \ge 0, \qquad \int G^n(\omega, \omega') d\mu(\omega') = c_0 = 1 = \int G^n(\omega, \omega') d\mu(\omega). \tag{17}$$

The theory of polynomial kernel approximation [18, 38] asserts that convolution kernels exist which satisfy the requirements (16)–(17). We quote [15] 3 possibilities: the Fejer kernel, the Jackson kernel and the modified Jackson kernel.

For example, considering the measure $d\mu(\omega) = \frac{d\omega}{\pi\sqrt{1-\omega^2}}$ with support in the interval $\omega \in I = (-1, 1)$ and the Tchebycheff orthonormal polynomials

$$T_i(\omega) = \cos(i \arccos \omega), \qquad -1 \leq \omega \leq 1.$$

The Fejer kernel G_F^n is defined by the coefficients $c_0 = 1$ and $c_i = 2\frac{n+1-i}{n+1}$ for $1 \leq i \leq n$. The Jackson kernels have better approximation properties than the Fejer kernel, with a slightly different definition of the coefficients c_i. An example of strong error bounds follows, see [15] for additional properties.

Proposition 3.1 *Consider the Jackson kernel. One has the inequalities*

$$\int |f_\varepsilon^n(t) - G^n *_\omega f_\varepsilon(t)| \, dx d\xi d\mu(\omega) \leq C \frac{t}{\varepsilon} \int \text{mod}_1(u^{\text{init}}, \frac{1}{n}) dx d\xi, \tag{18}$$

$$\int |f_\varepsilon^n(t) - f_\varepsilon(t)| \, dx d\xi d\mu(\omega) \leq C \left(1 + \frac{t}{\varepsilon}\right) \int \text{mod}_1(u^{\text{init}}, \frac{1}{n}) dx d\xi, \tag{19}$$

$$\int |f_\varepsilon^n(t) - M(u; \xi)| \, dx d\xi d\mu(\omega) \leq c\sqrt{\varepsilon} + C \left(1 + \frac{t}{\varepsilon}\right) \int \text{mod}_1(u^{\text{init}}, \frac{1}{n}) dx d\xi. \tag{20}$$

Similar bounds are derived for $u_\varepsilon^n - u$.

However these estimates do not allow to pass to the limit ε independently of N. It is instructive to write the formal limit in the regime $\varepsilon n = O(1)$. The unknowns of the resulting moment system are the quantities

$$u_{\varepsilon,i}^n(x, t) = \int f_{\varepsilon,i}^n(x, \omega, t) d\xi, \quad f_{\varepsilon,i}^n(x, \omega, t) = \int f_\varepsilon^n(x, \xi, \omega, t) T_i(\omega) d\mu(\omega).$$

An artificial damping phenomenon arises. Set for convenience $n + 1 = \frac{1}{\varepsilon}$. The projected equation for the modified Jackson kernel are

$$\partial_t u_{\varepsilon,i}^n + \text{div} \int a(\xi) f_{\varepsilon,i}^n d\xi = \frac{1}{\varepsilon} \left[c_i^{modJ} u_{\varepsilon,i}^n - u_{\varepsilon,i}^n \right]$$

$$= (n + 1) \left(\frac{(n+1-i)\cos\frac{\pi i}{n+1} + \sin\frac{\pi i}{n+1} \cot\frac{\pi}{n+1}}{n+1} - 1 \right) u_{\varepsilon,i}^n$$

$$= \left((n+1-i)\cos\frac{\pi i}{n+1} + \sin\frac{\pi i}{n+1} \cot\frac{\pi}{n+1} - n - 1 \right) u_{\varepsilon,i}^n = -h_n(i) u_{\varepsilon,i}^n.$$

Elementary calculations show that $h_n(0) = 0$, and that $h_n(x) > 0$ for $0 < x < n$ with $h_n(x) \to 0$ for all x as $n \to 0$. One also has that $0 < h_n(i) < i$ for $0 < i \leq n$. It implies after integration in x

$$\partial_t \int u_{\varepsilon,i}^n dx = -h_n(i) \int u_{\varepsilon,i}^n dx \implies \int u_{\varepsilon,i}^n dx(t) \tag{21}$$

$$= e^{-h_n(i)t} \int u_{\varepsilon,i}^n dx(0) \implies \lim_{t \to \infty} \int u_{\varepsilon,i}^n dx(t) = 0.$$

A similar damping phenomenon of the moments $i \neq 0$ also shows up if one uses the Jackson kernel, and is even stronger starting from the Fejer kernel. This seems the price to pay for the good convergence properties of Proposition 3.1.

3.2 Minimization Techniques

The initial purpose of the method proposed below was precisely to obtain a polynomial modeling of uncertainties with good properties, such as the maximum principle and no damping (by comparison with (21, it means $h_n(i) = 0$ for all i). Quite fortunately the universal entropy principle (14) can be generalized in this direction. It yields powerful tools with many good properties (even if some of them are still under studies).

Let us take one entropy S and $u_n \in P^n$ with $u_n \geq 0$ for all ω. Define

$$K^n(u^n) = \left\{ g^n(\cdot, \cdot) \in P_\omega^n, \ u^n(\omega) = \int g^n(\xi, \omega) d\xi, \ 0 \leq g^n \leq 1 \right\}.$$

For any $n \geq 0$, one tries to construct an equilibrium $M^n(u^n; \xi, \omega)$ as a minimizer

$$M^n(u^n) = \underset{g^n \in K^n(u^n)}{\operatorname{argmin}} \int S'(\xi) g^n d\xi d\mu(\omega). \tag{22}$$

For $n = 0$ this is a Brenier inequality [4, 5], it yields a unique minimizer. For general $n > 0$, let us assume for a while that M^n exists and is unique. One has the following a priori properties: under the assumption that a solution exists to the maximization problem (27), then the solution of the kinetic equation

$$\begin{cases} \partial_t f_\varepsilon^n + a(\xi).\nabla f_\varepsilon^n + \frac{1}{\varepsilon} f_\varepsilon^n = \frac{1}{\varepsilon} M^n(u_\varepsilon^n; \xi, \omega), \\[2mm] u_\varepsilon^n(x, \omega, t) = \int f_\varepsilon^n(x, \xi, \omega, t) d\xi, \\[2mm] f_\varepsilon^n(t = 0) = M^n(u^{\text{init},n}; \xi), \end{cases} \tag{23}$$

satisfies the entropy principle under the form

$$\partial_t \int S'(\xi) f_\varepsilon^n(x, \xi, \omega, t) d\xi d\mu(\omega) + \operatorname{div} \int a(\xi) S'(\xi) f_\varepsilon^n(x, \xi, \omega, t) d\xi d\mu(\omega) \leq 0.$$

Moreover under the same assumption, if u_ε^n converges strongly to some u^n, then one passes to the limit $\varepsilon \to 0$ in (23) and obtains the system of conservation laws for $0 \le i \le n$

$$\partial_t u_i^n + \text{div } \mathscr{F}_i^n[u^n] = 0, \qquad \mathscr{F}_i^n[u^n] := \int a(\xi) M^n(u^n; \xi, \omega) p_i(\omega) d\xi d\mu(\omega),$$
(24)

with the entropy inequality

$$\partial_t \mathscr{S}^n[u^n] + \text{div } \mathscr{G}^n[u^n] \le 0,$$

where the entropy and entropy fluxes are defined by

$$\mathscr{S}^n[u^n] := \int S'(\xi) M^n(u^n; \xi, \omega) d\xi d\mu(\omega)$$

and

$$\mathscr{G}^n[u^n] := \int S'(\xi) a(\xi) M^n(u^n; \xi, \omega) d\xi d\mu(\omega).$$

However a difficult question is to construct the solution of (22).

3.2.1 Quasi-Solution

A quasi-solution or feasible solution to the minimization problem (22) is proposed. This construction has two goals. The first one is to establish M^n is a quasi-minimizer (22) but for all S. The second one is to propose an implementable algorithm, at least for small n.

Let us remark that

$$S'(\xi) = \int_0^\infty S''(s) a_s(\xi) ds, \quad a_s(\xi) = \mathbf{I}_{\{0 < s < \xi\}},$$
(25)

meaning that any function S' such that $S'' \ge 0$ and $S'(0) = 0$ is a non-negative integral of functions $a_s(\xi)$ which also satisfy $a_s' \ge 0$ and $a_s(0) = 0$. That is the family of functions $(a_s)_{s>0}$ constitutes a generating family (actually the function $\xi \mapsto a_s(\xi)$ is the derivative of a branch of a Kruzkov entropy). Let us replace (22) with a family of similar problems

$$M^n(u^n) = \underset{g^n \in K^n(u^n)}{\text{argmin}} \int_\xi^\infty g^n ds d\mu(\omega), \quad \forall \xi.$$
(26)

More precisely any solution of (26) (if it exists) is also a solution of (22) (with the same restriction concerning the existence). Since the mass is preserved, that is $\int g^n(s,\omega)dsd\mu(\omega) = \int u^n(\omega)dsd\mu(\omega)$, this problem can be rewritten with the alternative formulation

$$M^n(u^n) = \operatorname*{argmax}_{g^n \in K^n(u^n)} \int_0^\xi g^n dsd\mu(\omega), \quad \forall \xi. \tag{27}$$

A quasi-solution is possible based on (27). The idea is to solve (27) progressively with respect to ξ, starting from $\xi = 0$ and then increasing its value until $u_+ = \max_I u^n(\omega)$. A constructive method (an algorithm) [15] shows that the quasi-solution writes

$$M^n(u^n; \xi, \omega) = \sum_{l \geq 0} h_l^n(\omega) \mathbf{I}_{\{\xi_l < \xi < \xi_{l+1}\}} \tag{28}$$

with $0 = \xi_0 < \xi_1 < \cdots < \xi_L < \xi_{L+1} = u_+$. The construction also shows the uniqueness of the feasible solution. The layer structure of this function is the key of the construction. The integral identity $\int_0^{u_+} M^n(u^n; \xi, \omega)d\xi = u^n(\omega)$ writes

$$\sum_{l \geq 0} (\xi_{l+1} - \xi_l)h_l^n(\omega) = u^n(\omega), \quad \omega \in I. \tag{29}$$

This function is constructed step by step, the first step for the bottom layer being trivial. The second step is the critical one where all the ideas of the method are carefully explained, in particular the role of the Bojavic-Devore theorem [3] for one sided approximation.

3.2.2 Discretization with Quasi-Solution

We discretize in time and space and implement the method issued from (28) under the form

$$\frac{\bar{u}_j^n - u_j^n}{\Delta t} + \frac{F^n[u_j^n] - F^n[u_{j-1}^n]}{\Delta x} = 0 \tag{30}$$

where $u_j^n \in P^n(\omega)$ is a polynomial in ω of degree n (fixed), in cell j and at the current time step. The generic flux $F^n[u_j^n]$ is constructed with (28). The value at next time step $t + \Delta t$ in cell j is denoted with a bar $\bar{u}_j^n \in P^n(\omega)$.

Let us assume the initial data is a positive and bounded polynomial

$$0 \leq U_{\min} \leq u_j^n(\omega) \leq U_{\max} < \infty, \quad \forall j \text{ and } \forall \omega \in I. \tag{31}$$

Consider the archetype of a convex flux which is the Burgers flux $F(\xi) = \frac{\xi^2}{2}$. The following result states that the explicit Euler scheme satisfies the maximum principle (this is a minimal stability requirement) under a CFL condition which is independent of n. The property is here checked directly on the scheme (30) but can also be derived as a consequence of the underlying kinetic formulation.

Theorem 3.1 *Assume the CFL condition* $U_{\max}\Delta t \leq \Delta x$. *Then*

$$U_{\min} \leq \overline{u}_j^n(\omega) \leq U_{\max}, \quad \forall j \text{ and } \forall \omega \in I. \tag{32}$$

3.3 More on Kinetic Polynomials

This section is based on the results recently announced in [16]. Not only it is proved that (22) is a well posed problem with existence and uniqueness of the minimizer, but the problem shows nice reformulation as an optimal control problem [36]. The minimization problem (22) concerns the variables (ξ, ω) but is independent of the variables (x, t). So we make for convenience a change of variables $(x, t) \leftarrow (\omega, \xi)$ together with a change of functions $q_n \leftarrow u_n$ and $u_n \leftarrow M^n$. It yields simpler notations, also better in terms of an optimal control problem.

Set $G = [0, 1]$ (which stands for the space of uncertain variables Ω). Let $T > 0$, $n \in \mathbb{N}$ and $q_n \in P_+^n$. Define $U_n = \{q_n \in P_+^n, \ 1 - q_n \in P_+^n\}$. Set

$$K_n(T, q_n) := \left\{ v_n \in L^\infty(\mathbb{R}^+ : U_n) : \int_0^T v_n(t)dt = q_n, \ v_n \equiv 0 \text{ for } t > T \right\}. \tag{33}$$

Take a strictly convex function denoted as $s = S$ and a Lebesgue integrable weight $w \geq 0$ with $\int_G w(x)dx > 0$ (with the correspondence $w(x)dx = d\mu(\omega)$ and $x = \omega$). Define the linear cost function

$$J(u_n) := \int_G \int_{\mathbb{R}^+} u_n(t, x)ds(t)w(x)dx. \tag{34}$$

Design of the polynomial Maxwellian (22) recasts as the following L^1 minimization problem.

Problem 1 Find $u_n \in K_n(T, q_n)$ such that

$$u_n = \underset{K_n(T,q_n)}{\operatorname{argmin}} J(v_n) \tag{35}$$

Theorem 3.2 *Assume the weight* $w \geq 0$ *satisfies* $\int_G w(x)dx > 0$. *Assume* s'' *is lower bounded from 0 and integrable. Assume* $T \geq \|q_n\|_{L^\infty(G)}$. *Then there exists a unique minimum to the problem (22).*

The proof is based on some convenient space-time comparison inequalities using ad-hoc tests functions. It is also proved that: (a) for T the solution u_n is vanishes for large time; (b) there exists $T_* > 0$ such that all solutions are the same for $T > T_*$.

A reformulation as an optimal control problem [35, 36] is appealing. Define

$$y_n(t, x) = \int_0^t u_n(t, x)dt \iff y_n'(t) = u_n(t) \text{ with } y_n(0) = 0. \tag{36}$$

In this context, the function $y_n(t) \in P_n$ is called the state and $u_n(t) \in U_n$ is called the control. The minimization problem (35) reformulates as follows.

Problem 2 Find an optimal control $u_n \in L^\infty(0, T : U_n)$ which minimizes the cost function and with the final state $y_n(T) = q_n \in P_n^+$.

Let us first remind the PMP maximum principle. Since the set of controls is discrete, convex and closed, one can invoke the PMP [35, 36]: for all optimal trajectories, there exists a Pontryagin multiplier $\lambda_n \in P_n$ such that

- the optimal control maximizes the criterion for almost all $t \in (0, T)$

$$\int_G (\lambda_n(x) - t)u_n(t, x)dx = \max_{v_n \in U_n} \int_G (\lambda_n(x) - t)v(x)dx. \tag{37}$$

This is called a normal trajectory, or a normal pair u_n, λ_n.
- or the optimal control maximizes the criterion for almost all $t \in (0, T)$

$$\int_G \lambda_n(x)u_n(t, x)dx = \max_{v_n \in U_n} \int_G \lambda_n(x)v_n(x)dx. \tag{38}$$

This is called a abnormal trajectory, or a abnormal pair u_n, λ_n. The abnormality or degeneracy comes from the fact that the criterion is independent of the time variable.

Abnormal trajectories are easy to construct if $q_n(x)$ vanishes at some point $x_\star \in [0, 1]$. In this case one can consider the polynomial $\lambda_n \in P_n$ with the quadrature property $\int_I \lambda_n(x)v_n(x)dx = -v_n(x_\star)$ for all $\forall v_n \in P_n$. Since $u_n(t, x_\star) = 0$ for all time t, it clear that λ_n satisfies (38).

Theorem 3.3 *Assume $q_n(x) \geq \varepsilon > 0$ over G. There exists an adjoint state $\lambda_n \in P_n$ such that the optimal solution of Problem 1–2 is solution of the PMP under the normal form*

$$u_n(t) = \underset{v_n \in U_n}{argmax} \left(\int_G (\lambda_n(x) - s'(t))v_n(x)dtw(x)dx \right) \text{ for almost all } t \in [0, T]. \tag{39}$$

A proof can be performed by showing that $\lambda \in P_n$ is a minimizer of a convenient cost function. Define the cost function as the integral in time of the criterion (37)

$$K(\lambda_n) := \int_0^\infty \int_G (\lambda_n(x) - t)u_n(t, x)dxdt \geq 0$$

where $u_n(t)$ satisfies the Pontryagin maximum principle. The cost is non negative by construction. It is well defined since u_n vanishes for large t. The cost function K is convex over P_n. The Danskin theorem yields that

$$dK(\lambda_n) = \left\langle \int_0^\infty u_n(t,x)dt, d\lambda_n \right\rangle.$$

The shooting method which is the essence of the study of normal trajectories is as follows.

Problem 3 (Shooting Method) Find $\lambda_n \in P_n$ such that $u_n(t)$ solution of (37) satisfies the endpoint condition $\int_0^\infty u_n(t)dt = q_n \in P_n^+$.

The shooting method is conveniently studied with the Lagrangian

$$L(\lambda_n) := K(\lambda_n) - \int_G \lambda_n(x)q_n(x)dx$$

where $q_n \in P_n^+$ is the given endpoint. The polynomial $\lambda_n \in P_n$ is solution of the shooting method if and only if it is an extremal point of the Lagrangian

$$dL(\lambda_n) = \left\langle \int_0^\infty u_n(t,x)dt - q_n(x), d\lambda_n \right\rangle = 0. \qquad (40)$$

Since L is convex and differentiable, a solution to (40) is also a minimum of the Lagrangian. The cornerstone of the proof is to show that L is infinite at infinity. Another interest of the PMP for our problem is the general principle.

Principle 3.2 *The Pontryagin multiplier is formally the adjoint entropic variable (in the sense of Godunov).*

The formal proof proceeds as follows. For $\lambda_n \in P_n$, consider $u_n(t)$ the minimizer of the cost function K and define

$$q_n = \int_0^\infty u_n(t)dt.$$

Define K^* the formal Legendre transform of K

$$K^*(\lambda_n) = \int \lambda_n(x)q_n(x)dx - K(\lambda_n).$$

The main difference between L and K^* is that q_n is given in L but is function of λ_n in K^*. One has

$$dK^*(\lambda_n) = \int (q_n d\lambda_n + \lambda_n dq_n)\, dx - dK(\lambda_n)$$

$$= \int \lambda_n dq_n dx + \int_0^1 \left(q_n - \int_0^\infty u_n(t)dt \right) d\lambda_n dx = \int \lambda_n dq_n dx. \qquad (41)$$

It can summarized as $dK = \langle q_n, d\lambda_n \rangle$ and $dK^* = \langle \lambda_n, dq_n \rangle$. If one assumes that the transformation $\lambda_n \mapsto q_n$ is a diffeomorphism, then K^* can be understood as a function of q_n. In this case K^* is a candidate to be an entropy. Let us now determine a candidate to be an entropy flux. We define

$$G(\lambda_n) := \int_0^\infty \int_G t(\lambda_n(x) - t)u_n(t, x)dxdt$$

which is well defined since u_n is defined in function of λ_n. Another use of the Danskin theorem yields

$$dG = \left\langle \int_0^\infty tu_n(t)dt, d\lambda_n \right\rangle.$$

The polar transform of G is

$$G^*(\lambda_n) = \int_0^1 \lambda_n(x) \left(\int_0^\infty tu_n(t)dt \right) dx - G(\lambda_n).$$

One has

$$dG^* = \int_0^1 d\lambda_n(x) \left(\int_0^\infty tu_n(t)dt \right) dx + \int_0^1 \lambda_n(x)d \left(\int_0^\infty tu_n(t)dt \right) dx - dG =$$

$$= \int_0^1 \lambda_n(x)d \left(\int_0^\infty tu_n(t)dt \right) dx. \tag{42}$$

One obtains the following formal result.

Lemma 3.3 *The system of projected equations*

$$\partial_t q_n(x, t) + \partial_x \int_0^\infty tu_n(x, t)dt = 0$$

admits the formal additional law

$$\partial_t K^*(\lambda_n) + \partial_x G^*(\lambda_n) = 0. \tag{43}$$

Proof Indeed one has

$$\int_0^1 \lambda_n(x)\partial_t q_n(x, t)dx + \int_0^1 \lambda_n(x)\partial_x \int_0^\infty tu_n(x, t)dtdx = 0.$$

Using (41) and (42), it is rewritten as (43) and the proof is ended.

3.4   Isentropic Euler System with $\gamma = 3$

An interesting question is to extend the polynomial modeling of uncertainties into systems of conservation laws with physical importance. A first example for the isentropic Euler system with $\gamma = 3$ in dimension one is as follows. Consider

$$\begin{cases} \partial_t \rho + \partial_x(\rho u) = 0, \\ \partial_t(\rho u) + \partial_x(\rho u^2 + p) = 0, \end{cases}$$

where $p = \frac{1}{12}\rho^3$. It admits the kinetic formulation [6, 29]

$$\partial_t f + v \partial_x f = \frac{1}{\varepsilon}\left(M_{\rho,\rho u} - f\right)$$

where $M_{\rho,\rho u}(v) \equiv 1$ for $u - \rho/2 < v < u + \rho/2$ and $M_{\rho,\rho u}(v) \equiv 0$ everywhere else. The Maxwellian $M_{\rho,\rho u}$ minimizes $\int_{\mathbb{R}} g(v)s'(v)dv$ over all functions $0 \le g \le 1$ such that $\int_{\mathbb{R}} g(v)dv = \rho > 0$ and $\int_{\mathbb{R}} g(v)v dv = \rho u \in \mathbb{R}$. A natural extension of the tools proposed previously would be to consider

$$M^n_{(\rho u)^n, \rho^n} = \underset{g^n \in \text{ admissible states}}{\text{argmin}} \int_{\mathbb{R}} \int_0^1 g^n(v, \omega)v^2 dv d\omega.$$

4   Numerical Methods

This section is devoted to provide elementary explanations and illustrations of some of the theoretical tools presented before and to explain advanced algorithms.

4.1   Regularity

A key feature is that weak solutions of a system of conservation laws with uncertainties (47) or (7) propagate in the uncertain space [25].

We consider the initial data

$$u^{\text{ini}}(x, \omega) = \begin{cases} 3 & \text{for } x < 1/2 \text{ and } -1 < \omega < 0, \\ 5 & \text{for } x < 1/2 \text{ and } 0 < \omega < 1, \\ 1 & \text{for } 1/2 < x \text{ and } -1 < \omega < 1. \end{cases} \tag{44}$$

The exact solution is a shock at velocity 2 for $\omega < 0$, and another shock at velocity 3 for $0 < \omega$

$$u(x, \omega, t) = \begin{cases} 3 & \text{for } x < 1/2 + 2t \text{ and } -1 < \omega < 0, \\ 5 & \text{for } x < 1/2 + 3t \text{ and } 0 < \omega < 1, \\ 1 & \text{elsewhere.} \end{cases} \tag{45}$$

This is visible in [15, p. 1010, Figure 4] where the numerical solution captured with a standard moment model is also represented.

4.2 Kinetic Polynomials

Kinetic polynomials can be used to design numerical methods with the preservation of the maximum principle. This is illustrated with an elementary implementation of the quasi-solutions.

We still consider the Burgers equation, but with a continuous initial data

$$u^{\text{ini}}(x, \omega) = \begin{cases} 12 & \text{for } x - \omega/5 < 1/2, \\ 1 & \text{for } x - \omega/5 < 3/2, \\ 12 - 11 (x - \omega/5 - 1/2) & \text{in between.} \end{cases} \tag{46}$$

The exact solution is a compressive ramp on all lines, and a shock at time $T = \frac{1}{11}$. So the exact solution is continuous in x and ω directions for $t < T$, and is discontinuous in the ω direction for $T < t$. The results are shown in [15, p. 1011, Figure 5].

4.3 Numerical Construction of Kinetic Polynomials

The construction of kinetic polynomials via optimal control theory brings the possibility to use many efficient numerical methods. For example it is proposed in [16] to use the AMPL language [20] to discretize and minimize (34)–(35). Note that L^1 minimization problems in combination with polynomial chaos expansions is pursued in [24].

An example of numerical implementation of the minimization problem (22) within the AMPL high level language is in Table 1 and a typical result is in Fig. 1.

Table 1 Script for an implementation of the solution of Problem 1 with the AMPL language [20]

```
##Parameters
param n := 3; # the polynomial degree
param T := 5;
param Nx := 40;
param Nt := 100;

## Variables
var y {k in 0..n, i in 0..Nt}; #y(t,x)=y_0(t)+y_1(t)x+...+y_n(t)x^n
var u {k in 0..n, i in 0..Nt}; #u(t,x)=u_0(t)+u_1(t)x+...+u_n(t)x^n
var utx {i in 0..Nt-1, j in 0..Nx} = sum\{k in 0..n\} u[k,i]*(j/Nx)^k;

## Cost
minimize cost: T/Nt*(sum{k in 0..n, i in 0..Nt-1}
(((i+1./2.)*T/Nt) *u[k,i]/(k+1.)));

## Constraints
subject to y_init {k in 0..n}: y[k,0] = 0;
subject to y_dyn {k in 0..n, i in 0..Nt-1}:
y[k,i+1] - y[k,i] - T/Nt*u[k,i]=0;
# q_n(x)=1+x+x**2+x**3
s.t. y_fin0: y[0,Nt]=1; s.t. y_fin1: y[1,Nt]=1;
s.t. y_fin2: y[2,Nt]=1; s.t. y_fin3: y[3,Nt]=1;
subject to cont {i in 0..Nt-1, j in 0..Nx}: 0 <= utx[i,j] <= 1;
## Inequalities in Un
data;

## Solver
option solver ipopt;
option ipopt_options " max_iter=10000 linear_solver=mumps
{halt_on_ampl_error yes}";
solve;
```

4.4 Connection with Polynomial Properties

Finally we evoke an axis of research [14] which is about a new way to construct polynomials with two bounds, one lower bound and one upper bound, in relation with a numerical implementation of kinetic polynomials. Some of the main results can be summarized as follows.

Start from

$$p_n \in P_n^+ := \{p_n \in P_n(x), \quad \text{such that } 0 \le p_n(x) \quad \forall x \in [0, 1]\}. \tag{47}$$

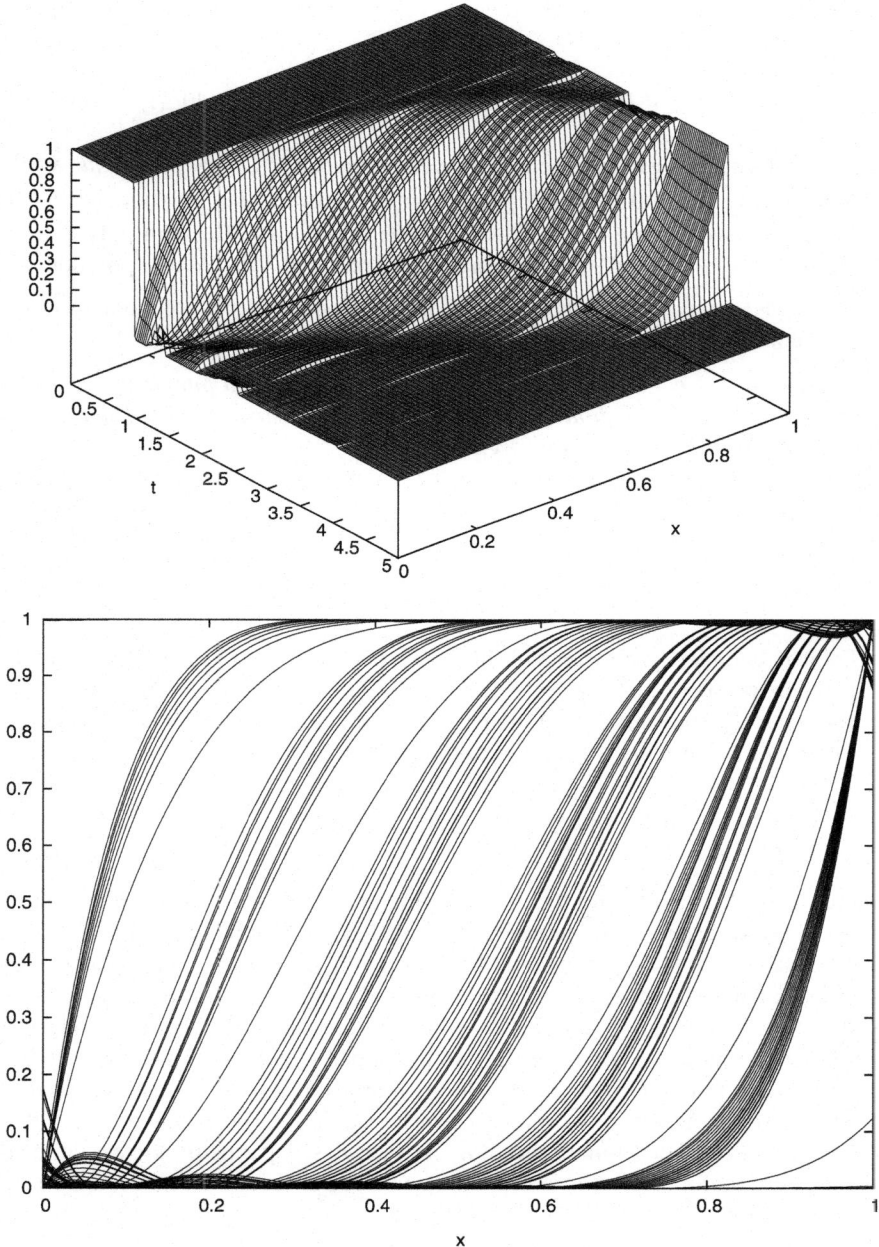

Fig. 1 Numerical computation of the polynomial Maxwellian-minimizer M^n (22), referred to as $u_n = M^n$ within this section. Numerical parameters: $N_x = 80$, $N_t = 200$, $n = 6$, $q_6 = 1 + x + x^2 + x^3$. The function $(x, t) \mapsto u_n(x, t)$ is represented on top as a surface, and is represented on bottom as many curves $x \mapsto u_n(x; t)$ parametrized by t

Define the set of polynomials which enters in the construction of kinetic polynomials as

$$p_n \in U_n := \{p_n \in P_n(x), \quad \text{such that } 0 \le p_n(x) \le 1 \quad \forall x \in [0,1]\}. \qquad (48)$$

Simpler subsets of U_n exist based on convex combinations $q_n = \sum_{j=0}^n \alpha_j u_j$ where the coefficients satisfy $0 \le \alpha_j$ and $\sum_{j=0}^n \alpha_j = 1$: the generating polynomials u_j can be either the basis of the monomials x^j, or the basis of the Berstein polynomials $B_{n,j}(x) = \frac{n!}{j!(n-j)!}x^j(1-x)^{n-j}$, or the basis of the rescaled Tchebycheff polynomials $\frac{T_j(2x-1)+1}{2}$. However none of these subsets is able to generate all polynomials in U_n only by convex combinations.

Theorem 4.1 *Let $n \in 2\mathbb{N}$ being even. There exists a smooth function from $\mathbb{R}^{3n/2}$ onto U_n. The smooth function is made explicit by a constructive algorithm and is 2π-periodic with respect to all its arguments.*

The norm of a uniform convergence is $\|f\| = \max_{0 \le x \le 1} |f(x)|$ for $f \in C^0[0,1]$.

Theorem 4.2 *Assume $f \in C^0[0,1]$ and $0 \le f(x) \le 1$ for $0 \le x \le 1$. Then*

$$\inf_{p_n \in U_n} \|f - p_n\| \le 2 \inf_{g_n \in P_n} \|f - g_n\|. \qquad (49)$$

Even if completely elementary, this is a remarkable result since the constant 2 is independent of n. The right hand side shows spectral convergence. This representation comes from quaternion algebras and the 4-squares Euler identity.

The next tests use this structure to minimize functionals like

$$J(p_n) := \int_0^1 (t - \lambda_n(x)) p_n(x)dx, \quad p_n \in U_n \qquad (50)$$

where $\lambda_n \in P_n$ is given and t may vary. This problem has interest in the context of this review paper. A reference is provided by a recent work [16] where a characterization of p_n is provided with the notion of a point of contact that comes from the seminal reference [3] is used. A point of contact of a function $f \in C^1[0,1]$ with $0 \le f \le 1$ is any point $0 \le y \le 1$ such that $f(y) = 0$ or $f(y) = 1$. The multiplicity order of the contact is the number of derivatives (+1) which vanish. For example if the point of contact is inside the interval, $0 < y < 1$, then necessarily $f'(y) = 0$, so the multiplicity order at y is necessarily ≥ 2. It is proved in [16] that p_n which realizes the minimum has not less than $n+1$ points of contact counted with order of multiplicity (this is similar to one-sided L^1 minimization for which we refer to [3]) for almost all t. We use this theoretical property to check the accuracy of the approximation. We remark that the optimal solution p_n (50) has the natural tendency to vanish where $t - \lambda_n(x) > 0$ and to be equal to 1 where $t - \lambda_n(x) < 0$, it is clearly a good strategy to minimize the cost function (50).

A numerical result representative of all the tests is the following. Take

$$\lambda_2(x) := T_2(2x-1) - t + x \text{ and } t = 0.3.$$

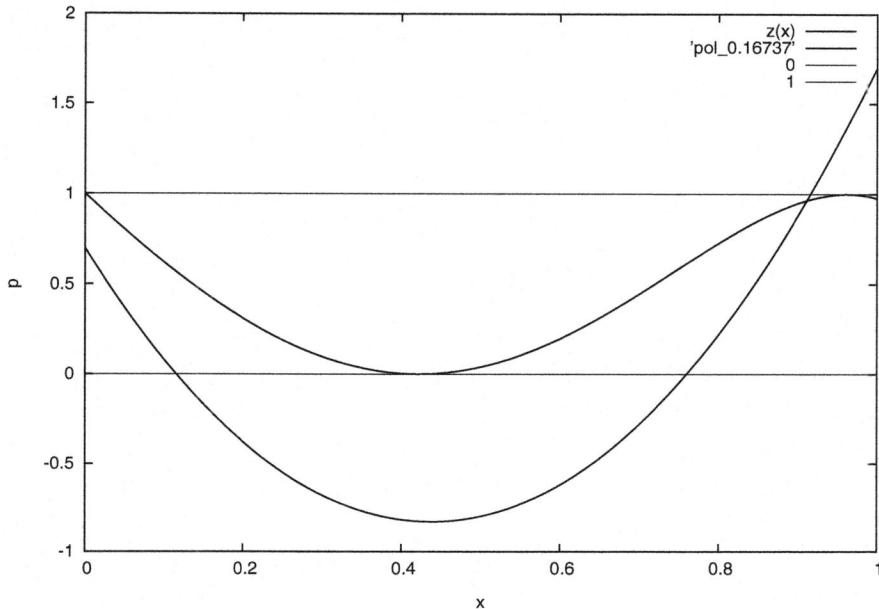

Fig. 2 Plot of $\lambda_2(x) - t$ and of a local minimum p_n^1 with $J(p_n^1) \approx -0.16737$. The total order of contact if $1 + 2 + 2 = 5$

A first numerical simulation yields the function displayed on Fig. 2, the numerical value of the cost function is $J(p_n^1) \approx -0.16737$. This function does not have the required number of contacts on the figure. But another minimum is captured by numerical simulations with another starting point, for which $J(p_n^2) \approx -0.188478 < J(p_n^1)$: its total order of contact is large enough (equal to $2n + 1 = 7$ since $n = 3$) and this is in accordance with the theory. No other minimum with lower value of the cost have been obtained by simulations, so it is the best candidate. Note that the exact calculation of the derivative $p_n'(x)$ is convenient to count without ambiguity the number of derivatives which vanish at points of contact (Fig. 3).

5   Conclusion

The examination of the challenges posed by polynomial modeling of uncertainties shows that alternatives to standard moment methods with chaos polynomials do exist. These new formulations try to introduce the polynomial structure used to model the uncertain variable ω into standard PDEs, but preserving at best the theoretical properties of the initial systems. Convolution techniques, kinetic formulations of conservation laws, minimization formulations and construction of quasi-solutions may have interest for non linear hyperbolic equations because they

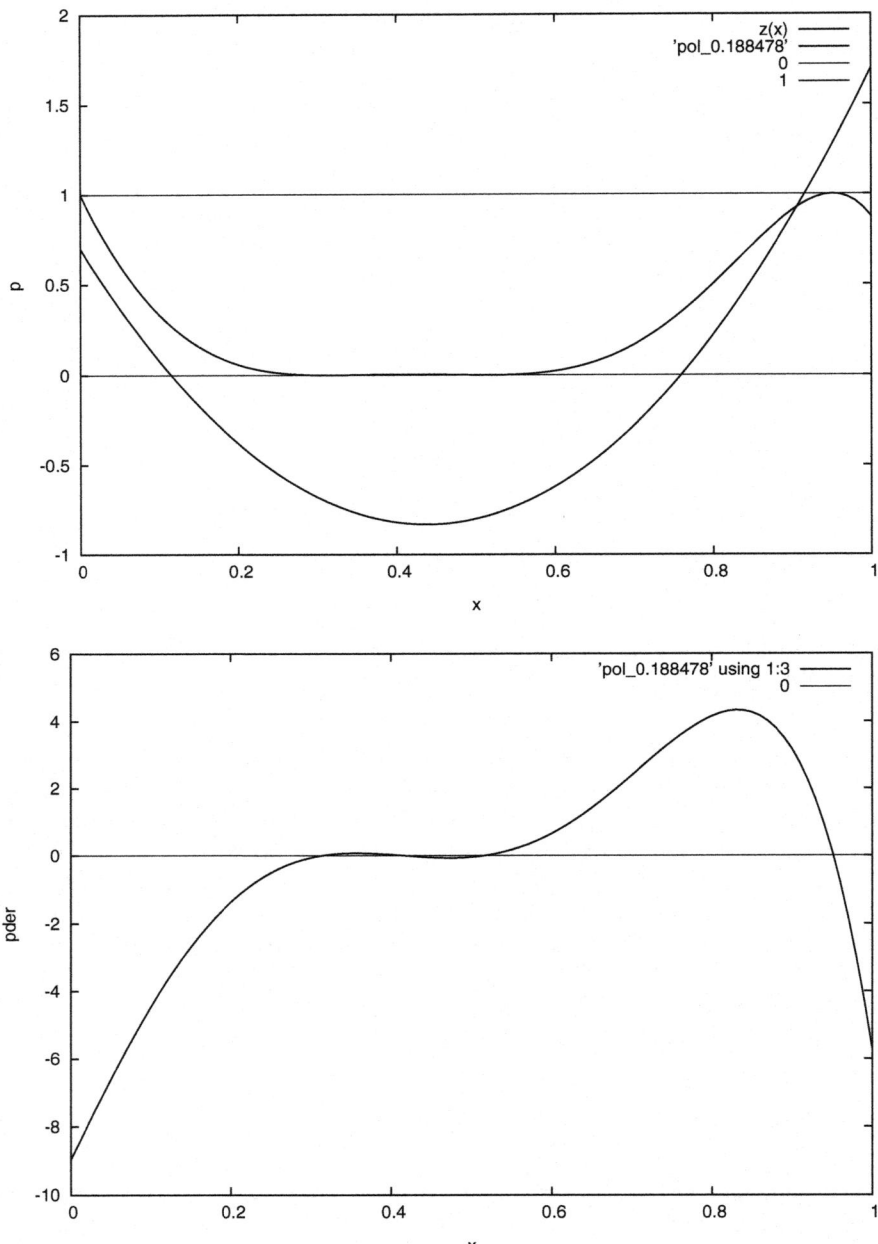

Fig. 3 On the top, plot of $\lambda_2(x) - t$ and of another local minimum p_n^2 with $J(p_n^2) \approx -0.188478$. The total order of contact if $1 + 2 + 2 + 2 = 7 = 2n + 1$, and so is the best candidate to be the global minimum. On the bottom, plot of the exact derivative $(p_n^2)'$

address the maximum principle and the preservation of entropies, and so they constitute an answer to the second question in the introduction. In certain cases, the preservation of mathematical structures yields proofs of convergence with respect to the parameters which control the polynomial degree in the uncertain space. However one loses the simplicity of the implementation provided by moment models [37], and so a clear path to the design of efficient, fast and multidimensional algorithms based on these structures is still to invent.

References

1. N.I. Akhiezer, *The Classical Moment Problem: And Some Related Questions in Analysis* (Olivery & Boyd, Edinburgh, 1965)
2. C. Bernardi, Y. Maday, Polynomial interpolation results in Sobolev space. J. Comput. Appl. Math. **43**, 53–80 (1992)
3. R. Bojanovic, R.A. Devore, On polynomials of best one side approximation. L'enseignement Math. **12**, 139–164 (1966)
4. Y. Brenier, Résolution d'équations d'évolution quasi-linéaires en dimension N d'espace à l'aide d'équations linéaires en dimension $N + 1$. J. Differ. Equ. **50**, 375–390 (1983)
5. Y. Brenier, L^2 formulation of multidimensional scalar conservation laws. Arch. Ration. Mech. Anal. **193**(1), 1–19 (2009)
6. Y. Brenier, L. Corrias, A kinetic formulation for multi-branch entropy solutions of scalar conservation laws. Ann. Inst. H. Poincaré Anal. Non Linéaire **15**, 169–190 (1998)
7. L. Caillabet, B. Canaud, G. Salin, S. Mazevet, P. Loubeyre, On the change in Inertial Confinement Fusion Implosions upon using an ab initio multiphase DT equation of state. Phys. Rev. Lett. **107**, 115004 (2011)
8. R.H. Cameron, W.T. Martin, The orthogonal development of non-linear functionals in series of Fourier-Hermite functionals. Ann. Math. **48**, 385–392 (1947)
9. E.J. Candès, J. Romberg, T. Tao, Robust uncertainty principles: exact signal reconstruction from highly incomplete frequency information. IEEE Trans. Inf. Theory **52**(2), 489–509 (2006)
10. C. Canuto, A. Quarteroni, Approximation results for orthogonal polynomials in Sobolev spaces. Math. Comput. **38**(157), 67–86 (1982)
11. C. Cercignani, *The Boltzmann Equation and Its Applications* (Springer, Berlin, 1988)
12. G.Q. Chen, C.D. Levermore, T.P. Liu, Hyperbolic conservation laws with stiff relaxation and entropy. Commun. Pure Appl. Math. **47**, 787–830 (1994)
13. B. Debusshere, H. Najim, P. Pebay, O. Knio, R. Ghanem, O. Le Maitre, Numerical challenges in the use of polynomial chaos representations for stochastic processes. SIAM J. Sci. Comput. **26**, 698–719 (2004)
14. B. Després, Polynomials with bounds and numerical approximation. Hal preprint server (2016). https://hal.archives-ouvertes.fr/hal-01307999
15. B. Després, B. Perthame, Uncertainty propagation; Intrusive kinetic formulations of scalar conservation laws. SIAM/ASA J. Uncertain. Quantif. **4**(1), 980–1013 (2016)
16. B. Després, E. Trélat, Two-sided space time L^1 approximation and optimal control of polynomial systems. HAL reprint server (2017). http://hal.upmc.fr/hal-01487186
17. B. Després, G. Poëtte, D. Lucor, Robust uncertainty propagation in systems of conservation laws with the entropy closure method, in *Uncertainty Quantification in Computational Fluid Dynamics*. Lecture Notes in Computational Science and Engineering, vol. 92 (Springer, Heidelberg, 2013), pp. 105–149
18. R.A. Devore, G.G. Lorenz, *Constructive Approximation* (Springer, Berlin, 1981)

19. U.S. Fjordholm, S. Lanthaler, S. Mishra, Statistical solutions of hyperbolic conservation laws I: foundations (2016). arxiv preprint server https://arxiv.org/abs/1605.05960
20. R. Fourer, D.G. Gay, B.W. Kernighan, A modeling language for mathematical programming. Manag. Sci. **36**, 519–554 (1990). AMPL: A Mathematical Programming Language, http://www.ampl.com/REFS/amplmod.pdf
21. R.G. Ghanem, P.D. Spanos, *Stochastic Finite Elements: A Spectral Approach* (Springer, New York, 1991)
22. D. Gottlieb, D. Xiu, Galerkin method for wave equations with uncertain coefficients. Commun. Comput. Phys. **3**(2), 505–518 (2008)
23. H. Grad, On the kinetic theory of rarefied gases. Commun. Pure Appl. Math. **2**, 331 (1949)
24. J.D. Jakeman, M.S. Eldred, K. Sargsyan, Enhancing l^1 minimization estimates of polynomial chaos expansions using basis selection. J. Comput. Phys. **289**, 18–34 (2015)
25. S. Jin, J.-G. Liu, Z. Ma, Uniform spectral convergence of the stochastic Galerkin method for the linear transport equations with random inputs in diffusive regime and a micro-macro decomposition based asymptotic preserving method (2016). Preprint 2016 online, http://www.math.wisc.edu/~jin/PS/Jin-Liu-Ma.pdf
26. M. Junk, Maximum entropy moment problems and extended Euler equations, in *Transport in Transition Regimes*. The IMA Volumes in Mathematics and Its Applications, vol. 135 (Springer, New York, 2004), pp. 189–198
27. P.D. Lax, *Hyperbolic Systems of Conservation Laws and the Theory of Shock Waves* (SIAM, Philadelphia, 1973)
28. G. Lin, C.-H. Su, G.E. Karniadakis, The stochastic piston problem. Proc. Natl. Acad. Sci. **101**, 15840–15845 (2004)
29. P.-L. Lions, B. Perthame, E. Tadmor, A kinetic formulation of multidimensional scalar conservation laws and related equations. J. Am. Math. Soc. **7**(1), 169–191 (1994)
30. I. Müller, T. Ruggeri, *Rational Extended Thermodynamics*. Springer Tract in Natural Philosophy, vol. 37 (Springer, New York, 1998)
31. B. Perthame, *Kinetic Formulation of Conservation Laws*. Oxford Lecture Series in Mathematics and Its Applications, vol. 21 (Oxford University Press, Oxford, 2002)
32. B. Perthame, E. Tadmor, A kinetic equation with kinetic entropy functions for scalar conservation laws. Commun. Math. Phys. **136**, 501–517 (1991)
33. V. Rana, H. Lim, J. Melvin, J. Glimm, B. Cheng, D.H. Sharp, Mixing with applications to inertial-confinement-fusion implosions. Phys. Rev. E **95**, 013203 (2017)
34. K. Sargsyan, C. Safta, B. Debusschere, H. Najm, Uncertainty quantification given discontinuous model response and a limited number of model runs. SIAM Int. J. Sci. Comput. **34**(1), B44–B64 (2012)
35. E.D. Sontag, *Mathematical Control Theory-Deterministic Finite Dimensional Systems*, 2nd edn. (Springer, Berlin, 1998)
36. E. Trélat, *Optimal Control and Applications* (Vuibert, Paris, 2008)
37. J. Tryoen, O. Le Maire, M. Ndjinga, A. Ern, Intrusive Galerkin methods with upwinding for uncertain nonlinear hyperbolic systems. J. Comput. Phys. **229**(18), 6485–6511 (2010)
38. A. Weisse, G. Wellein, A. Alvermann, H. Fehske, The kernel polynomial method. Rev. Mod. Phys. **78**, 275 (2006)
39. N. Wiener, The homogeneous chaos. Am. J. Math. **60**(4), 897–936 (1938)
40. K. Wu, H. Tang, D. Xiu, A stochastic Galerkin method for general system of quasilinear hyperbolic conservation laws with uncertainty (2016). arxiv preprint server, https://arxiv.org/abs/1601.04121
41. D. Xiu, G.E. Karniadakis, Modeling uncertainty in flow simulations via generalized polynomial chaos. J. Comput. Phys. **187**, 137–167 (2003)

Uncertainty Quantification for Kinetic Models in Socio–Economic and Life Sciences

Giacomo Dimarco, Lorenzo Pareschi, and Mattia Zanella

Abstract Kinetic equations play a major rule in modeling large systems of interacting particles. Recently the legacy of classical kinetic theory found novel applications in socio-economic and life sciences, where processes characterized by large groups of agents exhibit spontaneous emergence of social structures. Well-known examples are the formation of clusters in opinion dynamics, the appearance of inequalities in wealth distributions, flocking and milling behaviors in swarming models, synchronization phenomena in biological systems and lane formation in pedestrian traffic. The construction of kinetic models describing the above processes, however, has to face the difficulty of the lack of fundamental principles since physical forces are replaced by empirical social forces. These empirical forces are typically constructed with the aim to reproduce qualitatively the observed system behaviors, like the emergence of social structures, and are at best known in terms of statistical information of the modeling parameters. For this reason the presence of random inputs characterizing the parameters uncertainty should be considered as an essential feature in the modeling process. In this survey we introduce several examples of such kinetic models, that are mathematically described by nonlinear Vlasov and Fokker–Planck equations, and present different numerical approaches for uncertainty quantification which preserve the main features of the kinetic solution.

1 Introduction

Kinetic models describing the collective behavior of a large group of interacting agents have attracted a lot of interest in the recent years in view of their potential applications to various fields, like sociology, economy, finance and biology

G. Dimarco · L. Pareschi (✉)
University of Ferrara, Ferrara, Italy
e-mail: giacomo.dimarco@unife.it; lorenzo.pareschi@unife.it

M. Zanella
Politecnico di Torino, Torino, Italy
e-mail: mattia.zanella@polito.it

© Springer International Publishing AG, part of Springer Nature 2017
S. Jin, L. Pareschi (eds.), *Uncertainty Quantification for Hyperbolic and Kinetic Equations*, SEMA SIMAI Springer Series 14,
https://doi.org/10.1007/978-3-319-67110-9_5

[1, 2, 4, 6, 7, 11, 12, 23–25, 33–35, 40, 41, 51, 72, 74]. One of the major difficulties in applying the classical toolbox of kinetic theory to these systems is the lack of fundamental principles which define the microscopic dynamic. In addition, experimental results are typically non reproducible and, as a consequence, the model construction is dictated by its ability to describe qualitatively the system behavior and the formation of emergent social structures. A degree of uncertainty is therefore implicitly embedded in such models, since most modeling parameters can be only as statistical information from experimental results [5, 8, 15, 52, 65].

From a mathematical viewpoint, the kinetic models we will consider in the present survey are characterized by nonlinear Vlasov–Fokker–Planck equations with random inputs taking into account uncertainties in the initial data, in the interaction terms and/or in the boundary conditions. The models describe the evolution of a distribution function $f = f(\theta, x, w, t)$, $t \geq 0$, $x \in \mathbb{R}^{d_x}$, $w \in \mathbb{R}^{d_w}$, $d_x, d_w \geq 1$, and $\theta \in \Omega \subseteq \mathbb{R}^{d_\theta}$ a *random field*, accordingly to

$$\partial_t f + \mathscr{L}[f] = \nabla_w \cdot [\mathscr{B}[f]f + \nabla_w(Df)], \tag{1}$$

where $\mathscr{L}[\cdot]$ is a linear operator describing the agents' dynamics with respect to the x-variable, typically $\mathscr{L}[f] = w \cdot \nabla_x f$, $\mathscr{B}[\cdot]$ is a non-local operator of the form

$$\mathscr{B}[f](\theta, x, w, t) = \int_{\mathbb{R}^{d_x}} \int_{\mathbb{R}^{d_w}} P(x, x_*; w, w_*, \theta)(w - w_*)f(\theta, x_*, w_*, t)dw_*dx_*, \tag{2}$$

and $D(\theta, w) \geq 0$, for all $w \in \mathbb{R}^{d_w}$, is a function describing the local relevance of the diffusion. We refer to [27, 39, 80, 86, 87] for an introduction to the subject in relation with kinetic theory. In the rest of the chapter, to avoid unnecessary difficulties, we will mainly restrict to the case of a one-dimensional random input $d_\theta = 1$ distributed as $p(\theta)$. In the homogeneous case $f = f(\theta, w, t)$, $\mathscr{L}[f] \equiv 0$ the kinetic models are characterized by nonlinear Fokker-Planck equations.

1.1 The Classic Fokker-Planck Equation with Uncertainties

The most classical example is represented by the linear Fokker-Planck model obtained for $P \equiv 1$ corresponding to

$$\mathscr{B}[f](\theta, w) = (w - u(\theta)), \qquad D(\theta) = T(\theta), \tag{3}$$

where

$$u(\theta) = \int_{\mathbb{R}^{d_w}} f(\theta, w, t)w\, dw, \quad T(\theta) = \frac{1}{d_w} \int_{\mathbb{R}^{d_w}} f(\theta, w, t)(w - u(\theta))^2 f(\theta, w, t)\, dw$$

are the (conserved) mean velocity and temperature of the particles. In the above expressions we assumed an uncertain initial data such that $\int_{\mathbb{R}^{d_w}} f(\theta, w, 0)\, dw = 1$ for all $\theta \in \Omega$. The stationary solution in this case is represented by a Maxwellian distribution with uncertain momentum and temperature given by

$$f^{\infty}(\theta, w) = \frac{1}{(2\pi T(\theta))^{d_w/2}} \exp\left\{ -\frac{|w - u(\theta)|^2}{2T(\theta)} \right\}. \tag{4}$$

1.2 Opinion Formation with Uncertain Interaction

A kinetic Fokker-Planck model of opinion formation for $w \in I = [-1, 1]$, where ± 1 denote the two extremal opinions, corresponds to the choices [74, 84]

$$\mathscr{B}[f](\theta, w, t) = \int_I P(\theta, w, w_*)(w - w_*) f(\theta, w_*, t) dw_*, \quad D(w) = \frac{\sigma^2}{2}(1 - w^2)^2. \tag{5}$$

In the above nonlocal interaction term $P(\theta, \cdot, \cdot) \in [0, 1]$ is a function taking into account uncertainties in the compromise propensity between the agents' opinions.

In the simple case $P(w, w_*, \theta) = P(\theta)$ and deterministic initial data, the model preserves the mean opinion $u = \int_I wf(\theta, w, t)dw$ and we can analytically compute the steady state distribution

$$f^{\infty}(\theta, w) = \frac{C}{(1 - w^2)^2} (1 + w)^{\frac{P(\theta)u}{2\sigma^2}} (1 - w)^{\frac{P(\theta)u}{2\sigma^2}} \exp\left\{ -\frac{P(\theta)(1 - uw)}{\sigma^2(1 - w^2)} \right\}, \tag{6}$$

with $C > 0$ a normalization constant.

1.3 Wealth Distribution with Uncertain Diffusion

If we now consider $w \in [0, \infty)$ a measure of the agents' wealth, a Fokker-Planck model describing the wealth evolution of agents is obtained taking [33, 74]

$$\mathscr{B}[f](\theta, w, t) = \int_{[0, \infty]} a(w, w_*)(w - w_*) f(\theta, w_*, t) dw_*, \quad D(\theta, w) = \frac{\sigma(\theta)^2}{2} w^2, \tag{7}$$

where the term $\sigma(\theta)$ characterizes the uncertain strength of diffusion. An explicit expression of the steady state distribution is given in the case $a(w, w_*) \equiv 1$

$$f^{\infty}(\theta, w) = \frac{(\mu(\theta) - 1)^{\mu(\theta)}}{\Gamma(\mu(\theta)) w^{1 + \mu(\theta)}} \exp\left\{ -\frac{\mu(\theta) - 1}{w} \right\}, \tag{8}$$

where $\Gamma(\cdot)$ is the Gamma function and $\mu(\theta) = 1 + 2/\sigma^2(\theta)$ is the so-called Pareto exponent, which is now dependent on the random input.

1.4 Swarming Models with Uncertainties

As a final example we consider a kinetic model for the swarming behavior [11, 23–25, 32, 38, 40, 48, 58, 59]. In particular we focus on a model with self-propulsion and uncertain diffusion, see [9, 10]. The dynamics for the density $f = f(\theta, x, w, t)$ of agents in position $x \in \mathbb{R}^{d_x}$ with velocity $w \in \mathbb{R}^{d_w}$ is described by the Vlasov-Fokker-Planck equation (1) characterized by

$$\mathscr{L}[f] = w \cdot \nabla_x f, \qquad \mathscr{B}[f](\theta, x, w, t) = \alpha w(1 - |w|^2) + (w - u_f(\theta, x, t)), \qquad (9)$$

where

$$u_f(\theta, x, t) = \frac{\int_{\mathbb{R}^{d_x} \times \mathbb{R}^{d_w}} K(x, y) w f(\theta, y, w, t) \, dw \, dy}{\int_{\mathbb{R}^{d_x} \times \mathbb{R}^{d_w}} K(x, y) f(\theta, y, w, t) \, dw \, dy}, \qquad (10)$$

with $K(x, y) > 0$ a localization kernel, $\alpha > 0$ a self-propulsion term and $D(\theta) > 0$ the uncertain noise intensity.

In the space-homogeneous case $f = f(\theta, w, t)$, stationary solutions have the form

$$f^\infty(w, \theta) = C \exp\left\{ -\frac{1}{D(\theta)} \left(\alpha \frac{|w|^4}{4} + (1 - \alpha) \frac{|w|^2}{2} - u_{f\infty}(\theta) \cdot w \right) \right\}, \qquad (11)$$

with $C > 0$ a normalization constant and

$$u_{f\infty}(\theta) = \frac{\int_{\mathbb{R}^{d_w}} w f^\infty(w, \theta) dw}{\int_{\mathbb{R}^{d_w}} f^\infty(w, \theta) dw}.$$

We stress that, in all the above reported examples, uncertainty may be present in other modeling parameters, like boundary conditions, external forces, etc., by further increasing the dimensionality and the complexity of the kinetic model.

The development of numerical methods for kinetic equations presents several difficulties due to the high dimensionality and the intrinsic structural properties of the solution. Non negativity of the distribution function, conservation of invariant quantities, entropy dissipation and steady states are essential in order to compute qualitatively correct solutions. Preservation of these structural properties is even more challenging in presence of uncertainties which contribute to the curse of dimensionality. We refer to [43, 63, 82] for recent surveys on numerical methods for kinetic equations in the deterministic case.

For this reason we will focus on the construction of numerical methods for uncertainty quantification (UQ) which preserves the structural properties of the kinetic equation and, in particular, which are able to capture the correct steady state of the problem with arbitrary accuracy. We will discuss different numerical approaches based on the major techniques used for uncertainty quantification. In the deterministic case, similar approaches for nonlinear Fokker-Planck equations were previously derived in [17, 18, 29, 66, 71, 81]. Related methods for the case of nonlinear degenerate diffusions equations were proposed in [14, 28] and with nonlocal terms in [19, 26]. We refer also to [3] for the development of methods based on stochastic approximations and to [56] for a recent survey on schemes which preserve steady states of balance laws and related problems.

The simplest class of methods for quantifying uncertainty in partial differential equations (PDEs) are the stochastic collocation methods. Stochastic collocation methods are non-intrusive, so they preserve all properties of the deterministic numerical scheme, and easy to parallelize. In Sect. 3 we describe the structure preserving methods recently developed in [75, 76] together with a collocation approach and show how the resulting schemes preserve non negativity, conservation and entropy dissipation. In addition they capture the steady states with arbitrary accuracy and may achieve high convergence rates in the random space (spectral convergence for smooth solutions). Next in Sect. 4, we consider the closely related class of statistical sampling methods, most notably Monte Carlo (MC) sampling. In order to address the slow convergence of MC methods, we discuss here the development of Monte Carlo methods based on a Micro–Macro decomposition approach introduced in [44]. These methods preserve the structural properties of the kinetic problem, are capable to significatively reduce the statistical fluctuations of standard Monte Carlo and increase their computational efficiency by reducing the number of statistical samples in time. Section 5 is devoted to stochastic Galerkin methods based on generalized Polynomial Chaos (gPC). Although these deterministic methods may achieve high convergence rates for smooth solutions, they suffer from the disadvantage that they are highly intrusive and that increase the computational complexity of the problem. As a consequence, the main physical properties of the solution are typically lost at a numerical level. For this class of methods we show how to construct generalized Polynomial Chaos schemes based on the Micro–Macro formalism which preserve the steady states of the system [46]. Finally, in Sect. 6 several numerical applications to problem in socio-economy and life sciences are presented.

2 Preliminaries

In this section we recall some analytical properties of the considered kinetic models which will be useful for the development of the different numerical methods. Except in some simple case, a precise analytic description of the global equilibria of Eq. (1) is very difficult [21, 22, 85]. A deeper insight into the large time behavior can be achieved by resorting to the asymptotic behavior of the corresponding space homogeneous models, characterized by nonlinear Fokker–Planck type equations [83].

2.1 Fokker-Planck Type Equations

In the space homogeneous case the distribution function reduces to $f = f(\theta, w, t)$, $w \in \mathbb{R}^{d_w}$, $\theta \in \mathbb{R}^{d_\theta}$, $t > 0$ and is solution of the following problem

$$\partial_t f(\theta, w, t) = \mathscr{I}(f, f)(\theta, w, t), \tag{12}$$

where

$$\mathscr{I}(f, f)(\theta, w, t) = \nabla_w \cdot \Big[\mathscr{B}[f](\theta, w, t) f(\theta, w, t) + \nabla_w D(\theta, w) f(\theta, w, t) \Big], \tag{13}$$

together with an initial datum $f(\theta, w, 0) = f_0(\theta, w)$ and suitable boundary conditions on $w \in \mathbb{R}^{d_w}$.

We review in the present stochastic setting the classical results for the trend to equilibrium of problem (12) in the one-dimensioal case $w \in I \subseteq \mathbb{R}$ with a linear drift term, i.e.

$$\partial_t f(\theta, w, t) = \partial_w \Big[(w - u) f(\theta, w, t) + \partial_w (D(\theta, w) f(\theta, w, t)) \Big]. \tag{14}$$

Conservation of mass is imposed on the previous equation by considering suitable boundary conditions [74]. The stochastic stationary solution $f^\infty(\theta, w)$ of Eq. (14) is given by the solution of

$$(w - u) f^\infty(\theta, w) + \partial_w D(\theta, w) f^\infty(\theta, w).$$

The stochastic Fokker–Planck equation (14) may be rewritten in the equivalent forms

$$\partial_t f(\theta, w, t) = \partial_w \Big[D(\theta, w) f(\theta, w, t) \partial_w \log \frac{f(\theta, w, t)}{f^\infty(\theta, w, t)} \Big], \tag{15}$$

which corresponds to the *stochastic Landau form*, whereas the stochastic non logarithmic Laundau form of the equation is the following

$$\partial_t f(\theta, w, t) = \partial_w \Big[D(\theta, w) f^\infty(\theta, w, t) \partial_w \frac{f(\theta, w, t)}{f^\infty(\theta, w)} \Big]. \tag{16}$$

Convergence to equilibrium is usually determined through estimates of the entropy production. We define the relative entropy for all positive functions $f, \tilde{f}$ as follows

$$\mathscr{H}[f, \tilde{f}](\theta, w, t) = \int_I f(\theta, w, t) \log \left(\frac{f(\theta, w, t)}{\tilde{f}(\theta, w, t)} \right) dw, \tag{17}$$

we have [54]

$$\frac{d}{dt}\mathscr{H}[f,f^\infty](\theta,w,t) = -\mathscr{I}_D[f,f^\infty](\theta,w,t), \tag{18}$$

where the dissipation functional $\mathscr{I}_D[\cdot,\cdot]$ is defined as

$$\mathscr{I}_D[f,f^\infty] = \int_{\mathscr{I}} D(\theta,w)f(\theta,w,t)\left(\partial_w \log\left(\frac{f(\theta,w,t)}{f^\infty(\theta,w)}\right)\right)^2 dw. \tag{19}$$

In the classical setting $w \in \mathbb{R}$ and $D(\theta,w) = T(\theta)$, where $T(\theta)$ is the temperature, the steady state is given by the Maxwellian density (4) with $d_w = 1$ and relation (18) coupled with the log-Sobolev inequality

$$\mathscr{H}[f,f^\infty](\theta,w,t) \le \frac{1}{2}\mathscr{I}_D[f,f^\infty](\theta,w,t), \tag{20}$$

leads to the exponential decay of the relative entropy as proved in the following result [83].

Theorem 1 *Let $f(\theta,w,t)$ be the solution to the initial value problem*

$$\partial_t f(\theta,w,t) = \partial_w(w - u(\theta))f(\theta,w,t) + T(\theta)\partial_w^2 f(\theta,w,t)$$

with the initial condition $f(\theta,w,0) = f_0(\theta,w)$ with finite entropy. Then $f(\theta,w,t)$ converges for all $\theta \in \Omega$ to $f^\infty(\theta,w)$ given by (4) and

$$\mathscr{H}[f,f^\infty] \le e^{-2t/T(\theta)}\mathscr{H}[f_0,f^\infty].$$

For more general diffusion functions $D(\theta,w)$ analogous log-Sobolev inequality are not available. A strategy to study the convergence to equilibrium is to investigate the relation of relative entropy with the relative weighted Fisher information, see [21, 54, 70, 83] for more details.

2.2 Micro–Macro Formulation

In this paragraph we describe the Micro–Macro approach to kinetic equations of the form (12). The approach is based on the classical Micro–Macro decomposition originally developed by Liu and Yu in [69] for the fluid limit of the Boltzmann equation. The method has been fruitfully employed for the development of numerical methods by several authors (see [13, 36, 37, 47, 67, 91] and the references therein). These techniques has been also recently developed in [45, 53, 73] to construct spectral methods for the collisional operator of the Boltzmann equation that preserves exactly the Maxwellian steady state of the system. Since under suitable regularity assumptions on the initial distribution the Fokker-Planck equation admits a unique steady state solution $f^\infty(\theta, w)$, the Micro–Macro formulation is obtained decomposing the solution of the differential problem into the equilibrium part f^∞ and the non-equilibrium part g as follows

$$f(\theta, w, t) = f^\infty(\theta, w) + g(\theta, w, t), \tag{21}$$

where $g(\theta, w, t)$ is a distribution function such that

$$\int_{\mathbb{R}^{d_w}} \phi(w) g(\theta, w, t) dw = 0$$

for some moments $\phi(w) = 1, w$. The above decomposition (21) applied to the Fokker-Planck problem (12)–(13) yields the following result.

Proposition 1 *If the nonlinear Fokker-Planck equations (12)–(13) admits the unique equilibrium state $f^\infty(\theta, w)$, the differential operator $\mathscr{J}(\cdot, \cdot)$ defined in (13) with $\mathscr{B}[f]$ given by (2) may be rewritten as*

$$\mathscr{J}(f, f)(\theta, w, t) = \mathscr{J}(g, g)(\theta, w, t) + \mathscr{N}(f^\infty, g)(\theta, w, t), \tag{22}$$

where $\mathscr{N}(\cdot, \cdot)$ is a linear operator defined as

$$\mathscr{N}(f^\infty, g)(\theta, w, t) = \nabla_w \Big[\mathscr{B}[f^\infty] g(\theta, w, t) + \mathscr{B}[g] f^\infty(\theta, w) \Big].$$

The only admissible steady state solution of the problem

$$\begin{cases} \partial_t g(\theta, w, t) = \mathscr{J}(g, g)(\theta, w, t) + \mathscr{N}(f^\infty, g)(\theta, w, t), \\ f(\theta, w, t) = f^\infty(\theta, w) + g(\theta, w, t) \end{cases} \tag{23}$$

is given by $g^\infty(\theta, w) \equiv 0$.
The proof is an immediate consequence of the fact that at the steady state we have $\mathscr{J}(f^\infty, f^\infty) = 0$. Note that the steady state solution of the reformulated problem (23) is therefore independent of the uncertainty.

Remark 1 Under suitable assumptions, see Theorem 1, one can show that $f(\theta, w, t)$ exponentially decays to the equilibrium solution. As a consequence, the non-equilibrium part of the Micro–Macro approximation $g(\theta, w, t)$ exponentially decays to $g^\infty(\theta, w) \equiv 0$ for all $\theta \in \Omega$.

3 Collocation Methods

One of the most popular computational approaches for UQ relies on the class of collocation methods [88, 89]. These methods are non intrusive and permit to couple existing deterministic solvers for the PDEs with techniques for the quantification of the uncertainty. Moreover, since the structure of the solution remains unchanged, numerical analysis of collocation methods is a straightforward consequence of the results obtained for the underlying method used for solving the original equation.

In the following, since the linear transport part in (1) can be discretized using standard approaches, see for instance [43], we concentrate on homogeneous Fokker-Planck problem of the form (12)–(13). Connections with the full problem are then recovered using splitting methods or other partitioned time discretization schemes, like additive Runge-Kutta methods [60].

Collocation methods consist in solving the problem in a finite set of nodes $\{\theta_k\}_{k=0}^M$ of the random field. In this class of methods belongs the usual Monte Carlo sampling which will be treated in Sect. 4. If the distribution of the random input $\theta \sim p(\theta)$ is known, an efficient way to treat the uncertainty is to select the nodes in the random space according to Gaussian quadrature rules related to such distribution. This is straightforward in the univariate case, whereas becomes more challenging in the multivariate case [50].

For each $k = 0, \ldots, M$ we obtain a totally deterministic and decoupled problem since the value of the random variable is fixed. Therefore, solving this system of equations poses no difficulty provided one has a well-established deterministic algorithm. The result is an ensemble of $M + 1$ deterministic solutions which can be post-processed to recover the statistical values of interest. For example, in the univariate case if $\{\omega_k\}_{k=0}^M$ are the Gaussian weights on $\Omega \subseteq \mathbb{R}$ corresponding to $p(\theta)$ we can use the approximations

$$\mathbb{E}[f](w, t) = \int_\Omega f(\theta, w, t) p(\theta) \, d\theta \approx \mathbb{E}_M[f](w, t) = \sum_{k=0}^M \omega_k f(\theta_k, w, t), \quad (24)$$

$$Var[f](w, t) = \int_\Omega (f(\theta, w, t) - \mathbb{E}[f](w, t))^2 p(\theta) \, d\theta$$

$$\approx Var_M[f](w, t) = \sum_{k=0}^M \omega_k (f(\theta_k, w, t) - \mathbb{E}_M[f](w, t))^2, \quad (25)$$

where $\mathbb{E}[\cdot]$ and $Var[\cdot]$ denote the mean and the variance respectively. In the following we concentrates on the construction of numerical schemes which preserve the structural properties of the solution, like non-negativity, entropy dissipation and accurate asymptotic behavior [75, 76].

3.1 Structure Preserving Methods

In the one-dimensional case $d_w = 1$ for all $k = 0, \ldots, M$ the Fokker-Planck equation (12)–(13) may be written as

$$\partial_t f(\theta_k, w, t) = \partial_w \mathscr{F}[f](\theta_k, w, t), \quad w \in I \subseteq \mathbb{R} \quad (26)$$

where now

$$\mathscr{F}[f](\theta_k, w, t) = (\mathscr{B}[f](\theta_k, w, t) + D'(w))f(\theta_k, w, t) + D(w)\partial_w f(\theta_k, w, t) \qquad (27)$$

using the compact notation $D'(w) = \partial_w D(w)$. Typically, when I is a finite size set the problem is complemented with no-flux boundary conditions at the extremal points. In the sequel we assume $D(w) > 0$ in the internal points of I.

We introduce an uniform spatial grid $w_i \in I$, such that $w_{i+1} - w_i = \Delta w$. We denote as usual $w_{i\pm 1/2} = w_i \pm \Delta/2$ and consider a conservative discretization of (26)

$$\frac{d}{dt}f_i(\theta_k, t) = \frac{\mathscr{F}_{i+1/2}[f](\theta_k, t) - \mathscr{F}_{i-1/2}[f](\theta_k, t)}{\Delta w}, \qquad (28)$$

where for each $t \geq 0$ $\mathscr{F}_{i\pm 1/2}[f](\theta_k, t)$ is the numerical flux function characterizing the discretization.

Let us set $\mathscr{C}[f](w, \theta_k, t) = \mathscr{B}[f](w, \theta_k, t) + D'(w)$ and adopt the notations $D_{i+1/2} = D(w_{i+1/2}), D'_{i+1/2} = D'(w_{i+1/2})$. We will consider a general flux function which is combination of the grid points $i + 1$ and i

$$\mathscr{F}_{i+1/2}[f] = \tilde{\mathscr{C}}^k_{i+1/2}\tilde{f}_{i+1/2}(\theta_k, t) + D_{i+1/2}\frac{f_{i+1}(\theta_k, t) - f_i(\theta_k, t)}{\Delta w}, \qquad (29)$$

where

$$\tilde{f}_{i+1/2}(\theta_k, t) = (1 - \delta^k_{i+1/2})f_{i+1}(\theta_k, t) + \delta^k_{i+1/2}f_i(\theta_k, t). \qquad (30)$$

For example, the standard approach based on central difference is obtained by considering for all i the quantities

$$\delta^k_{i+1/2} = 1/2, \qquad \tilde{\mathscr{C}}^k_{i+1/2} = \mathscr{C}[f](w_{i+1/2}, \theta_k, t).$$

It is well-known, however, that such a discretization method is subject to restrictive conditions over the mesh size Δw in order to keep non negativity of the solution.

Here, we aim at deriving suitable expressions for the family of weight functions $\delta^k_{i+1/2}$ and for $\tilde{\mathscr{C}}^k_{i+1/2}$ in such a way that the method yields nonnegative solutions, without restriction on Δw, and preserves the steady state of the system with arbitrary order of accuracy.

First, observe that at the steady state the numerical flux should vanish. From (29) we get

$$\frac{f_{i+1}(\theta_k, t)}{f_i(\theta_k, t)} = \frac{-\delta^k_{i+1/2}\tilde{\mathscr{C}}^k_{i+1/2} + \dfrac{D_{i+1/2}}{\Delta w}}{(1 - \delta^k_{i+1/2})\tilde{\mathscr{C}}^k_{i+1/2} + \dfrac{D_{i+1/2}}{\Delta w}}. \qquad (31)$$

Similarly, if we consider the analytical flux imposing $\mathscr{F}[f](\theta_k, w, t) \equiv 0$, we have

$$D(w)\partial_w f(\theta_k, w, t) = -(\mathscr{B}[f](\theta_k, w, t) + D'(w))f(\theta_k, w, t), \tag{32}$$

which is in general not solvable, except in some special cases due to the nonlinearity on the right hand side. We may overcome this difficulty in the quasi steady-state approximation integrating equation (32) on the cell $[w_i, w_{i+1}]$

$$\int_{w_i}^{w_{i+1}} \frac{1}{f(\theta_k, w, t)}\partial_w f(\theta_k, w, t)dw = \\ -\int_{w_i}^{w_{i+1}} \frac{1}{D(w)}(\mathscr{B}[f](\theta_k, w, t) + D'(w))dw, \tag{33}$$

which gives

$$\frac{f(\theta_k, w_{i+1}, t)}{f(\theta_k, w_i, t)} = \exp\left\{-\int_{w_i}^{w_{i+1}} \frac{1}{D(w)}(\mathscr{B}[f](\theta_k, w, t) + D'(w))dw\right\}. \tag{34}$$

Now, by equating the ratio $f_{i+1}(\theta_k, t)/f_i(\theta_k, t)$ and $f(\theta_k, w_{i+1}, t)/f(\theta_k, w_i, t)$ in (31)–(34) for the numerical and exact flux respectively, and setting

$$\widetilde{\mathscr{C}}_{i+1/2}^k = \frac{D_{i+1/2}}{\Delta w}\int_{w_i}^{w_{i+1}} \frac{\mathscr{B}[f](\theta_k, w, t) + D'(w)}{D(w)}dw \tag{35}$$

we recover

$$\delta_{i+1/2}^k = \frac{1}{\lambda_{i+1/2}^k} + \frac{1}{1 - \exp(\lambda_{i+1/2}^k)}, \tag{36}$$

where

$$\lambda_{i+1/2}^k = \int_{w_i}^{w_{i+1}} \frac{\mathscr{B}[f](\theta_k, w, t) + D'(w)}{D(w)}dw = \frac{\Delta w\, \widetilde{\mathscr{C}}_{i+1/2}^k}{D_{i+1/2}}. \tag{37}$$

We have the following result [76]

Proposition 2 *The numerical flux function (29)–(30) with $\widetilde{\mathscr{C}}_{i+1/2}^k$ and $\delta_{i+1/2}^k$ defined by (35) and (36)–(37) vanishes when the corresponding flux (27) is equal to zero over the cell $[w_i, w_{i+1}]$. Moreover the nonlinear weight functions $\delta_{i+1/2}^k$ defined by (36)–(37) are such that $\delta_{i+1/2}^k \in (0, 1)$.*
By discretizing (37) through the midpoint rule

$$\int_{w_i}^{w_{i+1}} \frac{\mathscr{B}[f](\theta_k, w, t) + D'(w)}{D(w)}dw = \frac{\Delta w(\mathscr{B}_{i+1/2}(\theta_k, t) + D'_{i+1/2})}{D_{i+1/2}} + O(\Delta w^3), \tag{38}$$

we obtain the second order method defined by

$$\lambda_{i+1/2}^{k,\text{mid}} = \frac{\Delta w(\mathscr{B}_{i+1/2}(\theta_k, t) + D'_{i+1/2})}{D_{i+1/2}} \tag{39}$$

and

$$\delta_{i+1/2}^{k,\text{mid}} = \frac{D_{i+1/2}}{\Delta w(\mathscr{B}_{i+1/2}(\theta_k, t) + D'_{i+1/2})} + \frac{1}{1 - \exp(\lambda_{i+1/2}^{k,\text{mid}})}. \tag{40}$$

Higher order accuracy of the steady state solution can be obtained using suitable higher order quadrature formulas for the integral (35). We will refer to this type of schemes as structure preserving Chang-Cooper (SP-CC) type schemes.

Some remarks are in order.

Remark 2

- If we consider the limit case $D_{i+1/2} \to 0$ in (39)–(40) we obtain the weights

$$\delta_{i+1/2}^{k} = \begin{cases} 0, & \mathscr{B}_{i+1/2}(\theta_k, t) > 0, \\ 1, & \mathscr{B}_{i+1/2}(\theta_k, t) < 0 \end{cases}$$

  and the scheme reduces to a first order upwind scheme for the corresponding aggregation equation.
- For linear problems of the form $\mathscr{B}[f](\theta_k, w, t) = \mathscr{B}(\theta_k, w)$ the exact stationary state $f^\infty(w, \theta_k)$ can be directly computed from the solution of

$$D(w)\partial_w f^\infty(\theta_k, w) = -(\mathscr{B}(\theta_k, w) + D'(w))f^\infty(\theta_k, w), \tag{41}$$

together with the boundary conditions. Explicit examples of stationary states will be reported in the last section. Using the knowledge of the stationary state we have

$$\frac{f_{i+1}^\infty(\theta_k)}{f_i^\infty(\theta_k)} = \exp\left\{-\int_{w_i}^{w_{i+1}} \frac{1}{D(w)}(\mathscr{B}(\theta_k, w) + D'(w))dw\right\}$$

$$= \exp\left(-\lambda_{i+1/2}^\infty(\theta_k)\right), \tag{42}$$

therefore

$$\lambda_{i+1/2}^\infty(\theta_k) = \log\left(\frac{f_i^\infty(\theta_k)}{f_{i+1}^\infty(\theta_k)}\right) \tag{43}$$

and

$$\delta_{i+1/2}^{\infty}(\theta_k) = \frac{1}{\log(f_i^{\infty}(\theta_k)) - \log(f_{i+1}^{\infty}(\theta_k))} + \frac{f_{i+1}^{\infty}(\theta_k)}{f_{i+1}^{\infty}(\theta_k) - f_i^{\infty}(\theta_k)}. \tag{44}$$

In this case, the numerical scheme preserves the steady state exactly.
- The cases of higher dimension $d \geq 2$ may be derived similarly using dimensional splitting (see [75] for details).

3.1.1 Main Properties

In the following we recall some results on the preservation of the structural properties, like non negativity and entropy dissipation.

Concerning non negativity, first we report a result for an explicit time discretization scheme [75]. We introduce a time discretization $t^n = n\Delta t$ with $\Delta t > 0$ and $n = 0, \ldots, T$ and consider the simple forward Euler method

$$f_i^{n+1}(\theta_k) = f_i^n(\theta_k) + \Delta t \frac{\mathscr{F}_{i+1/2}^n(\theta_k) - \mathscr{F}_{i-1/2}^n(\theta_k)}{\Delta w}, \tag{45}$$

for all $k = 0, \ldots, M$.

Proposition 3 *Under the time step restriction*

$$\Delta t \leq \frac{\Delta w^2}{2(U\Delta w + D)}, \quad U = \max_{i,k} |\tilde{\mathscr{C}}_{i+1/2}^n(\theta_k)|, \tag{46}$$

the explicit scheme (45) with flux defined by (36)–(37) preserves nonnegativity for all $k = 0, \ldots, M$, i.e $f_i^{n+1}(\theta_k) \geq 0$ if $f_i^n(\theta_k) \geq 0$, $i = 0, \ldots, N$, $k = 0, \ldots, M$.
Higher order strong stability preserving (SSP) methods [57] are obtained by considering a convex combination of forward Euler methods. Therefore, the non negativity result can be extended to general SSP methods.

In practical applications, it is desirable to avoid the parabolic restriction $\Delta t = O(\Delta w^2)$ of explicit schemes. Unfortunately, fully implicit methods originate a nonlinear system of equations due to the nonlinearity of $\mathscr{B}[f]$ and the dependence of the weights $\delta_{i\pm1/2}^k$ from the solution. However, we have the following nonnegativity result for the semi-implicit case

$$f_i^{n+1}(\theta_k) = f_i^n(\theta_k) + \Delta t \frac{\hat{\mathscr{F}}_{i+1/2}^{n+1}(\theta_k) - \hat{\mathscr{F}}_{i-1/2}^{n+1}(\theta_k)}{\Delta w}, \tag{47}$$

where

$$
\begin{aligned}
\hat{\mathscr{F}}_{i+1/2}^{n+1}(\theta_k) = \tilde{\mathscr{C}}_{i+1/2}^{k,n} & \left[(1 - \delta_{i+1/2}^{k,n}) f_{i+1}^{n+1}(\theta_k) + \delta_{i+1/2}^{k,n} f_i^{n+1}(\theta_k) \right] \\
& + D_{i+1/2} \frac{f_{i+1}^{n+1}(\theta_k) - f_i^{n+1}(\theta_k)}{\Delta w}.
\end{aligned}
\tag{48}
$$

We have

Proposition 4 *Under the time step restriction*

$$
\Delta t < \frac{\Delta w}{2U}, \qquad U = \max_{i,k} |\tilde{\mathscr{C}}_{i+1/2}^{k,n}|
\tag{49}
$$

the semi-implicit scheme (47) preserves nonnegativity, i.e $f_i^{n+1}(\theta_k) \geq 0$ *if* $f_i^n(\theta_k) \geq 0$, $i = 0, \ldots, N$ *for all* $k = 0, \ldots, M$.

We refer to [75] for a detailed proof. Higher order semi-implicit approximations can be constructed following [16].

In order to discuss the entropy property we consider the prototype equation for all $k = 0, \ldots, M$

$$
\partial_t f(\theta_k, w, t) = \partial_w \left[P(\theta_k)(w - u) f(\theta_k, w, t) + \partial_w (D(w) f(\theta_k, w, t)) \right],
\tag{50}
$$

with $w \in I = [-1, 1]$ equipped with deterministic initial distribution $f(w, 0) = f_0(w)$, $u = \int_I w f_0(w) dw \in (-1, 1)$ and boundary conditions

$$
\partial_w (D(w) f(\theta_k, w, t)) + P(\theta_k)(w - u) f(\theta_k, w, t) = 0, \qquad w = \pm 1.
\tag{51}
$$

It can be shown that the introduced structure preserving scheme dissipates the numerical entropy [75]

Theorem 2 *Let us consider* $\mathscr{B}[f](\theta_k, w, t) = P(\theta_k)(w - u)$ *as in Eq. (50). The numerical flux (29)–(30) with* $\tilde{\mathscr{C}}_{i+1/2}^k$ *and* $\delta_{i+1/2}^k$ *given by (35)–(36) satisfies the discrete entropy dissipation for all* $k = 0, \ldots, M$

$$
\frac{d}{dt} \mathscr{H}_\Delta (f(\theta_k, w, t), f^\infty(\theta_k, w)) = -\mathscr{I}_\Delta (f(\theta_k, w, t), f^\infty(\theta_k, w)),
\tag{52}
$$

where

$$
\mathscr{H}_{\Delta w}(f(\theta_k, w, t), f^\infty(\theta_k, w)) = \Delta w \sum_{i=0}^{N} f_i \log \left(\frac{f_i(\theta_k, t)}{f_i^\infty(w, \theta_k)} \right)
\tag{53}
$$

and $\mathscr{I}_\Delta$ is the positive discrete dissipation function

$$\mathscr{I}_\Delta(f(w,\theta_k,t),f^\infty(w,\theta_k)) = \sum_{i=0}^N \left[\log\left(\frac{f_{i+1}(\theta_k,t)}{f_{i+1}^\infty(\theta_k,t)}\right) - \log\left(\frac{f_i(\theta_k,t)}{f_i^\infty(\theta_k)}\right)\right]$$

$$\cdot \left(\frac{f_{i+1}(\theta_k,t)}{f_{i+1}^\infty(\theta_k)} - \frac{f_i(\theta_k,t)}{f_i^\infty(\theta_k)}\right)\bar{f}_{i+1/2}^\infty(\theta_k)D_{i+1/2} \geq 0. \tag{54}$$

For more general equations the above approach does not permit to prove the entropy dissipation, see [75]. In the following, we introduce a different class of structure preserving schemes that, in addition to preservation of the steady state of the problem, ensure the entropy dissipation.

3.2 Entropic Average Schemes

Let us consider the general class of nonlinear Fokker-Planck equation with gradient flow structure [10, 22, 26]

$$\partial_t f(\theta_k,w,t) = \nabla_w \cdot [f(\theta_k,w,t)\nabla_w\xi(\theta_k,w,t)], \qquad w \in I \subseteq \mathbb{R}^{d_w}, \tag{55}$$

with $\{\theta_k\}_{k=0}^M$ the collocation nodes of the random field, and no-flux boundary conditions, where

$$\nabla_w\xi(\theta_k,w,t) = \mathscr{B}[f](\theta_k,w,t) + D\nabla_w \log f(\theta_k,w,t),$$
$$\mathscr{B}[f](\theta_k,w,t) = \nabla_w(U * f)(\theta_k,w,t), \tag{56}$$

with $U(\theta_k,\cdot)$ an uncertain interaction potential. A stochastic free energy functional is defined as follows

$$\mathscr{E}(\theta_k,t) = \frac{1}{2}\int_{\mathbb{R}^d}(U * f)(\theta_k,w,t)f(\theta_k,w,t)dw + D\int_{\mathbb{R}^d}\log f(\theta_k,w,t)f(\theta_k,w,t)dw,$$

which is dissipated along solutions as

$$\frac{d}{dt}\mathscr{E}(\theta_k,t) = -\int_{\mathbb{R}^d}|\nabla_w\xi|^2 f(\theta_k,w,t)dw = -\mathscr{I}(\theta_k,t), \tag{57}$$

where $\mathscr{I}(\theta_k,\cdot)$ is the entropy dissipation function. The corresponding discrete free energy is given by

$$\mathscr{E}_\Delta(\theta_k,t) = \Delta w \sum_{j=0}^N \left[\frac{1}{2}\Delta w \sum_{i=0}^N U_{j-i}(\theta_k)f_i(\theta_k,t)f_j(\theta_k,t)\right.$$

$$\left. + Df_j(\theta_k,t)\log f_j(\theta_k,t)\right] \tag{58}$$

In this case it is not possible to prove that the discrete entropy functional (58) is dissipated by the SP–CC type schemes developed in the previous sections, see [75]. For this reason we introduce the new entropic family of flux function

$$\tilde{f}^E_{i+1/2}(\theta_k, t) = \begin{cases} \dfrac{f_{i+1}(\theta_k, t) - f_i(\theta_k, t)}{\log f_{i+1}(\theta_k, t) - \log f_i(\theta_k, t)} & f_{i+1}(\theta_k, t) \neq f_i(\theta_k, t), \\ f_{i+1}(\theta_k, t) & f_{i+1}(\theta_k, t) = f_i(\theta_k, t), \end{cases} \tag{59}$$

for all $k = 0, \ldots, M$. We will refer to the above approximation of the solution at the grid point $i + 1/2$ as *entropic average* of the grid points i and $i + 1$. In the general case of the flux function (27) with non constant diffusion the resulting numerical flux reads

$$\mathscr{F}^E_{i+1/2}(\theta_k, t) = D_{i+1/2} \left(\frac{\tilde{\mathscr{C}}_{i+1/2}(\theta_k, t)}{D_{i+1/2}} \right.$$

$$\left. + \frac{\log f_{i+1}(\theta_k, t) - \log f_i(\theta_k, t)}{\Delta w} \right) \tilde{f}^E_{i+1/2}(\theta_k, t). \tag{60}$$

Concerning the stationary state, we obtain immediately by imposing the numerical flux equal to zero

$$\frac{\tilde{\mathscr{C}}_{i+1/2}(\theta_k, t)}{D_{i+1/2}} + \frac{\log f_{i+1}(\theta_k, t) - \log f_i(\theta_k, t)}{\Delta w} = 0,$$

and therefore we get

$$\frac{f_{i+1}(\theta_k, t)}{f_i(\theta_k, t)} = \exp \left(-\frac{\Delta w \, \tilde{\mathscr{C}}_{i+1/2}(\theta_k, t)}{D_{i+1/2}} \right). \tag{61}$$

By equating the above ratio with the quasi-stationary approximation (34) we get the same expression for $\tilde{\mathscr{C}}_{i+1/2}(\theta_k, t)$ for all $k = 0, \ldots, M$ as in (35)

$$\tilde{\mathscr{C}}_{i+1/2}(\theta_k, t) = \frac{D_{i+1/2}}{\Delta w} \int_{w_i}^{w_{i+1}} \frac{\mathscr{B}[f](w, \theta_k, t) + D'(w)}{D(w)} dw. \tag{62}$$

A fundamental result concerning the entropic average (59) is the following

Lemma 1 *The entropy average defined in* (59) *may be written as a convex combination with nonlinear weights*

$$\tilde{f}^E_{i+1/2}(\theta_k, t) = \delta^{k,E}_{i+1/2} f_i(\theta_k, t) + (1 - \delta^{k,E}_{i+1/2}) f_{i+1}(\theta_k, t), \tag{63}$$

where

$$\delta^{k,E}_{i+1/2} = \frac{f_{i+1}(\theta_k, t)}{f_{i+1}(\theta_k, t) - f_i(\theta_k, t)} + \frac{1}{\log f_i(\theta_k, t) - \log f_{i+1}(\theta_k, t)} \in (0, 1). \qquad (64)$$

On the contrary to the Chang-Cooper average the restrictions for the non negativity property of the solution are stronger. Similar to central differences, we have a restriction on the mesh size which becomes prohibitive for small values of the diffusion function $D(w)$.

Concerning the entropy dissipation we can summarize the main results in the following [75]

Theorem 3 *The numerical flux (60)–(59) for a constant diffusion D satisfies the discrete entropy dissipation*

$$\frac{d}{dt}\mathscr{E}_\Delta(\theta_k, t) = -\mathscr{I}_\Delta(\theta_k, t), \qquad (65)$$

where $\mathscr{E}_\Delta(\theta_k, t)$ is given by (58) and $I_\Delta(\theta_k, t)$ is the discrete entropy dissipation function

$$\mathscr{I}_\Delta(\theta_k, t) = \Delta w \sum_{j=0}^{N}(\xi_{j+1}(\theta_k, t) - \xi_j(\theta_k, t))^2 \tilde{f}^E_{i+1/2}(\theta_k, t) \geq 0, \qquad (66)$$

with $\xi_{j+1}(\theta_k, t) - \xi_j(\theta_k, t)$ the discrete version of (56).
Further, we can state the following entropy dissipation results for problem (50) in the nonlogarithmic Landau form (16).

Theorem 4 *Let us consider $\mathscr{B}[f](w, \theta_k, t) = P(\theta_k)(w - u)$ as in Eq. (50). The numerical flux (60)–(59) with $\tilde{\mathscr{C}}^k_{i+1/2}$ given by (35) satisfies the discrete entropy dissipation*

$$\frac{d}{dt}\mathscr{H}_\Delta(f(\theta_k, t), f^\infty(\theta_k, t)) = -\mathscr{I}^E_\Delta(f(\theta_k, t), f^\infty(\theta_k, t)), \qquad (67)$$

where $\mathscr{H}_{\Delta w}(f(\theta_k, t), f^\infty(\theta_k, t))$ is given by (53) and $\mathscr{I}^E_\Delta(\theta_k, t)$ is the positive discrete dissipation function

$$\mathscr{I}^E_\Delta(f(\theta_k, t), f^\infty(\theta_k, t)) = \sum_{i=0}^{N}\left[\log\left(\frac{f_{i+1}(\theta_k, t)}{f^\infty_{i+1}(\theta_k, t)}\right) - \log\left(\frac{f_i(\theta_k, t)}{f^\infty_i(\theta_k, t)}\right)\right]^2$$
$$\cdot D_{i+1/2}\tilde{f}^E_{i+1/2}(\theta_k, t) \geq 0. \qquad (68)$$

3.3   Numerical Results

We consider a stochastic Fokker–Planck equation with uncertainty in the initial distribution, i.e.

$$\begin{cases} \partial_t f(\theta, w, t) = \partial_w \Big[wf(\theta, w, t) + T(\theta)\partial_w^2 f(\theta, w, t) \Big], \\ f(\theta, w, 0) = f_0(\theta, w), \end{cases} \tag{69}$$

for all $w \in \mathbb{R}$ with

$$f_0(\theta, w) = \frac{1}{2} \left\{ \frac{1}{\sqrt{2\pi\sigma^2(\theta)}} e^{-\frac{(w-c)^2}{2\sigma^2(\theta)}} + \frac{1}{\sqrt{2\pi\sigma^2(\theta)}} e^{-\frac{(w+c)^2}{2\sigma^2(\theta)}} \right\}, \qquad c = 1/10 \tag{70}$$

and $\sigma^2(\theta) = 1/10 + \epsilon\theta$, $\theta \sim U([-1, 1])$, $\epsilon = 5 \times 10^{-3}$. In (69) the diffusion coefficient is the temperature

$$T(\theta) = \int_{\mathbb{R}} w^2 f_0(\theta, w) dw.$$

It is well-known that the steady–state solution of this problem is the Maxwellian distribution (4).

In the previous paragraphs we showed how an essential aspect for the accurate description of the stochastic steady state relies in the approximation of the family of integrals $\delta_{i+1/2}^k$, $\lambda_{i+1/2}^k$, see (36)–(37). In this case, however, since the steady state is known we can evaluate exactly these weight functions as in (43)–(44). We postpone to the last section of the present contribution the discussion on numerical results obtained with more general weight functions for which no exact formulation are given. In Fig. 1 (right) we report the relative L^1 error for the expectation of

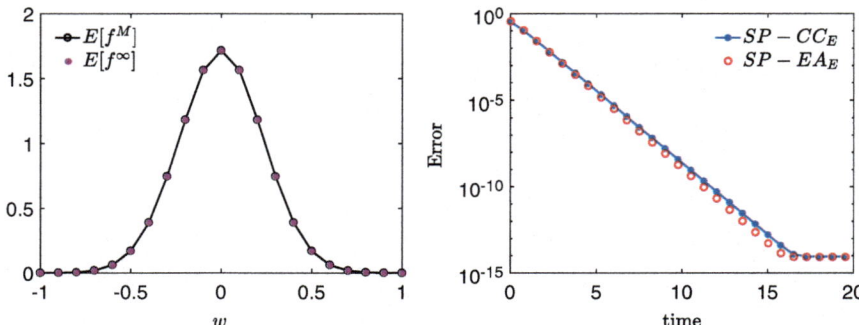

Fig. 1 Left: exact and numerical approximation of the expected steady state distribution. Right: evolution of the L^1 relative error for the expected solution calculated for both $SP - CC_E$ and $SP - EA_E$ methods. In both figures we considered a grid on $[-1, 1]$ with $N = 21$ points and $M = 10$ nodes in the random field, the final time $T = 20$ and $\Delta t = \Delta w^2/2$. The nodes of the random field have been chosen with Gauss–Legendre polynomials

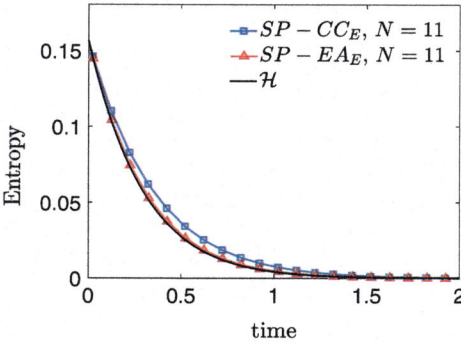

Fig. 2 Dissipation of the numerical expected entropy for $SP - CC_E$ and $SP - EA_E$ schemes on a coarse grid with $N = 11$ points

the solution in time. As expected the schemes are capable to capture the stochastic steady state exactly. Next in Fig. 2 the evolution of the expectation for the numerical entropy is given.

4 Variance Reduction Monte Carlo Methods

Among the different type of techniques used in UQ, certainly the Monte Carlo methods represent one of the most popular and important classes [20, 44, 55, 74]. They show all their potential when the dimension of the uncertainty space becomes very large. In addition, Monte Carlo methods are effective when the probability distribution of the random inputs is not known analytically or lacks of regularity since other approaches based on orthogonal stochastic polynomials may be impossible to use of may produce poor results.

In this section, we first describe a standard Monte Carlo approach which deals with random initial data and then we describe a modification of this algorithm which permits to strongly decrease the computational costs and increase the accuracy close to the steady state.

4.1 The Standard Monte Carlo Method

We describe the method when applied to the solution of a Vlasov-Fokker-Planck type equation (1) with deterministic parameters $P = P(x, x_*, w, w_*)$ and $D = D(w, t)$ and random initial data $f(\theta, x, w, 0) = f_0(\theta, x, w)$. First we assume that the kinetic equation has been discretized by a deterministic solver in the variables w, x

and t. In this setting, the simplest Monte Carlo (MC) method for UQ is based on the following steps.

Algorithm 1 (Standard Monte Carlo (MC) Method)

1. **Sampling**: Sample M independent identically distributed (i.i.d.) initial data f_0^k, $k = 1, \ldots, M$ from the random field f_0 and approximate these over the grid (for example by piece-wise constant cell averages).
2. **Solving**: For each realization f_0^k the underlying kinetic equation (1) is solved numerically by the deterministic solver. We denote the solutions at time t^n by $f_{\Delta w, \Delta x}^{k,n}, k = 1, \ldots, M$, where Δw and Δx characterize the discretizations in w and x.
3. **Estimating**: Estimate the expected value of the random solution field with the sample mean of the approximate solution

$$E_M[f_{\Delta w}^n] = \frac{1}{M} \sum_{k=1}^{M} f_{\Delta w, \Delta x}^{k,n}. \tag{71}$$

The above algorithm is straightforward to implement in any existing code for the Vlasov-Fokker-Planck equations. Furthermore, the only (data) interaction between different samples is in step 3, when ensemble averages are computed. Thus, the MC algorithms for UQ are non-intrusive and easily parallelizable as well.

The typical error estimate that one obtains using such an approach is of the type

$$\|E[f(\cdot, t^n)] - E_M[f_{\Delta w}^n]\| \leq C_1 M^{-1/2} + C_2(\Delta w)^q + C_3(\Delta x)^p + C_4(\Delta t)^r \tag{72}$$

where $\| \cdot \|$ is a suitable norm, C_1, C_2, C_3 and C_4 are positive constants depending only on the second moments of the initial data and the interaction term, and q, p and r characterize the accuracy of the discretizations in the phase-space. Clearly, it is possible to equilibrate the discretization and the sampling errors in the a-priori estimate taking $M = O(\Delta w^{-2q})$, $\Delta x = O(\Delta w^{q/p})$ and $\Delta t = O(\Delta w^{q/r})$. This means that in order to have comparable errors the number of samples should be extremely large, especially when dealing with high order deterministic discretizations. This may make the Monte Carlo approach very expensive in practical applications.

4.2   The Micro-Macro Monte Carlo Method

In order to improve the performances of standard MC methods, we introduce a novel class of variance reduction Monte Carlo methods [44]. The key idea is to take advantage of the knowledge of the steady state in order to reduce both the variance and the computational cost of the Monte Carlo estimate. The method is essentially a control variate strategy based on a suitable microscopic-macroscopic decomposition of the distribution function.

We describe the method in the space homogeneous case, an example of such technique in the non homogeneous case is reported in Sect. 6, while we refer to [44] for a detailed discussion and extensions of such method to more general kinetic equations. Following Sect. 2.2 we introduce the Micro–Macro decomposition

$$f(\theta, w, t) = f^\infty(\theta, w) + g(\theta, w, t),$$

where $f^\infty(\theta, w)$ is the steady state solution of the problem considered. Then, the method consists in using the Monte Carlo estimation procedure only on the non equilibrium part $g(\theta, w, t)$ solution of (23).

The crucial aspect is that the equilibrium state $g^\infty(\theta, w)$ is zero and therefore, independent from θ. More precisely, we can decompose the expected value of the distribution function in an equilibrium and non equilibrium part

$$
\begin{aligned}
\mathbb{E}[f](w, t)] &= \int_\Omega f(\theta, w, t) p(\theta) d\theta \\
&= \int_\Omega f^\infty(\theta, w) p(\theta) d\theta + \int_\Omega g(\theta, w, t) p(\theta) d\theta,
\end{aligned}
\tag{73}
$$

and then exploit the fact that $f^\infty(\theta, w)$ is known to have an estimate of the error committed by the Monte Carlo integration of type

$$e_M[f] \simeq \sigma_g M^{-1/2} \tag{74}$$

instead of

$$e_M[f] \simeq \sigma_f M^{-1/2}, \tag{75}$$

where σ_g and σ_f are the variances of respectively the perturbation and the distribution function and where we have supposed for simplicity that expected value of the equilibrium part is computed with a negligible error. Now, since it is known that the perturbation g goes to zero in time exponentially fast, then also its variance goes to zero, which means that at the steady state the Monte Carlo integration becomes only dependent on the way in which the expected value of the equilibrium part is computed.

The simplest version of the algorithm consists of the following steps:

Algorithm 2 (Micro-Macro Monte Carlo (M^3C) Method)

1. **Small scale sampling**: Sample M_E independent identically distributed (i.i.d.) initial data f_0^k, $k = 1, \ldots, M_E$ from the random field f_0. For each sample compute the corresponding equilibrium state $f_{\Delta w}^{k,\infty}$ from its moments evaluated through suitable quadrature rules in w based on the discretization parameter Δw.
2. **Large scale sampling**: Select $M \ll M_E$ samples f_0^k, $k = 1, \ldots, M$ and compute $g_0^k = f_0^k - f^{k,\infty}$ and approximate these over the grid (for example by piece-wise constant cell averages).

3. **Solving**: For each realization g_0^k the underlying kinetic equation (23) is solved numerically by the deterministic solver. We denote the solutions at time t^n by $g_{\Delta w}^{k,n}$, $k = 1, \ldots, M$.
4. **Estimating**: We estimate the expected value of the random solution field

$$f_{\Delta w}^{k,n} = f_{\Delta w}^{k,\infty} + g_{\Delta w}^{k,n},$$

with the sample mean of the approximate solution

$$E_{M,M_E}[f_{\Delta w}^n] = \frac{1}{M_E} \sum_{k=1}^{M_E} f_{\Delta w}^{k,\infty} + \frac{1}{M} \sum_{k=1}^{M} g_{\Delta w}^{k,n}. \tag{76}$$

Using such an approach one obtains an error estimate of the type

$$\|E[f(\cdot, t^n)] - E_{M,M_E}[f_{\Delta w}^n]\| \leq C_E M_E^{-1/2} + C_1^n M^{-1/2} + C_2(\Delta w)^q + C_3(\Delta t)^r \tag{77}$$

where now the constant C_1^n depends on time and on the second moment of the solution $g(\theta, w, t^n)$ which will vanish for large times. In fact, independently of θ we have that $g(\theta, w, t^n) \to 0$ as $n \to \infty$. Therefore, the method reduces the variance of the estimator in time and asymptotically, since $C_1^n \to 0$ as $n \to \infty$, depends only on the fine scale sampling which does not affect the overall computational cost.

The efficiency of the M³C can be further improved in the case of monotonic convergence to equilibrium of the distribution function f by introducing a strategy of sampling reduction at each time step. The resulting algorithm is the following

Algorithm 3 (Fast Micro-Macro Monte Carlo (FM³C) Method)

1. **Small scale sampling**: Sample M_E independent identically distributed (i.i.d.) initial data f_0^k, $k = 1, \ldots, M_E$ from the random field f_0. For each sample compute the corresponding equilibrium state $f_{\Delta w}^{k,\infty}$ from its moments evaluated through suitable quadrature rules in w based on the discretization parameter Δw.
2. **Large scale sampling**: Select $M_0 \ll M_E$ samples f_0^k, $k = 1, \ldots, M_0$ and compute $g_0^k = f_0^k - f^{k,\infty}$ and approximate these over the grid (for example by piece-wise constant cell averages).
3. **Solving**: For each realization g_0^k the underlying kinetic equation (23) is solved numerically by the deterministic solver. This is realized at each time step $n = 0, 1, 2, \ldots$ as follows.

   a. **Advance in time**: Starting from $g_{\Delta w}^{k,n}$, $k = 1, \ldots, M_n$ compute the solution $g_{\Delta w}^{k,n+1}$ with one time step of the deterministic solver.
   b. **Discard samples**: At each time step we compute the variance of $g_{\Delta w}^{k,n+1}$ as

$$Var_{M_n}[g_{\Delta w}^{n+1}] = \frac{1}{M_n} \sum_{k=1}^{M_n} (g_{\Delta w}^{k,n+1} - E_{M_n}[g_{\Delta w}^{k,n+1}])^2 \leq Var_{M_n}[g_{\Delta w}^n].$$

Set $M_{n+1} = \llbracket M_n \left(Var_{M_n}[g_{\Delta w}^{n+1}]/Var_{M_n}[g_{\Delta w}^n] \right) \rrbracket$ where $\llbracket \cdot \rrbracket$ denotes the integer part and discard uniformly $M_n - M_{n+1}$ samples.

4. **Estimating**: We estimate the expected value of the random solution field

$$f_{\Delta w}^{k,n} = f_{\Delta w}^{k,\infty} + g_{\Delta w}^{k,n},$$

with the sample mean of the approximate solution

$$E_{M_n,M_E}[f_{\Delta w}^n] = \frac{1}{M_E} \sum_{k=1}^{M_E} f_{\Delta w}^{k,\infty} + \frac{1}{M_n} \sum_{k=1}^{M_n} g_{\Delta w}^{k,n}. \tag{78}$$

The algorithm preserves the advantages of the simple M^3C method but with a greater computational efficiency since the number of samples, and therefore the number of deterministic equations that we have to solve, decreases in time and asymptotically vanishes.

Remark 3

- In the case the underlying uncertainty probability density function $p(\theta)$ is known, the M^3C method can be applied without any small scale sampling since the estimate of the expected value reduces to

$$E_{M_n}[f_{\Delta w}^n] = \int_\Omega f_{\Delta w}^\infty(\theta, w)p(\theta)d\theta + \frac{1}{M_n} \sum_{k=1}^{M_n} g_{\Delta w}^{k,n}. \tag{79}$$

In this case M^3C methods achieves arbitrary accuracy for large times.
- In contrast with Multi Level Monte Carlo (MLMC) methods [55], which can produce non monotone estimators, the estimators produced by the M^3C method are monotonic, i.e. mean estimator of positive quantities (such as density) is also positive, the same holds true for the entropy property.
- The extension of the M^3C method to the non homogeneous case is not straightforward, and depends on the type of problem considered. Some applications are reported in Sect. 6. For more general cases we refer to [44] where a detailed discussion is done.

4.3 Numerical Results

In this section we show some results concerning the Micro-Macro Monte Carlo methods by comparing them to the standard MC method for UQ. In particular, we study the behaviors of our approach in solving the stochastic Fokker–Planck equation with uncertainty in the initial distribution (69)–(70).

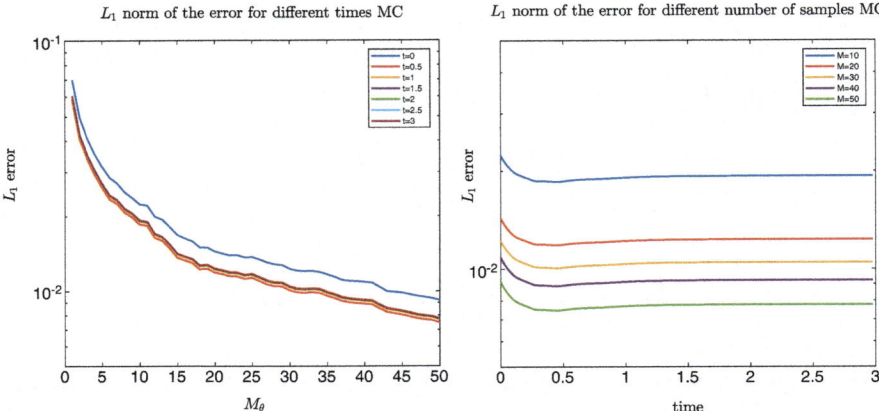

Fig. 3 Monte Carlo method. Left: evolution of the L_1 norm of the error for the expected distribution over an increasing number of stochastic inputs. Different lines represent the error at different times. Right: evolution of the L^1 norm of the error for the expected distribution over time. Different lines represent a different number of stochastic inputs over which the expected value is computed. Grid $[-1, 1]$ with $N = 100$ points, final time $T = 3$ and $\Delta t = \Delta w^2/2$. The solution has been averaged over 100 different realizations

In Fig. 3 we report the L_1 norm of the error for the expected solution with a standard MC method. Left image shows the error for an increasing number of samples for different times, while right image shows the trend of the error over time for a different number of random inputs. The final time is set to $T_f = 3$, the number of cells in velocity is 100, while the stability condition gives $\Delta t = \Delta w^2/2$. The maximum number of samples which furnishes the set of initial conditions is $M_\theta = 50$, while the solution is averaged over 100 realizations. One can clearly see the $M_\theta^{-1/2}$ slope for the error in the left picture.

In Fig. 4, the L_1 norm of the error is reported in the same setting for the M^{3}C method. The same number of averages and stochastic initial condition have been employed. We can see how the error decreases as a function of time in an exponential fashion on the contrary of the MC case for which the error is almost independent on time.

Finally, in Fig. 5 we show the behavior of the fast M^{3}C method. The number of samples for which the time evolution of the perturbation g is considered is reported on the right and it diminishes exponentially with time. The corresponding L_1 norm of the error is shown on the left. For this case, we increased the initial number of random samples to 1000 to highlight the behavior of the fast approach.

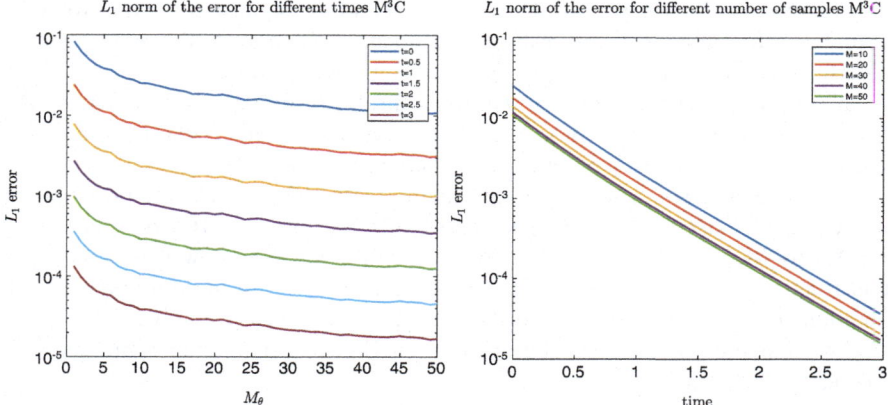

Fig. 4 M^3C method. Left: evolution of the L_1 norm of the error for the expected distribution over an increasing number of stochastic inputs. Different lines represent the error at different times. Right: evolution of the L^1 norm of the error for the expected distribution over time. Different lines represent a different number of stochastic inputs over which the expected value is computed. Grid $[-1, 1]$ with $N = 100$ points, final time $T = 3$ and $\Delta t = \Delta w^2/2$. The solution has been averaged over 100 different realizations

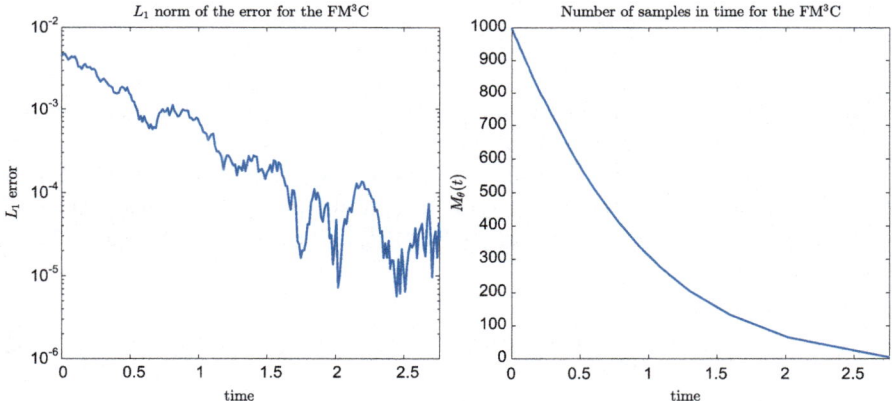

Fig. 5 Fast M^3C method. Left: evolution of the L_1 norm of the error for the expected distribution over time. Different lines represent a different number of stochastic inputs over which the expected value is computed. Right: number of random nodes over time. Grid on $[-1, 1]$ with $N = 100$ points, final time $T = 3$ and $\Delta t = \Delta w^2/2$

5 Stochastic Galerkin Methods

Among the various methods for UQ in PDEs, stochastic Galerkin (SG) methods based on generalized polynomial chaos (gPC) expansions are very attractive thanks to the spectral convergence property with respect to the random input

[31, 62, 64, 68, 78, 79, 88, 90]. On the other hand, their intrusive nature forces a complete reformulation of the problem and standard schemes for the corresponding deterministic problem cannot be used in a straightforward way.

In particular, it is well known that, this intrusive formulation may lead to the loss of important structural properties of the original problem, like hyperbolicity, positivity and preservation of large time behavior [30, 42, 77].

In this section we analyze gPC-SG methods for the numerical approximation of stochastic Vlasov-Fokker-Planck equations in the form (1). In particular, using the Micro–Macro approach in the gPC-SG setting we show how it is possible to construct methods which preserve the asymptotic behavior of the solution [46]. We mention here related approaches for kinetic equations developed in [61, 92].

We recall first some basic notions on Galerkin approximation techniques for stochastic computations.

5.1 Preliminaries on gPC-SG Techniques

Let us consider the function $f(\theta, x, w, t), f \in L^2$ in the random variable $\theta \in \Omega \subseteq \mathbb{R}$, solution of the differential problem

$$\partial_t f(\theta, x, w, t) = \mathscr{J}(f, f)(\theta, x, w, t), \tag{80}$$

with $\mathscr{J}$ given by (13). The present setup of the problem may be naturally extended to a r-dimensional vector of random variables.

We consider the space $\mathbb{P}_M$ of polynomials of degree up to M, generated by a family of orthogonal polynomials with respect to the probability density function $p(\theta)$ of the random variable θ, namely $\{\Phi_h(\theta)\}_{h=0}^M$. They form an orthogonal basis of $L^2(\Omega)$, i.e.

$$\mathbb{E}\Big[\Phi_h(\theta)\Phi_k(\theta)\Big] = \int_\Omega \Phi_h(\theta)\Phi_k(\theta)p(\theta)\,d\theta = \mathbb{E}\Big[\Phi_h^2(\theta)\Big]\delta_{hk} \tag{81}$$

where δ_{hk} is the Kronecker delta function. Let us assume that $p(\theta)$ has finite second order moment, we can represent the function $f(x, w, \theta, t)$ through the complete polynomial chaos expansion as follows

$$f(\theta, x, w, t) = \sum_{m \in \mathbb{N}} \hat{f}_m(x, w, t)\Phi_m(\theta), \tag{82}$$

where $\hat{f}_m(x, t)$ is given by

$$\hat{f}_m(x, w, t) = \mathbb{E}\Big[f(\theta, x, w, t)\Phi_m(\theta)\Big], \qquad m \in \mathbb{N}. \tag{83}$$

The generalized polynomial chaos expansion approximates the solution $f(\theta, x, w, t)$ of (80) with its M-th order truncation $f^M(\theta, x, w, t)$ and considers the Galerkin projections of the differential problem for each $h = 0, \ldots, M$

$$\partial_t \mathbb{E}[f^M(\theta, x, w, t)\Phi_h(\theta)] = \mathbb{E}[\mathscr{J}(f^M, f^M)(\theta, x, w, t)\Phi_h(\theta)]. \tag{84}$$

Thanks to the orthogonality of the polynomial basis of the space $\mathbb{P}_M$ we obtain a coupled system of $M + 1$ purely deterministic equations

$$\partial_t \hat{f}_h(x, w, t) = \hat{\mathscr{J}}_h(\hat{f}, \hat{f})(x, w, t), \qquad h = 0, \ldots, M, \tag{85}$$

where $\hat{f} = \hat{f}_{kk=0}{}^M$.

These subproblems must then be solved through suitable numerical techniques. The approximation of the statistical quantities of interest are defined in terms of the introduced projections. From (83) being $\Phi_0 \equiv 1$ we have

$$\mathbb{E}[f(\theta, x, w, t)] = \hat{f}_0(x, w, t), \tag{86}$$

and thanks to the orthogonality it is possible to show that

$$\begin{aligned}
Var[f(\theta, x, w, t)] &= \mathbb{E}\left[\left(\sum_{h=0}^{M} \hat{f}_h \Phi_h(\theta) - \hat{f}_0 \right)^2 \right] \\
&= \sum_{h=0}^{M} \hat{f}_h^2(x, w, t)\mathbb{E}[\Phi_h^2(\theta)] - \hat{f}_0^2(x, w, t).
\end{aligned} \tag{87}$$

5.1.1 gPC-SG Methods for Vlasov–Fokker–Planck Equations

Let us consider the stochastic Vlasov–Fokker–Planck equation (1) with a nonlocal drift $\mathscr{B}[\cdot]$ of the form (2).

The gPC-SG approximation is given by the following system of deterministic differential equations

$$\partial_t \hat{f}_h(x, w, t) + \mathscr{L}[\hat{f}_h(x, w, t)] =$$
$$\nabla_w \cdot \left[\sum_{k=0}^{M} b_{hk}[\hat{f}](x, w, t)\hat{f}_k(x, w, t) + \nabla_w D(x, w)\hat{f}_h(x, w, t) \right], \tag{88}$$

where

$$b_{hk}[\hat{f}](x, w, t) = \frac{1}{\|\Phi_h\|_{L^2}^2} \sum_{m=0}^{M} \int_{\Omega} \mathscr{B}[\hat{f}_m]\Phi_k(\theta)\Phi_m(\theta)dp(\theta). \tag{89}$$

Note that, due to the nonlinearity of Fokker–Planck problems, we obtain a coupled system of deterministic Vlasov–Fokker–Planck equations describing the evolution of each projection. In vector notations we have

$$\partial_t \hat{\mathbf{f}}(x, w, t) + \mathscr{L}[\hat{\mathbf{f}}](x, w, t) = \nabla_w \cdot [\mathbf{B}[\hat{\mathbf{f}}](x, w, t)\hat{\mathbf{f}}(x, w, t) + \nabla_w \mathbf{D}(x, w)\hat{\mathbf{f}}(x, w, t)], \tag{90}$$

where $\hat{\mathbf{f}} = (\hat{f}_0, \ldots, \hat{f}_M)^T$ and the component of the $(M + 1) \times (M + 1)$ matrix $\mathbf{B}[\hat{\mathbf{f}}](x, w, t)$ are given by (89).

In a similar way, we can derive the gPC-SG formulation of stochastic Vlasov–Fokker–Planck equations with uncertain diffusion terms.

Remark 4 In the case that uncertainty is present only in the initial data, and therefore $\mathscr{B}[f](x, w, \theta, t) = \mathscr{B}(x, w, t)$, the matrix $\mathbf{B}$ is diagonal and we need to solve the decoupled system of Vlasov type equations

$$\partial_t \hat{f}_h(x, w, t) + \mathscr{L}[\hat{f}_h](x, w, t) = \nabla_w \cdot [b_{hh}\hat{f}_h(x, w, t) + \nabla_w D(x, w)\hat{f}_h(x, w, t)], \tag{91}$$

$h = 0, \ldots, M$. Hence, a structure preserving approach as in Sect. 3.1 may be introduced in order to preserve the large time behavior of the collision step of each projection by defining a family of weight functions

$$\lambda_{i+1/2}^h = \frac{D_{i+1/2}}{\Delta w} \int_{w_i}^{w_{i+1}} \frac{b_{hh}(x, w, t) + D'(x, w)}{D(x, w)} dw,$$

$$\delta_{i+1/2}^h = \frac{1}{\lambda_{i+1/2}^h} + \frac{1}{1 - \exp(\lambda_{i+1/2}^h)}. \tag{92}$$

In this setting the scheme capture with arbitrary accuracy the steady state and the expected value of the numerical solution is kept nonnegative. However, for more general nonlocal type operators $\mathscr{B}[\cdot]$ this approach cannot be applied for the construction of a stochastic Galerkin expansion which preserves the steady state solution and nonnegativity of the mean.

5.2   A Micro–Macro gPC Approach

We discussed in the previous section how the gPC-SG method for stochastic Fokker–Planck equations generates a coupled system of partial differential equations. Although gPC-SG guarantees spectral convergence on the random field under suitable regularity assumptions, its accuracy in describing the long-time solutions of the problems depends on the particular scheme emploied for solving the coupled system.

Let us consider suitable regularity assumptions on the initial distribution such that the stochastic Fokker–Planck problem admits the unique steady state solution

$f^\infty(\theta, w)$. With the aim of preserving the steady states of the problem in the Galerkin setting we introduce a Micro–Macro gPC-SG scheme. Thanks to the formalism introduced in Sect. 5.1 and by analogy with (21) the Micro–Macro gPC decomposition for all $M \geq 0$ reads [46]

$$f^M(\theta, w, t) = f^{\infty,M}(\theta, w) + g^M(\theta, w, t), \qquad w \in \mathbb{R}^{d_w}, t \geq 0, \tag{93}$$

where

$$f^{\infty,M}(w, \theta) = \sum_{h=0}^{M} \widehat{f^\infty}_h(w) \Phi_h(\theta), \qquad \widehat{f^\infty}_h(w) = \int_\Omega f^\infty(\theta, w) \Phi_h(\theta) dp(\theta).$$

Since Eq. (93) is equivalent to $\hat{\mathbf{f}} = \widehat{\mathbf{f}^\infty} + \hat{\mathbf{g}}$, we can reformulate the original problem in terms of $\hat{\mathbf{g}}$. Equation (88) may be reformulated for all $h = 0, \ldots, M$ in terms of the nonequilibrium part of the Micro–Macro gPC decomposition $\hat{g}_h$ as follows

$$\begin{cases} \partial_t \hat{g}_h(w, t) &= \hat{\mathscr{I}}_h(\hat{g}, \hat{g})(w, t) + \hat{\mathscr{N}}_h(\widehat{f^\infty}, \hat{g})(w, t), \\ f^M(w, \theta, t) &= f^{\infty,M}(w, \theta) + g^M(w, \theta, t), \end{cases} \tag{94}$$

where the operator $\hat{\mathscr{I}}_h$ is the Galerkin projection of the quadratic operators of the collisional type defined in (13) and $\hat{\mathscr{N}}_h$ is a linear operator defined as

$$\hat{\mathscr{I}}_h(\hat{g}, \hat{g})(w, t) = \nabla_w \cdot \left[\sum_{k=0}^{M} b_{hk}[\hat{g}] \hat{g}_k(w, t) + \nabla_w D(w) \hat{g}_h(w, t) \right],$$

$$\hat{\mathscr{N}}_h(\widehat{f^\infty}, \hat{g})(w, t) = \nabla_w \cdot \left[\sum_{k=0}^{M} b_{hk}[\widehat{f^\infty}] \hat{g}_k(w, t) + b_{hk}[\hat{g}] \widehat{f^\infty}_k(w) \right]. \tag{95}$$

Now, the equilibrium state of each gPC projection is $\hat{g}_h \equiv 0$ and any consistent schemes for the numerical approximations of the differential terms in (95) admits $\hat{g}_h \equiv 0$ as equilibrium state for all $h = 0, \ldots, M$. For example, we can use a standard central difference approximation scheme for the differential terms in (95) to achieve second order accuracy for transient times and exact preservation of the steady state asymptotically.

5.3 Numerical Results

We consider the evolution of the Fokker–Planck equation (69) with the uncertain initial condition (70). Following the approach introduced in the previous section, we obtain the SG system of equations

$$\partial_t \hat{f}_h(w, t) = \partial_w \left[w \hat{f}_h(w, t) + \partial_w \sum_{k=0}^{M} d_{hk} \hat{f}_k(w, t) \right], \tag{96}$$

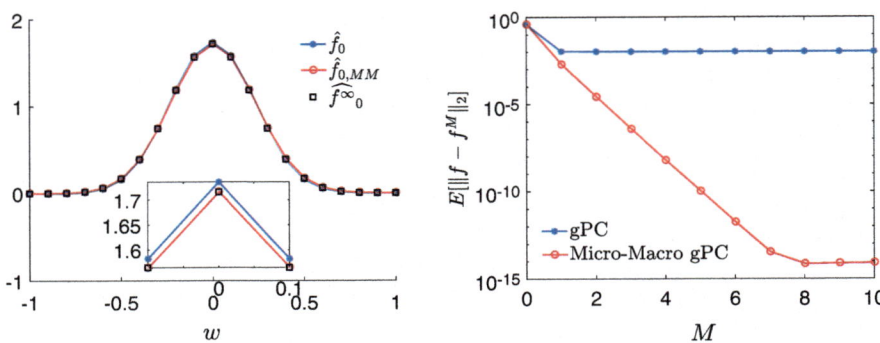

Fig. 6 Left: expected stationary solution of the uncertain Fokker–Planck equation obtained through standard SG and MM-SG methods with central differences and $M = 10$. Right: expected L^2 error for standard SG and MM-SG. In both cases we considered a discretization of the interval $[-1, 1]$ with $N = 21$ gridpoints and $\Delta t = \Delta w/2$

with

$$d_{hk} = \frac{1}{\|\Phi_h\|_2^2} \int_\Omega T(\theta)\Phi_h(\theta)\Phi_k(\theta)dp(\theta).$$

In order to build the Micro–Macro gPC decomposition of the SG system we take advantage of the analytical solution given by the Maxwellian distribution (4), which can be approximated by its M-order truncation as in Sect. 5.2. Therefore we aim at solving the modified problem for all $h = 0, \ldots, M$

$$\begin{cases} \partial_t \hat{g}_h(w, t) = \partial_w \Big[w\hat{g}_h(w, t) + \partial_w \sum_{k=0}^{M} d_{hk}\hat{g}_k(w, t) \Big], \\ f^M(\theta, w, t) = g^M(\theta, w, t) + f^{\infty,M}(\theta, w). \end{cases} \tag{97}$$

In all our numerical examples we use second order central difference approximations of the derivatives in w. In Fig. 6 we compare the numerical long time solution obtained through a standard SG system (96) and the Micro–Macro SG system (MM). We can observe how the Micro–Macro gPC-SG method gives an accurate description of the expected steady state of the problem, on the contrary the error of the standard gPC-SG method saturates at the accuracy obtained with the central differences.

6 Other Applications

In this section we present several numerical examples of stochastic Fokker–Planck and Vlasov–Fokker-Planck equations solved with the schemes introduced in the previous sections. In particular we focus on some recent models in socio–economic and life sciences as discussed in the Introduction.

6.1 Example 1: Opinion Model with Uncertain Interactions

Let us consider a distribution function $f = f(\theta, w, t)$ describing the density of agents with opinion $w \in I = [-1, 1]$ whose evolution is given in terms of a stochastic Fokker-Planck equation characterized by the nonlocal term (5) with uncertain compromise propensity function $P(\theta, w, {}_*) \in [0, 1]$. In the following we will solve the problem both in the collocation and in the Galerkin setting.

We consider as deterministic initial distribution $f(\theta_k, w, 0) = f_0(w)$ for all $k = 1, \dots, M$, with

$$f_0(w, 0) = \beta \left[\exp(-c(w + 1/2)^2) + \exp(-c(w - 1/2)^2) \right], \qquad c = 30, \qquad (98)$$

with $\beta > 0$ a normalization constant and let $u = \int_{-1}^{1} w f_0(w) dw$ the mean opinion. We choose a uniformly distributed random input $\theta \sim U([-1, 1])$ and a random interaction function of the form $P(\theta) = 0.75 + \theta/4$.

We discretize the random variable by considering $M > 1$ Gauss–Legendre collocation nodes. In Fig. 7 we compute the relative L_1 error for mean and variance with respect to the exact steady state (6) using $N = 80$ points for the $SP-CC$ scheme with various quadrature rules adopted for the evaluation of the weights function in (37). Singularities at the boundaries in the integration of (37) can be avoided using open Newton–Cotes methods. In the sequel, we will adopt the notation $SP - CC_k$, $k = 2, 4, 6, G$, to denote the structure preserving schemes with Chang–Cooper flux when (37) is approximated with second, fourth, sixth order open Newton–Cotes or Gaussian quadrature, respectively.

In Fig. 8 the time evolution of the expected solution and variance are given. We can observe from the estimation of the variance the regions of higher variability of

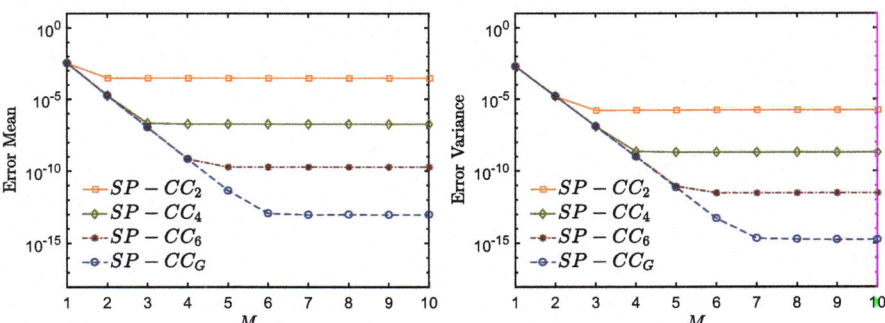

Fig. 7 Example 1. Values of L^1 error in the estimation of the expected solution (left) and its variance (right) for $T = 20$ and for an increasing number of collocation nodes. The numerical error has been computed with respect to the expected analytical solution (left) and its variance (right), see (6). We compare the error for the SP–CC scheme with different quadrature methods in case of random interaction $P(\theta) = 0.75 + \theta/4$, $\theta \sim U([-1, 1])$. Initial distribution (98), $\sigma^2/2 = 0.1$, $N = 80$, $\Delta t = \Delta w^2/(2\sigma^2)$

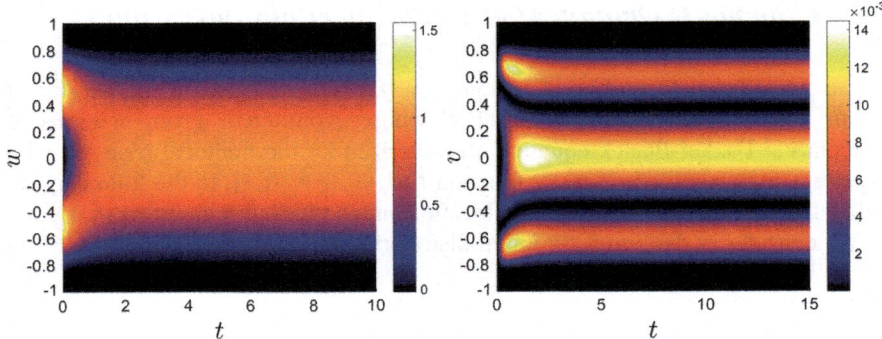

Fig. 8 Example 1. Evolution of $\mathbb{E}^M[f(w,\theta,t)]$ (left) and $Var^M[f(w,\theta,t)]$ (right) for the opinion model obtained with $M = 10$ collocation points and the $SP - CC_G$ scheme over the time interval $[0, 10]$

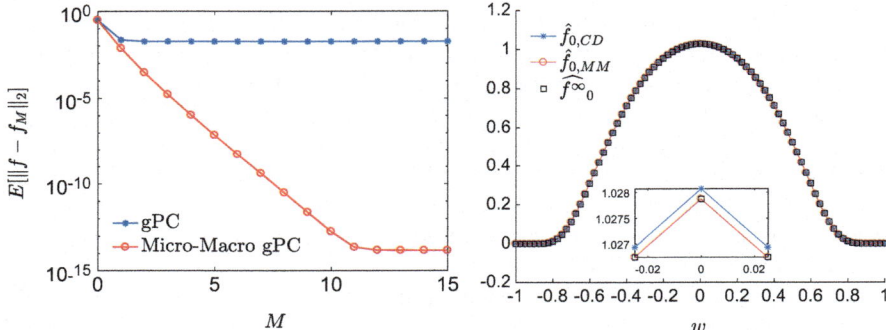

Fig. 9 Example 1. Left: Estimation of $\mathbb{E}[\|f - f^M\|_{L^2}]$ computed at time $t = 20$ and for an increasing number of $M \geq 0$, we compare the errors computed through a standard method for the solution of the system of coupled PDEs of the Fokker–Planck type, with the Micro–Macro gPC method. We used $N = 80$ gridpoints, $\sigma^2/2 = 0.1$, $\Delta t^2 = \Delta w^2/(2\sigma^2)$. Right: Large time behavior for the estimated expected solution of the opinion model.

the expected solution due to uncertain interactions. The evolutions of the statistical quantities have been computed through a collocation $SP - CC_G$ method with 20 quadrature points for the evaluation of (37).

Finally, as in Sect. 5.2 we consider a Micro-Macro gPC Galerkin setting based on the knowledge of the stationary solution (6).

In Fig. 9 we present the behavior numerical error $\mathbb{E}[\|f^\infty - f^M\|_2]$ for large time where the differential terms in w are solved by central differences. We report also the large time behavior for the expected solution in both schemes, where it is possible to observe how the Micro–Macro gPC is able to capture with high accuracy the steady state of the problem.

6.2 Example 2: Wealth Evolution with Uncertain Diffusion

We consider the Fokker-Planck equation defined by (7) where now $f = f(\theta, w, t)$ with $w \in \mathbb{R}^+$ representing the wealth of the agents and the uncertainty acts on the diffusion parameter. We consider the deterministic initial distribution $f(\theta_k, w, 0) = f_0(w)$ for all $k = 1, \ldots, M$ with

$$f_0(w, 0) = \beta \exp\left\{ -c(w - \tilde{u})^2 \right\}, \qquad c = 20, \quad \tilde{u} = 2, \tag{99}$$

where $\beta > 0$ is a normalization constant. To deal with the truncation of the computational domain in the interval $[0, L]$, following [75], after introducing N grid points we consider the quasi stationary boundary condition in order to evaluate $f_N(\theta_k, t)$, i.e.

$$\frac{f_N(\theta_k, t)}{f_{N-1}(\theta_k, t)} = \exp\left\{ -\int_{w_{N-1}}^{w_N} \frac{\mathscr{B}[f](\theta_k, w, t) + D'(\theta_k, w)}{D(\theta_k, w)} dw \right\}, \tag{100}$$

for all $k = 1, \ldots, M$. In Fig. 10 we report in a semilog scale the relative L_1 error for mean and variance with respect to the exact steady state introduced in (8) of the semi-implicit SP–CC scheme for several integration methods with $N = 200$ points over with $L = 10$, and an increasing number of collocation nodes $M = 1, \ldots, 15$. The time step is chosen in such a way that the CFL condition for the positivity of the semi-implicit scheme is satisfied, i.e. $\Delta t = O(\Delta w)$ see Sect. 3.1.1. For the tests we considered $\sigma^2(\theta) = 0.1 + 5 \times 10^{-2}\theta$, where $\theta \sim U([-1, 1])$. We can observe how the error decays exponentially for an increasing number of collocation nodes. The time evolution of the mean and the variance is plotted in Fig. 11.

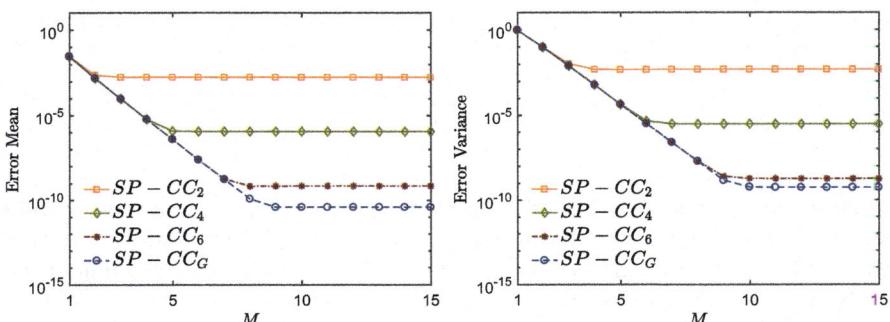

Fig. 10 Example 2. Error for the SP–CC scheme with different quadrature methods in case of random diffusion constant $\sigma^2(\theta) = 0.1 + \theta/200$, $\theta \sim U([-1, 1])$. We report the L^1 relative error in the estimation of the expected solution (left) and its variance (right) for $T = 20$ and for an increasing number of nodes in the random space. The numerical error has been computed with respect to the expected analytical solution (left) and its variance (right) obtained from (8). The initial distribution $f_0(w)$ is (99), we consider the domain $[0, L]$, $L = 10$ with $N = 200$ points and $\Delta t = \Delta w/L$ with a semi-implicit approximation

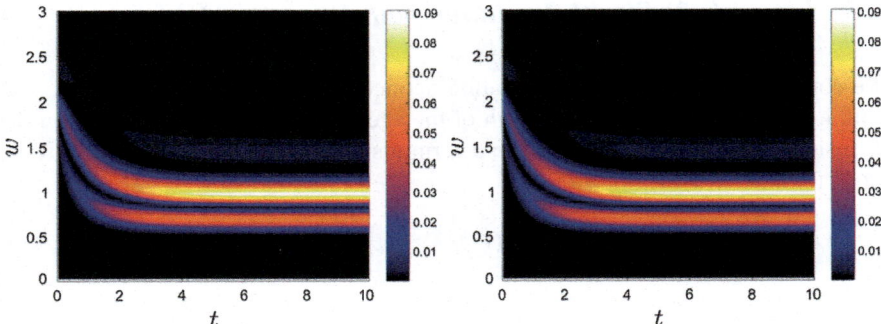

Fig. 11 Example 2. Evolution of expected solution $\mathbb{E}^M[f(\theta, w, t)]$ (left) and its variance $\mathrm{Var}^M[f(\theta, w, t)]$ (right) for the wealth evolution model. The evolution is computed through $M = 10$ collocation points and the $SP - CC_G$ scheme over the time interval $[0, 10]$, $\Delta t = \Delta w/L$ with $w \in [0, L]$, $L = 10$

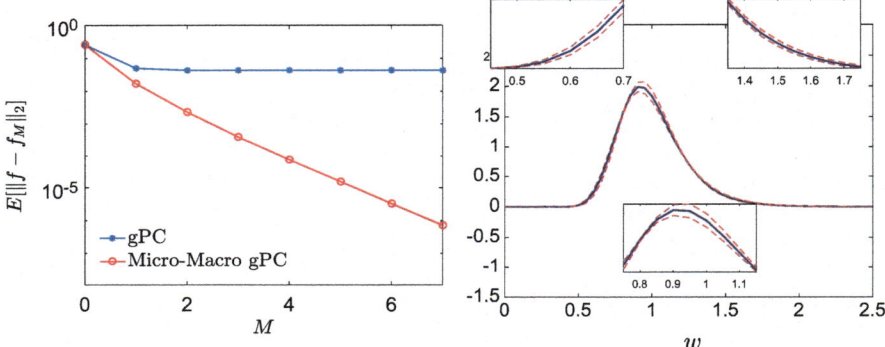

Fig. 12 Example 2. Left: Estimation of $\mathbb{E}[\|f - f^M\|_{L^2(\Omega)}]$ computed at time $T = 20$ and for an increasing number of $M \geq 0$, we compare the errors computed through a standard gPC-SG method and the Micro–Macro gPC-SG method. We used $N = 200$ gridpoints, $\sigma^2 = 0.1 + \theta/200$. Right: Statistical dispersion of the expected asymptotic solution of the wealth distribution model calculated with the Micro–Macro gPC-SG method

Next we consider the SG-gPC formulation of the equation for the wealth evolution. Since in this case the uncertainty enters in the definition of the diffusion variable $\sigma^2 = \sigma^2(\theta)$ taking $a(\cdot, \cdot) \equiv 1$ the analytical steady state solution of the problem is given in (8) and we can consider the Micro–Macro gPC scheme as in Sect. 5.2.

In Fig. 12 we compare the error for a standard gPC approximation and the Micro–Macro gPC. In both cases central differences have been used for the differential terms in w. We can see how $\mathbb{E}[\|f - f^M\|_2]$ computed at time $T = 20$, close to the stationary solution, decreases in relation to the number of terms of the gPC approximation whereas the standard gPC show a limited accuracy given by the error in approximating the large time behavior of the problem.

6.3 Example 3: Swarming Model with Uncertainties

Finally, the last example is devoted to a Vlasov-Fokker-Planck equation describing the swarming behavior of large group of agents. It is worth to observe how for this problem one steady state solution is provided by the global Maxwellian, which is a locally stable pattern, see [24, 49]. We compare the numerical solution of the problem making use of MC and M^3C scheme analyzed in Sect. 4.

We consider an uncertain self-propelled swarming model described by the Vlasov-Fokker-Planck equation (1) characterized by (9). This describes the time evolution of a distribution function $f(x, w, \theta, t)$ which represents the density of individuals in position $x \in \mathbb{R}^{d_x}$ having velocity $w \in \mathbb{R}^{d_w}$ at time $t > 0$. The initial data consists in a bivariate normal distribution of the form

$$f_0(x, w) = C(f_0^A(x, w) + f_0^B(x, w)), \tag{101}$$

where

$$f_0^A(x, w) = \frac{1}{2\pi\sqrt{\sigma_x^2 \sigma_w^2}} \exp\left\{-\frac{1}{2}\left(\frac{(x - \mu_x)^2}{\sigma_x^2} + \frac{(w - \mu_{w,A})^2}{\sigma_w^2}\right)\right\} \tag{102}$$

and

$$f_0^B(x, w) = \frac{1}{2\pi\sqrt{\sigma_x^2 \sigma_w^2}} \exp\left\{-\frac{1}{2}\left(\frac{(x - \mu_x)^2}{\sigma_x^2} + \frac{(w - \mu_{w,B})^2}{\sigma_w^2}\right)\right\} \tag{103}$$

with $\mu_x = 0$, $\sigma_x = 0.25$, $\mu_{w,A} = -\mu_{w,B} = 1.5$, $\sigma_w^2 = 0.25$ and $C > 0$ is a normalization constant. The uncertainty is present in the diffusion coefficient, i.e. $D = D(\theta) = 0.2 + 0.1\theta$, and it is distributed accordingly to $\theta \sim U([-0.1, 0.1])$.

We compute the solution by using the structure preserving scheme discussed in Sect. 3.1 for solving the homogeneous Fokker-Planck equation and we combine this method with a WENO scheme for the linear transport part. A second order time splitting approach combines the two discretization in space and velocity. More in details, we compare a Monte Carlo collocation with the Micro-Macro collocation discussed in Sect. 3. The number of cells in space is fixed to $N_x = 100$, in velocity space to $N_v = 100$ while the number of random inputs is fixed to $M = 50$. The solution is averaged over 10 different realization and the final time is fixed to $T = 250$. The size of the domain is $[0, L]$ with $L = 10$ in space and $[-L_v, L_v] = [-3, 3]$ in velocity space. In Fig. 13 the time evolution of the expected distribution with respect to the uncertain variable is reported for different times computed by the MC approach together with the time evolution of the expected perturbation g from the steady state solution computed with the M^3C method. In Fig. 14, the variance of the distribution over time and the variance of the perturbation g over time are reported, the firsts computed by the MC method, the seconds with the M^3C one. Finally, in Fig. 15, the L_1 norm of the error for the MC and the M^3C methods are

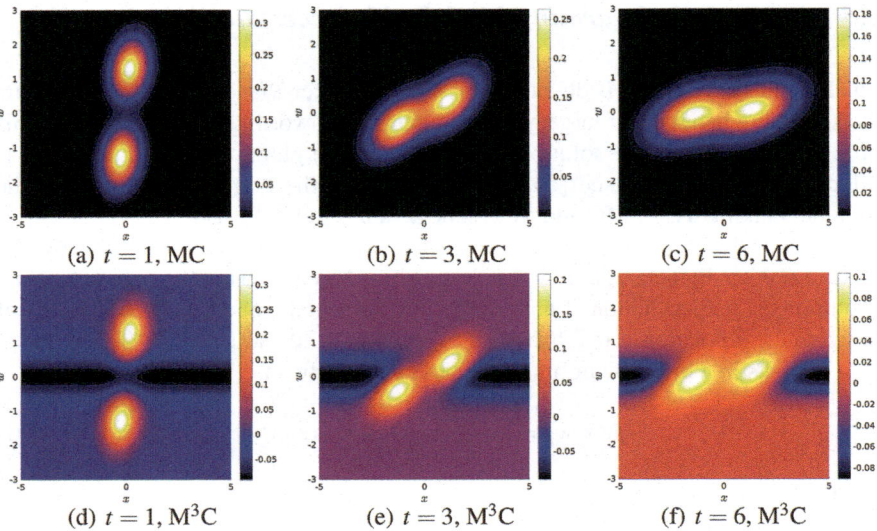

Fig. 13 Example 3. Time evolution of the distribution function, expected solution over time for the MC and the M³C methods. The top images report the expected solution computed with MC for $t = 1$, $t = 3$ and $t = 6$. The bottom images report the expected perturbation from the steady state equilibrium computed with the M³C method for $t = 1$, $t = 3$ and $t = 6$

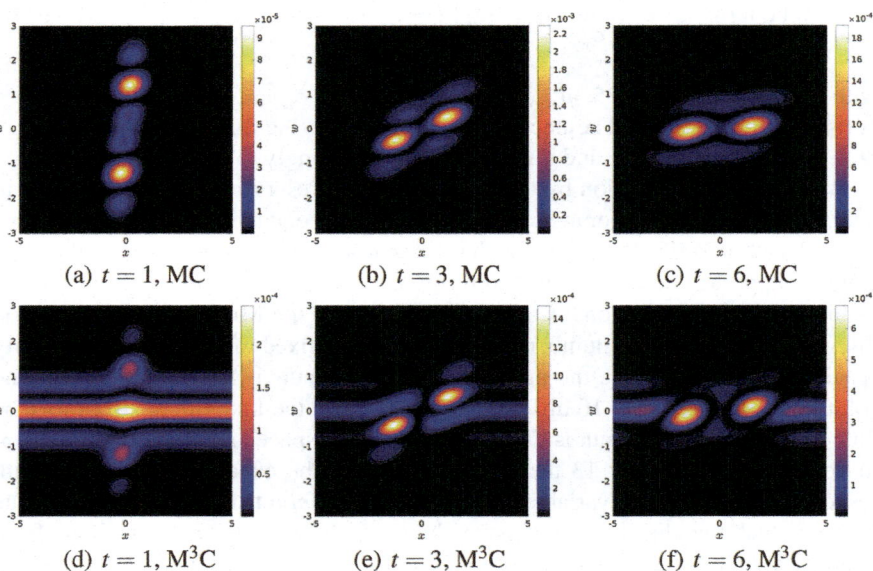

Fig. 14 Example 3. Time evolution of the variance of the asymptotic solution over time for the MC and the M³C methods. The top images report the variance computed with MC for $t = 1$, $t = 3$ and $t = 6$. The bottom images report the variance of the perturbation from the steady state equilibrium computed with the M³C method for $t = 1$, $t = 3$ and $t = 6$

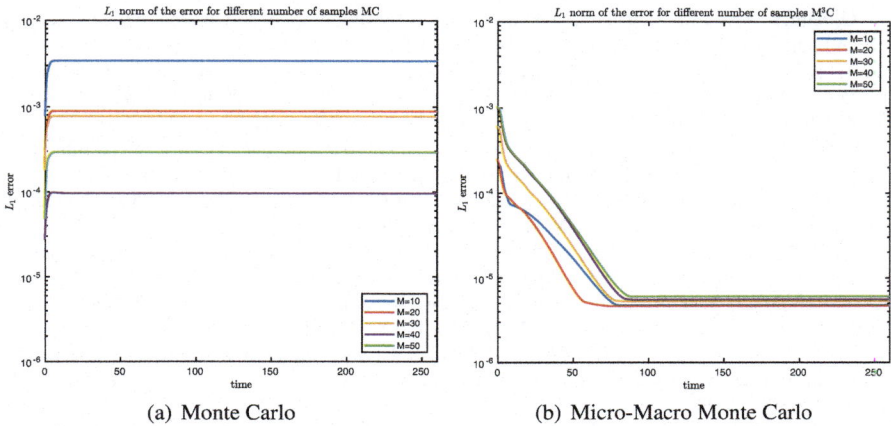

(a) Monte Carlo (b) Micro-Macro Monte Carlo

Fig. 15 Example 3. Left: Estimation of the L_1 error of the expected distribution over time computed for an increasing number of random inputs M for the MC method. Right: Estimation of the L_1 error of the expected distribution over time computed for an increasing number of random inputs M for the M^{3}C method. The error of the *MC* method remains constant in time while for M^{3}C method the error decreases

reported as a function of time for different number of random inputs. The gain in computational accuracy of the M^{3}C method is clearly evident for large times. The reference solution has been computed by a collocation method which employs the Gauss nodes as quadrature nodes with $M = 100$ random inputs.

Acknowledgements The research that led to the present survey was partially supported by the research grant *Numerical methods for uncertainty quantification in hyperbolic and kinetic equations* of the group GNCS of INdAM. MZ acknowledges support from GNCS and "Compagnia di San Paolo" (Torino, Italy).

References

1. S.M. Ahn, S.Y. Ha, Stochastic flocking dynamics of the Cucker–Smale model with multiplicative white noises. J. Math. Phys. **51**(10), 103–301 (2010)
2. G. Ajmone Marsan, N. Bellomo, M. Egidi, Towards a mathematical theory of complex socio–economical systems by functional subsystems representation. Kinet. Relat. Model. **1**(2), 249–278 (2008)
3. G. Albi, L. Pareschi, Binary interaction algorithms for the simulation of flocking and swarming dynamics. Multiscale Model. Simul. **11**(1), 1–29 (2013)
4. G. Albi, M. Herty, L. Pareschi, Kinetic description of optimal control problems and applications to opinion consensus. Commun. Math. Sci. **13**(6), 1407–1429 (2015)
5. G. Albi, L. Pareschi, M. Zanella, Uncertainty quantification in control problems for flocking models. Math. Probl. Eng. **2015**, 1–14 (2015)
6. G. Albi, L. Pareschi, G. Toscani, M. Zanella, Recent advances in opinion modeling: control and social influence, in *Active Particles Vol.1: Theory, Methods, and Applications*, ed. by N. Bellomo, P. Degond, E. Tadmor (Birkhäuser–Springer, Berlin, 2017), pp. 49–98

7. G. Albi, L. Pareschi, M. Zanella, Opinion dynamics over complex networks: kinetic modelling and numerical methods. Kinet. Relat. Model. **10**(1), 1–32 (2017)
8. M. Ballerini, N. Cabibbo, R. Candelier, A. Cavagna, E. Cisbani, I. Giardina, A. Orlandi, G. Parisi, A. Procaccini, M. Viale, V. Zdravkovic, Empirical investigation of starling flocks: a benchmark study in collective animal behavior. Anim. Behav. **76**(1), 201–215 (2008)
9. A.B.T. Barbaro, P. Degond, Phase transition and diffusion among socially interacting self-propelled agents. Discrete Continuous Dyn. Syst. Ser. B **19**, 1249–1278 (2014)
10. A.B.T. Barbaro, J.A. Cañizo, J.A. Carrillo, P. Degond, Phase transitions in a kinetic model of Cucker–Smale type. Multiscale Model. Simul. **14**(3), 1063–1088 (2016)
11. N. Bellomo, J. Soler, On the mathematical theory of the dynamics of swarms viewed as complex systems. Math. Models Methods Appl. Sci. **22**(1), 1140006 (2012)
12. N. Bellomo, B. Piccoli, A. Tosin, Modeling crowd dynamics from a complex system viewpoint. Math. Models Methods Appl. Sci. **22**(suppl 2), 1230004 (2012)
13. M. Bennoune, M. Lemou, L. Mieussens, Uniformly stable numerical schemes for the Boltzmann equation preserving the compressible Navier–Stokes asymptotics. J. Comput. Phys. **227**, 3781–3803 (2008)
14. M. Bessemoulin-Chatard, F. Filbet, A finite volume scheme for nonlinear degenerate parabolic equations. SIAM J. Sci. Comput. **34**, 559–583 (2012)
15. M. Bongini, M. Fornasier, M. Hansen, M. Maggioni, Inferring interaction rules from observations of evolutive systems I: the variational approach. Math. Models Methods Appl. Sci. **27**, 909 (2017)
16. S. Boscarino, F. Filbet, G. Russo, High order semi-implicit schemes for time dependent partial differential equations. J. Sci. Comput. **68**, 975–1001 (2016)
17. C. Buet, S. Dellacherie, On the Chang and Cooper numerical scheme applied to a linear Fokker-Planck equation. Commun. Math. Sci. **8**(4), 1079–1090 (2010)
18. C. Buet, S. Cordier, V. Dos Santos, A conservative and entropy scheme for a simplified model of granular media. Transp. Theory Stat. Phys. **33**(2), 125–155 (2004)
19. M. Burger, J.A. Carrillo, M.-T. Wolfram, A mixed finite element method for nonlinear diffusion equations. Kinet. Relat. Models **3**, 59–83 (2010)
20. R.E. Caflisch, Monte Carlo and Quasi Monte Carlo methods. Acta Numer. **7**, 1–49 (1998)
21. J.A. Carrillo, G. Toscani, Exponential convergence toward equilibrium for homogeneous Fokker–Planck–type equations. Math. Methods Appl. Sci. **21**, 1269–1286 (1998)
22. J.A. Carrillo, R.J. McCann, C. Villani, Kinetic equilibration rates for granular media and related equations: entropy dissipation and mass transportation estimates. Revista Matemática Iberoamericana **19**, 971–1018 (2003)
23. J.A. Carrillo, M. Fornasier, J. Rosado, G. Toscani, Asymptotic flocking dynamics for the kinetic Cucker–Smale model. SIAM J. Math. Anal. **42**(1), 218–236 (2010)
24. J.A. Carrillo, M. Fornasier, G. Toscani, F. Vecil, Particle, kinetic and hydrodynamic models of swarming, in *Mathematical Modeling of Collective Behavior in Socio–Economic and Life Sciences* (Birkhauser, Boston, 2010), pp. 297–336
25. J.A. Carrillo, Y.-P. Choi, M. Hauray, The derivation of swarming models: mean-field limit and Wasserstein distances, in *Collective Dynamics from Bacteria to Crowds*, vol. 553, CISM International Centre for Mechanical Sciences (Springer, Heidelberg, 2014), pp. 1–46
26. J.A. Carrillo, A. Chertock, Y. Huang, A finite-volume method for nonlinear nonlocal equations with a gradient flow structure. Commun. Comput. Phys. **17**, 233–258 (2015)
27. C. Cercignani, *The Boltzmann Equation and its Applications* (Springer, New York, 1988)
28. C. Chainais-Hillairet, A. Jüngel, S. Schuchnigg, Entropy-dissipative discretization of nonlinear diffusion equations and discrete Beckner inequalities. ESAIM Math. Model. Numer. Anal. **50**(1), 135–162 (2016)
29. J.S. Chang, G. Cooper, A practical difference scheme for Fokker–Planck equations. J. Comput. Phys. **6**(1), 1–16 (1970)
30. A. Chertock, S. Jin, A. Kurganov, An operator splitting based stochastic Galerkin method for the one–dimensional compressible Euler equations with uncertainty (Preprint, 2016)

31. H. Cho, D. Venturi, G.E. Karniadakis, Numerical methods for high–dimensional probability density function equations. J. Comput. Phys. **305**(15), 817–837 (2016)
32. Y.-P. Choi, S.-Y. Ha, Z. Li, Emergent dynamics of the Cucker–Smale flocking model and its variants, in *Active Particles, Volume 1*, eds. by N. Bellomo, P. Degond, E. Tadmor. Modeling and Simulation in Science, Engineering and Technology (Birkhäuser, Cham, 2017), pp. 299–331
33. S. Cordier, L. Pareschi, G. Toscani, On a kinetic model for a simple market economy. J. Stat. Phys. **120**(1), 253–277 (2005)
34. E. Cristiani, B. Piccoli, A. Tosin, Modeling self–organization in pedestrian and animal groups from macroscopic and microscopic viewpoints, in *Mathematical Modeling of Collective Behavior in Socio–Economic and Life Sciences*, ed. by G. Naldi, L. Pareschi, G. Toscani. Modeling and Simulation in Science, Engineering and Technology (Birkhäuser, Boston, 2010), pp. 337–364
35. E. Cristiani, B. Piccoli, A. Tosin, Multiscale modeling of granular flows with application to crowd dynamics. Multiscale Model. Simul. **9**(1), 155–182 (2011)
36. N. Crouseilles, M. Lemou, An asymptotic preserving scheme based on a micro–macro decomposition for collisional Vlasov equation: diffusion and high–field scaling limits. Kinet. Relat. Model. **4**(2), 441–477 (2011)
37. N. Crouseilles, G. Dimarco, M. Lemou, Asymptotic preserving and time diminishing schemes for rarefied gas dynamic. Kinet. Relat. Model. **10**, 643–668 (2017)
38. F. Cucker, S. Smale, Emergent behavior in flocks. IEEE Trans. Autom. Control **52**(5), 852–862 (2007)
39. P. Degond, L. Pareschi, G. Russo, (eds.), *Modeling and Computational Methods for Kinetic Equations*, Modeling and Simulation in Science, Engineering and Technology (Birkhäuser Boston Inc., Boston, 2004)
40. P. Degond, J.-G. Liu, S. Motsch, V. Panferov, Hydrodynamic models of self-organized dynamics: derivation and existence theory. Methods Appl. Anal. **20**(2), 89–114 (2013)
41. P. Degond, J.-G. Liu, C. Ringhofer, Evolution in a non–conservative economy driven by local Nash equilibria. Philos. Trans. A Math. Phys. Eng. Sci. **372**(2028), 20130394 (2014)
42. B. Després, G. Poëtte, D. Lucor, Robust uncertainty propagation in systems of conservation laws with the entropy closure method, in *Uncertainty Quantification in Computational Fluid Dynamics*. Lecture Notes in Computational Science and Engineering, vol. 92 (Springer, Berlin, 2010), pp. 105–149
43. G. Dimarco, L. Pareschi, Numerical methods for kinetic equations. Acta Numer. **23**, 369–520 (2014)
44. G. Dimarco, L. Pareschi, Variance reduction Monte Carlo methods for uncertainty quantification in the Boltzmann equation and related problems (Preprint, 2018)
45. G. Dimarco, Q. Li, B. Yan, L. Pareschi, Numerical methods for plasma physics in collisional regimes. J. Plasma Phys. **81**(1), 305810106 (2015)
46. G. Dimarco, L. Pareschi, M. Zanella, Micro-Macro generalized polynomial chaos techniques for kinetic equations. (Preprint, 2018)
47. A. Dimits, W. Lee, Partially linearized algorithms in gyrokinetic particle simulation. J. Comput. Phys. **107**(2), 309–323 (1993)
48. M.R. D'Orsogna, Y.L. Chuang, A.L. Bertozzi, L. Chayes, Self-propelled particles with soft-core interactions: patterns, stability and collapse. Phys. Rev. Lett. **96**, 104302 (2006)
49. R. Duan, M. Fornasier, G. Toscani, A kinetic flocking model with diffusion. Commun. Math. Phys. **300**, 95–145 (2010)
50. D.A. Dunavant, High degree efficient symmetrical Gaussian quadrature rules for the triangle. Int. J. Numer. Methods Eng. **21**, 1129–1148 (1985)
51. B. Düring, M.-T. Wolfram, Opinion dynamics: inhomogeneous Boltzmann-type equations modelling opinion leadership and political segregation. Proc. R. Soc. Lond. A Math. Phys. Eng. Sci. **471**, 20150345 (2015)
52. B. Düring, P. Markowich, J.-F. Pietschmann, M.-T. Wolfram, Boltzmann and Fokker–Planck equations modelling opinion formation in the presence of strong leaders. Proc. R. Soc. A Math. Phys. Eng. Sci. **465**(2112), 3687–3708 (2009)

53. F. Filbet, L. Pareschi, T. Rey, On steady–state preserving spectral methods for the homogeneous Boltzmann equation. Comptes Rendus Mathematique **353**(4), 309–314 (2015)
54. G. Furioli, A. Pulvirenti, E. Terraneo, G. Toscani, Fokker-Planck equations in the modeling of socio–economic phenomena. Math. Models Methods Appl. Sci. **27**(1), 115–158 (2017)
55. M.B. Giles, Multilevel Monte Carlo methods. Acta Numer. **24**, 259–328 (2015)
56. L. Gosse, *Computing Qualitatively Correct Approximations of Balance Laws. Exponential-Fit, Well-Balanced and Asymptotic-Preserving*. SEMA SIMAI Springer Series (Springer, Berlin, 2013)
57. S. Gottlieb, C.W. Shu, E. Tadmor, Strong stability-preserving high-order time discretization methods. SIAM Rev. **43**(1), 89–112 (2001)
58. S.-Y. Ha, E. Tadmor, From particle to kinetic and hydrodynamic descriptions of flocking. Kinet. Relat. Models **3**(1), 415–435 (2008)
59. S.-Y. Ha, K. Lee, D. Levy, Emergence of time–asymptotic flocking in a stochastic Cucker–Smale system. Commun. Math. Sci. **7**(2), 453–469 (2009)
60. E. Hairer, S.P. Norsett, G. Wanner, Solving Ordinary Differential Equation I: Nonstiff Problems. Springer Series in Comput. Mathematics, Vol. 8, Springer-Verlag 1987, Second revised edition 1993.
61. J. Hu, S. Jin, A stochastic Galerkin method for the Boltzmann equation with uncertainty. J. Comput. Phys. **315**, 150–168 (2016)
62. J. Hu, S. Jin, D. Xiu, A stochastic Galerkin method for Hamilton–Jacobi equations with uncertainty. SIAM J. Sci. Comput. **37**(5), A2246–A2269 (2015)
63. S. Jin, Asymptotic preserving (AP) schemes for multiscale kinetic and hyperbolic equations: a review, in *Lecture Notes for Summer School on Methods and Models of Kinetic Theory, (M&MKT), Porto Ercole (Grosseto, Italy)* Riv. Mat. Univ. Parma. **3**(2), 177–216 (2012)
64. S. Jin, D. Xiu, X. Zhu, A well-balanced stochastic Galerkin method for scalar hyperbolic balance laws with random inputs. J. Sci. Comput. **67**, 1198–1218 (2016)
65. Y. Katz, K. Tunstrøm, C.C. Ioannou, C. Huepe, I.D. Couzin, Inferring the structure and dynamics of interactions in schooling fish. Proc. Natl. Acad. Sci. U. S. A. **108**(46), 18720–18725 (2011)
66. E.W. Larsen, C.D. Levermore, G.C. Pomraning, J.G. Sanderson, Discretization methods for one-dimensional Fokker–Planck operators. J. Comput. Phys. **61**(3), 359–390 (1985)
67. M. Lemou, L. Mieussens, A new asymptotic preserving scheme based on micro–macro formulation for linear kinetic equations in the diffusion limit. SIAM J. Sci. Comput. **31**(1), 334–368 (2008)
68. O. Le Maitre, O.M. Knio, *Spectral Methods for Uncertainty Quantification: with Applications to Computational Fluid Dynamics*. Scientific Computation (Springer, Dordrechat, 2010)
69. T.-P. Liu, S.-H. Yu, Boltzmann equation: micro–macro decomposition and positivity of shock profiles. Commun. Math. Phys. **246**(1), 133–179 (2004)
70. D. Matthes, A. Jüngel, G. Toscani, Convex Sobolev inequalities derived from entropy dissipation. Arch. Ration. Mech. Anal. **199**(2), 563–596 (2011)
71. M. Mohammadi, A. Borzì, Analysis of the Chang–Cooper discretization scheme for a class of Fokker-Planck equations. J. Numer. Math. **23**(3), 271–288 (2015)
72. G. Naldi, L. Pareschi, G. Toscani, (eds.), *Mathematical Modeling of Collective Behavior in Socio-Economic and Life Sciences* (Birkhäuser, Boston, 2010)
73. L. Pareschi, T. Rey, Residual equilibrium schemes for time dependent partial differential equations. Computers & Fluids **156**, 329–342 (2017)
74. L. Pareschi, G. Toscani, *Interacting Multiagent Systems: Kinetic Equations and Monte Carlo Methods* (Oxford University Press, Oxford, 2013)
75. L. Pareschi, M. Zanella, Structure–preserving schemes for nonlinear Fokker–Planck equations and applications. J. Sci. Comput. 1–26 (2017)
76. L. Pareschi, M. Zanella, Structure–preserving schemes for mean–field equations of collective behavior. *Proceedings of the 16th International Conference on Hyperbolic Problems: Theory, Numerics, Applications*, Aachen 2016, to appear

77. P. Pettersson, G. Iaccarino, J. Nordström, A stochastic Galerkin method for the Euler equations with Roe variable transformation. J. Comput. Phys. **257**, 481–500 (2014)
78. P. Pettersson, G. Iaccarino, J. Nordström, *Polynomial Chaos Methods for Hyperbolic Partial Differential Equations: Numerical Techniques for Fluid Dynamics Problems in the Presence of Uncertainties*. Mathematical Engineering (Springer, Berlin, 2015)
79. G. Poëtte, B. Després, D. Lucor, Uncertainty quantification for systems of conservation laws. J. Comput. Phys. **228**(7), 2443–2467 (2009)
80. H. Risken, *The Fokker–Planck Equation. Methods of Solution and Applications*, 2nd edn. (Springer, Berlin, 1989)
81. H.L. Scharfetter, H.K. Gummel, Large signal analysis of a silicon Read diode oscillator. IEEE Trans. Electron Devices **16**, 64–77 (1969)
82. E. Sonnendrucker, Numerical methods for Vlasov equations. Technical report, MPI TU Munich, 2013
83. G. Toscani, Entropy production and the rate of convergence to equilibrium for the Fokker–Planck equation. Q. Appl. Math. **LVII**(3), 521–541 (1999)
84. G. Toscani, Kinetic models of opinion formation. Commun. Math. Sci. **4**(3), 481–496 (2006)
85. G. Toscani, C. Villani, Sharp entropy dissipation bounds and explicit rate of trend to equilibrium for the spatially homogeneous Boltzmann equation. Commun. Math. Phys. **203**(3), 667–706 (1999)
86. C. Villani, A review of mathematical topics in collisional kinetic theory, in *Handbook of Mathematical Fluid Mechanics*, ed. by S. Friedlander, D. Serre, vol. I (North–Holland, Amsterdam, 2002), pp. 71–305
87. A.A. Vlasov, *Many–Particle Theory and its Application to Plasma*. Russian Monographs and Text on Advanced Mathematics and Physics, vol. VII (Gordon and Breach, Science Publishers, Inc., New York, 1961)
88. D. Xiu, *Numerical Methods for Stochastic Computations* (Princeton University Press, Princeton, 2010)
89. D. Xiu, J.S. Hesthaven, High–order collocation methods for differential equations with random inputs. SIAM J. Sci. Comput. **27**(3), 1118–1139 (2005)
90. D. Xiu, G.E. Karniadakis, The Wiener–Askey polynomial chaos for stochastic differential equations. SIAM J. Sci. Comput. **24**(2), 614–644 (2002)
91. B. Yan, A hybrid method with deviational particles for spatial inhomogeneous plasma. J. Comput. Phys. **309**, 18–36 (2016)
92. Y. Zhu, S. Jin, The Vlasov-Poisson-Fokker-Planck system with uncertainty and a one-dimensional asymptotic-preserving method. SIAM Multiscale Model. Simul. **15**(4), pp. 1502–1529.

Uncertainty Quantification for Kinetic Equations

Jingwei Hu and Shi Jin

Abstract Kinetic equations contain uncertainties in their collision kernels or scattering coefficients, initial or boundary data, forcing terms, geometry, etc. Quantifying the uncertainties in kinetic models have important engineering and industrial applications. In this article we survey recent efforts in the study of kinetic equations with random inputs, including their mathematical properties such as regularity and long-time behavior in the random space, construction of efficient stochastic Galerkin methods, and handling of multiple scales by stochastic asymptotic-preserving schemes. The examples used to illustrate the main ideas include the random linear and nonlinear Boltzmann equations, linear transport equation and the Vlasov-Poisson-Fokker-Planck equations.

1 Introduction

Kinetic equations describe the non-equilibrium dynamics of a gas or system comprised of a large number of particles using a probability density function. In multiscale modeling hierarchy, they serve as a basic building block that bridges atomistic and continuum models. On one hand, they are more efficient (requiring fewer degrees of freedom) than molecular dynamics; on the other hand, they provide reliable information at the mesoscopic level when the macroscopic fluid mechanics laws of Navier-Stokes and Fourier become inadequate. The most fundamental (and the very first) kinetic equation is the Boltzmann equation, an integro-differential

J. Hu
Department of Mathematics, Purdue University, West Lafayette, IN, USA
e-mail: jingweihu@purdue.edu

S. Jin (✉)
Department of Mathematics, University of Wisconsin-Madison, Madison, WI, USA

Department of Mathematics, Institute of Natural Sciences, MOE-LSEC and SHL-MAC, Shanghai Jiao Tong University, Shanghai, China
e-mail: sjin@wisc.edu

© Springer International Publishing AG, part of Springer Nature 2017
S. Jin, L. Pareschi (eds.), *Uncertainty Quantification for Hyperbolic and Kinetic Equations*, SEMA SIMAI Springer Series 14,
https://doi.org/10.1007/978-3-319-67110-9_6

equation describing particle transport and binary collisions [11, 16]. Proposed by Ludwig Boltzmann in 1872, the equation is considered as the basis of the modern kinetic theory. During the past decades, there have been enormous studies on the Boltzmann and related kinetic models, both theoretically and numerically (cf. [13, 20, 74]). This trend is ever-growing as the application of the kinetic theory has already gone beyond traditional fields like rarefied gas dynamics [12], radiative transfer [15], and branched out to microfabrication technology [48, 61], biological and even social sciences [63].

In spite of the vast amount of existing research, the study of kinetic equations has mostly remained deterministic and ignored *uncertainty*. In reality, however, there are many sources of uncertainties that can arise in these equations. They may be due to

- *Incomplete knowledge* of the interaction mechanism between particles. Kinetic equations typically contain an integral operator modeling particle interactions. Inside this integral, there is a term called collision or scattering kernel describing the transition rate during particle collisions. Ideally, the collision kernel should be calculated from first principles using scattering theory [11]. This, if not impossible, is extremely complicated for complex particle systems. Therefore, empirical collision kernels are often used in practice with the aim to reproduce correct viscosity and diffusion coefficients [8, 9, 33, 50]. Specifically, these kernels contain adjustable parameters whose values are determined by matching with available experimental data for various kinds of particles.
- *Imprecise measurement* of the boundary data. A commonly used boundary for kinetic equations is the so-called Maxwell boundary condition [11, 12], which assumes part of the particles are bounced back specularly and part of them are absorbed by the wall and re-emitted according to a special Gaussian distribution. This distribution depends on the (measured) macroscopic properties of the wall such as temperature and bulk velocity.

The uncertainties are of course not limited to the aforementioned examples: they may also come from inaccurate measurement of the initial data, our lack of knowledge of gas-surface interactions, forcing and geometry, etc. Understanding the impact of these uncertainties is critical to the simulations of the complex kinetic systems to validate the kinetic models, and will allow scientists and engineers to obtain more reliable predictions and perform better risk assessment.

Despite tremendous amount of research activities in uncertainty quantification (UQ) in recent decades in many areas of sciences and engineering, the study of uncertainty in kinetic models, albeit important and necessary, has remained mostly untouched territory until very recently. It is the goal of this survey to review recent development of UQ for kinetic equations. Here the uncertainty is introduced through random inputs, and we adopt the generalized polynomial chaos based stochastic Galerkin (gPC-sG) approximation, which has been successfully applied to many physical and engineering problems, see for instance, the overviews in [27, 55, 66, 77]. Due to the high-dimensionality and intrinsic physical properties of kinetic equations, the construction of stochastic methods represents a great challenge. We will use some prototype equations including the classical Boltzmann equation, linear Boltzmann equations, and Vlasov-Poisson-Fokker-Planck system to illustrate the main strategy.

It is well-known that the gPC-sG approach is intrusive, requiring more coding efforts compared with non-intrusive methods such as the stochastic collocation [32, 78]. The reason of our choice is twofold: (1) Due to its Galerkin formulation, mathematical analysis of these methods can be conducted more conveniently. Indeed many of the analytical methods well-established in kinetic theory can be conveniently adopted or extended to study the stochastic Galerkin system of the random kinetic equations; (2) Kinetic equations often contain small parameters such as the mean free path/time which asymptotically lead to hyperbolic/diffusion equations. We are interested in developing the stochastic analogue of the asymptotic-preserving (AP) scheme, a scheme designed to capture the asymptotic limit at the discrete level. The stochastic Galerkin method yields systems of deterministic equations that *resemble the deterministic kinetic equations*, although in vector forms. Thus it allows one to easily use the deterministic AP framework for the random problems, allowing minimum "intrusivity" to the legacy deterministic codes. The stochastic Galerkin method can ensure the desired convergence in the weak sense. The resulting stochastic Asymptotic-Preserving (sAP) [46] sG methods will allow all numerical parameters, such as mesh size, time-step and the number of gPC modes chosen *independently* of the (possibly small) mean free path/time.

On the other hand, the study of regularity, coercivity and hypocoercivity on the random kinetic equations, which will be reviewed in this article as well, provides theoretical foundation for not only the stochastic Galerkin methods, but also the stochastic collocation methods.

The rest of this paper is organized as follows. In the next section, we give a brief review of some kinetic equations with random inputs and their basic properties. Section 3 discusses the theoretical issues such as coercivity, hypocoercivity, regularity, and long-time behavior for random kinetic equations. We then introduce in Sect. 4 the gPC-sG method. Special emphasis is given to the unique issues arising in kinetic equations such as property of the collision operator under gPC-sG approximation and efficient treatment of the nonlinear collision integral. Spectral accuracy of the gPC-sG method is also established. In Sect. 5, we consider the kinetic equations in diffusive scalings and construct the stochastic AP scheme following its deterministic counterpart. We conclude in Sect. 6 and list a few open problems in this field.

2 Preliminaries on Kinetic Equations with Random Inputs

In this section, we review some kinetic equations and their basic properties that will be used in this article. Due to the large variety of kinetic models, it is impossible to give a thorough description of all of them. Therefore, we will concentrate on several prototype models: the linear neutron transport equation, the semiconductor Boltzmann equation, the Vlasov-Poisson-Fokker-Planck equation, and the classical nonlinear Boltzmann equation. Other related kinetic models will be briefly mentioned at the end of the section.

As mentioned in Sect. 1, for real-world problems, the collision/scattering kernel, initial/boundary data, source, or other physical parameters in the kinetic equations may contain uncertainties that propagate into the solution and affect its property substantially. To characterize these random inputs, we assume certain quantities depend on a random vector $\mathbf{z} \in \mathbb{R}^n$ in a properly defined probability space $(\Sigma, \mathscr{A}, \mathbb{P})$, whose event space is Σ and is equipped with σ-algebra $\mathscr{A}$ and probability measure $\mathbb{P}$. We also assume the components of $\mathbf{z}$ are mutually independent random variables with known probability $\omega(\mathbf{z}) : I_{\mathbf{z}} \longrightarrow \mathbb{R}^+$, obtained already through some dimension reduction technique, e.g., Karhunen-Loève (KL) expansion [60], and do not pursue further the issue of random input parameterization. We treat $\mathbf{z}$ as a parameter and the properties given in this section hold for every given $\mathbf{z}$.

2.1 The Linear Transport Equation with Isotropic Scattering

We first introduce the linear transport equation in one dimensional slab geometry:

$$\varepsilon \partial_t f + v \partial_x f = \frac{\sigma}{\varepsilon} \mathscr{L} f - \varepsilon \sigma^a f + \varepsilon S, \quad t > 0, \ x \in [0, 1], \ v \in [-1, 1], \ \mathbf{z} \in I_{\mathbf{z}}, \tag{1}$$

$$\mathscr{L} f(t, x, v, \mathbf{z}) = \frac{1}{2} \int_{-1}^{1} f(t, x, v', \mathbf{z}) \, dv' - f(t, x, v, \mathbf{z}), \tag{2}$$

with the initial condition

$$f(0, x, v, \mathbf{z}) = f^0(x, v, \mathbf{z}). \tag{3}$$

This equation arises in neutron transport, radiative transfer, etc. and describes particles (for example neutrons) transport in a background media (for example nuclei). $f(t, x, v, \mathbf{z})$ is the density distribution of particles at time t, position x, and $v = \Omega \cdot e_x = \cos \theta$ where θ is the angle between the moving direction and x-axis. $\sigma(x, \mathbf{z})$, $\sigma^a(x, \mathbf{z})$ are total and absorption cross-sections respectively. $S(x, \mathbf{z})$ is the source term. For $\sigma(x, \mathbf{z})$, we assume

$$\sigma(x, \mathbf{z}) \geq \sigma_{\min} > 0. \tag{4}$$

ε is the dimensionless Knudsen number, the ratio between particle mean free path and the characteristic length (such as the length of the domain). The equation is scaled in long time with strong scattering.

We are interested in problems that contain *uncertainties* in the collision cross-section, source, initial or boundary data. Thus in our problem f, σ, σ^a and S all depend on $\mathbf{z}$.

Denote

$$[\phi] = \frac{1}{2} \int_{-1}^{1} \phi(v)\, dv \tag{5}$$

as the average of a velocity dependent function ϕ.

Define in the Hilbert space $L^2([-1, 1];\ \phi^{-1}\, dv)$ the inner product and norm

$$\langle f, g \rangle_\phi = \int_{-1}^{1} f(v)g(v)\phi^{-1}\, dv, \qquad \|f\|_\phi^2 = \langle f, f \rangle_\phi. \tag{6}$$

The linear operator $\mathscr{L}$ satisfies the following properties [6]:

- $[\mathscr{L}f] = 0$, for every $f \in L^2([-1, 1])$;
- The null space of f is $\mathscr{N}(\mathscr{L}) = \text{Span}\{\phi \mid \phi = [\phi]\}$;
- The range of f is $\mathscr{R}(\mathscr{L}) = \mathscr{N}(\mathscr{L})^\perp = \{f \mid [f] = 0\}$;
- *Coercivity:* $\mathscr{L}$ is non-positive self-adjoint in $L^2([-1, 1]; \phi^{-1}\, dv)$, i.e., there is a positive constant s_m such that

$$\langle f, \mathscr{L}f \rangle_\phi \leq -2s_m \|f\|_\phi^2, \qquad \forall f \in \mathscr{N}(\mathscr{L})^\perp; \tag{7}$$

- $\mathscr{L}$ admits a pseudo-inverse, denoted by $\mathscr{L}^{-1}$, from $\mathscr{R}(\mathscr{L})$ to $\mathscr{R}(\mathscr{L})$.

Let $\rho = [f]$. For each fixed $\mathbf{z}$, the classical diffusion limit theory of linear transport equation [6, 7, 52] gives that, as $\varepsilon \to 0$, ρ solves the following diffusion equation:

$$\partial_t \rho = \partial_x(\kappa(x, \mathbf{z})\partial_x \rho) - \sigma^a(x, \mathbf{z})\rho + S(x, \mathbf{z}), \tag{8}$$

where the diffusion coefficient

$$\kappa(x, \mathbf{z}) = \frac{1}{3}\sigma(x, \mathbf{z})^{-1}. \tag{9}$$

When $\mathbf{z}$ is random, (8) is a random diffusion equation.

2.2 The Semiconductor Boltzmann Equation

The semiconductor Boltzmann equation describes the electron transport in a semiconductor device [61]:

$$\varepsilon \partial_t f + \mathbf{v} \cdot \nabla_{\mathbf{x}} f + \nabla_{\mathbf{x}} \phi \cdot \nabla_{\mathbf{v}} f = \frac{1}{\varepsilon} \mathscr{Q}^s(f), \quad t > 0, \ \mathbf{x} \in \Omega \subset \mathbb{R}^d, \ \mathbf{v} \in \mathbb{R}^d, \ \mathbf{z} \in I_{\mathbf{z}}, \tag{10}$$

where $f(t, \mathbf{x}, \mathbf{v}, \mathbf{z})$ is again the particle distribution function, $\phi(t, \mathbf{x}, \mathbf{z})$ is the electric potential given a priori or produced self-consistently by f through the Poisson equation:

$$\Delta_{\mathbf{x}}\phi = \rho - h,$$

where $\rho(t, \mathbf{x}, \mathbf{z}) = \int f \, d\mathbf{v}$, and $h(\mathbf{x}, \mathbf{z})$ is the doping profile (some physical parameters such as the material permittivity are omitted for brevity). The collision operator $\mathcal{Q}^s(f)$ is a linear approximation of the electron-phonon interaction:

$$\mathcal{Q}^s(f)(\mathbf{v}, \mathbf{z}) = \int_{\mathbb{R}^d} [s(\mathbf{v}_*, \mathbf{v}, \mathbf{z})f(\mathbf{v}_*, \mathbf{z}) - s(\mathbf{v}, \mathbf{v}_*, \mathbf{z})f(\mathbf{v}, \mathbf{z})] \, d\mathbf{v}_*, \qquad (11)$$

where $s(\mathbf{v}, \mathbf{v}_*, \mathbf{z})$ describes the transition rate from $\mathbf{v}$ to $\mathbf{v}_*$ and may take various forms depending on the approximation. Here we assume

$$s(\mathbf{v}, \mathbf{v}_*, \mathbf{z}) = \sigma(\mathbf{v}, \mathbf{v}_*, \mathbf{z})M^s(\mathbf{v}_*),$$

with M^s being the normalized Maxwellian:

$$M^s(\mathbf{v}) = \frac{1}{\pi^{d/2}}e^{-|\mathbf{v}|^2};$$

the scattering kernel σ being rotationally invariant, symmetric and bounded:

$$\sigma(\mathbf{v}, \mathbf{v}_*, \mathbf{z}) = \sigma(|\mathbf{v}|, |\mathbf{v}_*|, \mathbf{z}), \quad 0 < \sigma_{\min} \leq \sigma(\mathbf{v}, \mathbf{v}_*, \mathbf{z}) = \sigma(\mathbf{v}_*, \mathbf{v}, \mathbf{z}) \leq \sigma_{\max}.$$

Define the collision frequency

$$\lambda(\mathbf{v}, \mathbf{z}) = \int_{\mathbb{R}^d} \sigma(\mathbf{v}, \mathbf{v}_*, \mathbf{z})M^s(\mathbf{v}_*) \, d\mathbf{v}_*, \qquad (12)$$

then it is easy to see $\sigma_0 \leq \lambda(\mathbf{v}, \mathbf{z}) \leq \sigma_1$. Therefore, (11) can be written as

$$\mathcal{Q}^s(f)(\mathbf{v}, \mathbf{z}) = \int_{\mathbb{R}^d} \sigma(\mathbf{v}, \mathbf{v}_*, \mathbf{z}) [M^s(\mathbf{v})f(\mathbf{v}_*, \mathbf{z}) - M^s(\mathbf{v}_*)f(\mathbf{v}, \mathbf{z})] \, d\mathbf{v}_*$$

$$= M^s(\mathbf{v}) \int_{\mathbb{R}^d} \sigma(\mathbf{v}, \mathbf{v}_*, \mathbf{z})f(\mathbf{v}_*, \mathbf{z}) \, d\mathbf{v}_* - \lambda(\mathbf{v}, \mathbf{z})f(\mathbf{v}, \mathbf{z}). \qquad (13)$$

It can be shown that the collision operator (13) satisfies

$$\int_{\mathbb{R}^d} \mathcal{Q}^s(f)(\mathbf{v}, \mathbf{z})f(\mathbf{v}, \mathbf{z})/M^s(\mathbf{v}) \, d\mathbf{v}$$

$$= -\frac{1}{2}\int_{\mathbb{R}^d}\int_{\mathbb{R}^d} \sigma(\mathbf{v}, \mathbf{v}_*, \mathbf{z})M^s(\mathbf{v})M^s(\mathbf{v}_*)\left(\frac{f(\mathbf{v})}{M^s(\mathbf{v})} - \frac{f(\mathbf{v}_*)}{M^s(\mathbf{v}_*)}\right)^2 d\mathbf{v}_*d\mathbf{v} \leq 0.$$

$$(14)$$

Furthermore, the followings are equivalent

$$\int_{\mathbb{R}^d} \mathscr{Q}^s(f) \frac{f}{M} \, d\mathbf{v} = 0 \iff \mathscr{Q}^s(f) = 0 \iff f = \rho(t, \mathbf{x}, \mathbf{z}) M^s(\mathbf{v}). \tag{15}$$

Then, as $\varepsilon \to 0$, (10) leads to the following *drift-diffusion* limit [67]:

$$\partial_t \rho = \nabla_{\mathbf{x}} \cdot (D (\nabla_{\mathbf{x}} \rho + 2\rho \mathbf{E})), \tag{16}$$

where $\mathbf{E} = -\nabla_{\mathbf{x}} \phi$ is the electric field, D is the diffusion coefficient matrix defined by

$$D = \int_{\mathbb{R}^d} \frac{\mathbf{v} \otimes \mathbf{v} M^s(\mathbf{v})}{\lambda(\mathbf{v}, \mathbf{z})} \, d\mathbf{v}.$$

2.3 The Vlasov-Poisson-Fokker-Planck System

The Vlasov-Poisson-Fokker-Planck (VPFP) system arises in the kinetic modeling of the Brownian motion of a large system of particles in a surrounding bath [14]. One application of such system is the electrostatic plasma, in which one considers the interactions between the electrons and a surrounding bath via the Coulomb force. In the dimensionless VPFP system with uncertainty, the time evolution of particle density distribution function $f(t, \mathbf{x}, \mathbf{v}, \mathbf{z})$ under the action of an electrical potential $\phi(t, \mathbf{x}, \mathbf{z})$ satisfies

$$\begin{cases} \partial_t f + \frac{1}{\delta} \mathbf{v} \cdot \nabla_{\mathbf{x}} f - \frac{1}{\varepsilon} \nabla_{\mathbf{x}} \phi \cdot \nabla_{\mathbf{v}} f = \frac{1}{\delta\varepsilon} \mathscr{F} f, \\ -\Delta_{\mathbf{x}} \phi = \rho - 1, \quad t > 0, \quad \mathbf{x} \in \Omega \subset \mathbb{R}^d, \ \mathbf{v} \in \mathbb{R}^d, \ \mathbf{z} \in I_{\mathbf{z}}, \end{cases} \tag{17}$$

with initial condition

$$f(0, \mathbf{x}, \mathbf{v}, \mathbf{z}) = f^0(\mathbf{x}, \mathbf{v}, \mathbf{z}). \tag{18}$$

Here, $\mathscr{F}$ is a collision operator describing the Brownian motion of the particles, which reads,

$$\mathscr{F} f = \nabla_{\mathbf{v}} \cdot \left(M^v \nabla_{\mathbf{v}} \left(\frac{f}{M^v} \right) \right), \tag{19}$$

where M^v is the *global equilibrium* or *global Maxwellian*,

$$M^v = \frac{1}{(2\pi)^{\frac{d}{2}}} e^{-\frac{|v|^2}{2}}. \tag{20}$$

δ is the reciprocal of the scaled thermal velocity, ε represents the scaled thermal mean free path. There are two different regimes for this system. One is the *high field regime*, where $\delta = 1$. As $\varepsilon \to 0$, f goes to the local Maxwellian $M_l^v = \frac{1}{(2\pi)^{\frac{d}{2}}}e^{-\frac{|v-\nabla_x\phi|^2}{2}}$, and the VPFP system converges to a hyperbolic limit [2, 31, 65]:

$$\begin{cases} \partial_t \rho + \nabla_x \cdot (\rho \nabla_x \phi) = 0, \\ -\Delta_x \phi = \rho - 1. \end{cases} \qquad (21)$$

Another regime is the *parabolic regime*, where $\delta = \varepsilon$. When $\varepsilon \to 0$, f goes to the global Maxwellian M^v, and the VPFP system converges to a parabolic limit [68]:

$$\begin{cases} \partial_t \rho - \nabla_x \cdot (\nabla_x \rho - \rho \nabla_x \phi) = 0, \\ -\Delta_x \phi = \rho - 1. \end{cases} \qquad (22)$$

2.4 The Classical Nonlinear Boltzmann Equation

We finally introduce the classical Boltzmann equation that describes the time evolution of a rarefied gas [11]:

$$\partial_t f + \mathbf{v} \cdot \nabla_x f = \mathcal{Q}^b(f,f), \quad t > 0, \ \mathbf{x} \in \Omega \subset \mathbb{R}^d, \ \mathbf{v} \in \mathbb{R}^d, \ \mathbf{z} \in I_z, \qquad (23)$$

where $\mathcal{Q}^b(f,f)$ is the bilinear collision operator modeling the binary interaction among particles:

$$\mathcal{Q}^b(f,f)(\mathbf{v}, \mathbf{z}) = \int_{\mathbb{R}^d} \int_{S^{d-1}} B(\mathbf{v}, \mathbf{v}_*, \eta, \mathbf{z}) \left[f(\mathbf{v}', \mathbf{z})f(\mathbf{v}'_*, \mathbf{z}) - f(\mathbf{v}, \mathbf{z})f(\mathbf{v}_*, \mathbf{z}) \right] d\eta \, d\mathbf{v}_*. \qquad (24)$$

Here $(\mathbf{v}, \mathbf{v}_*)$ and $(\mathbf{v}', \mathbf{v}'_*)$ are the velocity pairs before and after a collision, during which the momentum and energy are conserved; hence $(\mathbf{v}', \mathbf{v}'_*)$ can be represented in terms of $(\mathbf{v}, \mathbf{v}_*)$ as

$$\begin{cases} \mathbf{v}' = \dfrac{\mathbf{v} + \mathbf{v}_*}{2} + \dfrac{|\mathbf{v} - \mathbf{v}_*|}{2}\eta, \\ \mathbf{v}'_* = \dfrac{\mathbf{v} + \mathbf{v}_*}{2} - \dfrac{|\mathbf{v} - \mathbf{v}_*|}{2}\eta, \end{cases}$$

with the parameter η varying on the unit sphere S^{d-1}. The collision kernel $B(\mathbf{v}, \mathbf{v}_*, \eta, \mathbf{z})$ is a non-negative function depending on $|\mathbf{v} - \mathbf{v}_*|$ and cosine of the deviation angle θ:

$$B(\mathbf{v}, \mathbf{v}_*, \eta, \mathbf{z}) = B(|\mathbf{v} - \mathbf{v}_*|, \cos\theta, \mathbf{z}), \quad \cos\theta = \frac{\eta \cdot (\mathbf{v} - \mathbf{v}_*)}{|\mathbf{v} - \mathbf{v}_*|}.$$

The specific form of B is determined from the intermolecular potential via the scattering theory. For numerical purpose, a commonly used model is the variable hard-sphere (VHS) model introduced by Bird [9]:

$$B(|\mathbf{v} - \mathbf{v}_*|, \cos\theta, \mathbf{z}) = b_\lambda(\mathbf{z})|\mathbf{v} - \mathbf{v}_*|^\lambda, \quad -d < \lambda \le 1, \tag{25}$$

where $\lambda > 0$ corresponds to the hard potentials, and $\lambda < 0$ to the soft potentials.

The collision operator (24) conserves mass, momentum, and energy:

$$\int_{\mathbb{R}^d} \mathcal{Q}^b(f,f) \, d\mathbf{v} = \int_{\mathbb{R}^d} \mathcal{Q}^b(f,f)\mathbf{v} \, d\mathbf{v} = \int_{\mathbb{R}^d} \mathcal{Q}^b(f,f)|\mathbf{v}|^2 \, d\mathbf{v} = 0. \tag{26}$$

It satisfies the celebrated Boltzmann's H-theorem:

$$-\int_{\mathbb{R}^d} \mathcal{Q}^b(f,f) \ln f \, d\mathbf{v} \ge 0,$$

which implies that the entropy is always non-decreasing. Furthermore, the following statements are equivalent

$$\int_{\mathbb{R}^d} \mathcal{Q}^b(f,f) \ln f \, d\mathbf{v} = 0 \iff \mathcal{Q}^b(f,f) = 0 \iff f = \mathcal{M}^b(\mathbf{v})_{(\rho(t,\mathbf{x},\mathbf{z}), \mathbf{u}(t,\mathbf{x},\mathbf{z}), T(t,\mathbf{x},\mathbf{z}))},$$

where $\mathcal{M}^b$ is the local equilibrium/Maxwellian defined by

$$\mathcal{M}^b = \frac{\rho}{(2\pi T)^{d/2}} e^{-\frac{(\mathbf{v}-\mathbf{u})^2}{2T}},$$

with ρ, $\mathbf{u}$, T being, respectively, the density, bulk velocity, and temperature:

$$\rho = \int_{\mathbb{R}^d} f \, d\mathbf{v}, \quad \mathbf{u} = \frac{1}{\rho} \int_{\mathbb{R}^d} f\mathbf{v} \, d\mathbf{v}, \quad T = \frac{1}{d\rho} \int_{\mathbb{R}^d} f|\mathbf{v} - \mathbf{u}|^2 \, d\mathbf{v}. \tag{27}$$

A widely used boundary condition for Boltzmann-like kinetic equations is the Maxwell boundary condition which is a linear combination of specular reflection and diffusion (particles are absorbed by the wall and then re-emitted according to a Maxwellian distribution of the wall). Specifically, for any boundary point $\mathbf{x} \in \partial\Omega$, let $n(\mathbf{x})$ be the unit normal vector to the boundary, pointed to the domain, then the in-flow boundary condition is given by

$$f(t, \mathbf{x}, \mathbf{v}, \mathbf{z}) = g(t, \mathbf{x}, \mathbf{v}, \mathbf{z}), \quad (\mathbf{v} - \mathbf{u}_w) \cdot n > 0,$$

with

$$g(t, \mathbf{x}, \mathbf{v}, \mathbf{z}) = (1 - \alpha)f(t, \mathbf{x}, \mathbf{v} - 2[(\mathbf{v} - \mathbf{u}_w) \cdot n]n, \mathbf{z})$$
$$+ \frac{\alpha}{(2\pi)^{\frac{d-1}{2}} T_w^{\frac{d+1}{2}}} e^{-\frac{|\mathbf{v} - \mathbf{u}_w|^2}{2T_w}} \int_{(\mathbf{v} - \mathbf{u}_w) \cdot n < 0} f(t, \mathbf{x}, \mathbf{v}, \mathbf{z})|(\mathbf{v} - \mathbf{u}_w) \cdot n| \, d\mathbf{v},$$

$$(28)$$

where $\mathbf{u}_w = \mathbf{u}_w(t, \mathbf{x}, \mathbf{z})$, $T_w = T_w(t, \mathbf{x}, \mathbf{z})$ are the velocity and temperature of the wall (boundary). The constant α ($0 \leq \alpha \leq 1$), which may depend on $\mathbf{z}$ as well, is the accommodation coefficient with $\alpha = 1$ corresponding to the purely diffusive boundary, and $\alpha = 0$ the purely specular reflective boundary.

2.5   Other Related Kinetic Models: A Glance

In addition to the above introduced equations, we mention a few related kinetic models. Interested readers may consult the survey papers [11, 20, 74] for details. First of all, the collision operator does not have to be the aforementioned forms: when the deviation angle θ is small, the Boltzmann collision integral (24) diverges and one has to consider its grazing collision limit—the Fokker-Planck-Landau operator [51], which is a diffusive operator relevant in the study of Coulomb interactions. When the quantum effect is non-negligible (particles behave as Bosons or Fermions), (11) or (24) needs to be modified to include an extra factor like $(1 \pm f)$, resulting in the so-called quantum or degenerate collision operators [19, 73]. Other generalizations such as the multi-species model [71] (system consists of more than one type of particles), inelastic model [75] (during collisions only the mass and momentum are conserved whereas the energy is dissipative, for example, in granular materials) are also possible. Secondly, the forcing term on the left hand side is not necessary as that shown in (10): generally one can couple the kinetic equation with the Maxwell equation where both electric and magnetic effects are present [72].

3   Coercivity, Hypocoercivity, Regularity and Long Time Behavior

Coercivity, or more generally hypocoercivity, describing the dissipative nature of the kinetic collision operators, plays important roles in the study of the solution of kinetic equations toward the local or global Maxwellian [74, 76]. For uncertain problems, one can extend such behavior to the random space, thus gives rise to regularity or long-time estimates in the random space of the solution, allowing one to quantify the long-time impact of the uncertainties for some statistical quantities of interest. In this section, we will review some of recent results in this direction, in particular, how such analysis can be used to understand the regularity and propagation of uncertainty for random kinetic equations.

In this section we will restrict our discussion to the one-dimensional random variable z with finite support I_z (e.g., uniform and beta distributions). Generalization to multi-dimensional random variables with finite support can be carried out in a similar fashion.

3.1 The Linear Transport Equation

To study the regularity and long-time behavior in the random space of the linear transport equation (1)–(3), we first recall the Hilbert space of the random variable

$$H(I_z;\ \omega\,\mathrm{d}z) = \left\{ f \mid I_z \rightarrow \mathbb{R}^+,\ \int_{I_z} f^2(z)\omega(z)\,\mathrm{d}z < +\infty \right\}, \qquad (29)$$

equipped with the inner product and norm defined as

$$\langle f, g \rangle_\omega = \int_{I_z} fg\,\omega(z)\,\mathrm{d}z, \quad \|f\|_\omega^2 = \langle f, f \rangle_\omega. \qquad (30)$$

We also define the kth order differential operator with respect to z as

$$D^k f(t, x, v, z) := \partial_z^k f(t, x, v, z), \qquad (31)$$

and the Sobolev norm in H as

$$\|f(t, x, v, \cdot)\|_{H^k}^2 := \sum_{\alpha \leq k} \|D^\alpha f(t, x, v, \cdot)\|_\omega^2. \qquad (32)$$

Finally, we introduce norms in space and velocity as follows,

$$\|f(t, \cdot, \cdot, \cdot)\|_\Gamma^2 := \int_Q \|f(t, x, v, \cdot)\|_\omega^2\,\mathrm{d}x\,\mathrm{d}v, \qquad t \geq 0, \qquad (33)$$

$$\|f(t, \cdot, \cdot, \cdot)\|_{\Gamma^k}^2 := \int_Q \|f(t, x, v, \cdot)\|_{H^k}^2\,\mathrm{d}x\,\mathrm{d}v, \qquad t \geq 0, \qquad (34)$$

where $Q = [0, 1] \times [-1, 1]$ denotes the domain in the phase space. For simplicity, we will suppress the dependence of t and just use $\|f\|_\Gamma$, $\|f\|_{\Gamma^k}$ in the following proof.

An important property of $\mathscr{L}$ is its coercivity, given in (7), based on which the following results were established in [47].

Theorem 1 (Uniform Regularity) *If for some integer $m \geq 0$,*

$$\|D^k \sigma(z)\|_{L^\infty} \leq C_\sigma, \qquad \|D^k f_0\|_\Gamma \leq C_0, \qquad k = 0, \ldots, m, \tag{35}$$

then the solution f to the linear transport equation (1)–(3), with $\sigma^a = S = 0$ and periodic boundary condition in x, satisfies,

$$\|D^k f\|_\Gamma \leq C, \qquad k = 0, \cdots, m, \qquad \forall t > 0, \tag{36}$$

where C_σ, C_0 and C are constants independent of ε.

The above theorem shows that, under some smoothness assumption on σ, the regularity of the initial data is preserved in time and the Sobolev norm of the solution is bounded uniformly in ε.

Theorem 2 (ε^2-Estimate on $[f] - f$) *With all the assumptions in Theorem 1 and furthermore, $\sigma \in W^{k,\infty} = \{\sigma \in L^\infty([0,1] \times I_z) | D^j \sigma \in L^\infty([0,1] \times I_z)$ for all $j \leq k\}$. For a given time $T > 0$, the following regularity result of $[f] - f$ holds:*

$$\|D^k([f] - f)\|_\Gamma^2 \leq e^{-\sigma_{\min} t / 2\varepsilon^2} \|D^k([f_0] - f_0)\|_\Gamma^2 + C' \varepsilon^2 \tag{37}$$

for any $t \in (0, T]$ and $0 \leq k \leq m_,$, where C' and C are constants independent of ε.

The first term on the right hand side of (37) is the behavior of the initial layer, which is damped exponentially in t/ε. After the initial layer, the high order derivatives in z of the difference between f and its local equilibrium $[f]$ is of $O(\varepsilon)$.

3.2 The Semiconductor Boltzmann Equation

The results in the previous subsection can be extended to the (linear) semiconductor Boltzmann equation by assuming $\phi = 0$ in (10).

Introduce the Hilbert space of the velocity variable $L_M^2 := L^2\left(\mathbb{R}^d; \dfrac{dv}{M^s(v)}\right)$, with the corresponding inner product $\langle \cdot, \cdot \rangle_{L_M^2}$ and norm $\|\cdot\|_{L_M^2}$. First, the collision operator $\mathscr{Q}^s$ has the following *coercivity* property for any $f \in L_M^2$ [69],

$$\langle \mathscr{Q}^s(f), f \rangle_{L_M^2} \leq -\sigma_{\min} \|f - \rho M^s\|_{L_M^2}^2, \tag{38}$$

Introduce the following norms

$$\|f(t, \cdot, \cdot, \cdot)\|_\Gamma^2 := \int_\Omega \int_{\mathbb{R}^d} \frac{\|f\|_\omega^2}{M(v)} \, dv \, dx,$$

$$\|f(t, \cdot, \cdot, \cdot)\|_{\Gamma^k}^2 := \int_\Omega \int_{\mathbb{R}^d} \frac{\|f\|_{H^k}^2}{M(v)} \, dv \, dx.$$

We assume a periodic boundary condition in space. The following results were proved in [59].

Theorem 3 (Uniform Regularity) *Assume for some integer $m \geq 0$,*

$$\|D^k\sigma\|_{L^\infty(v,z)} \leq C_\sigma, \qquad \|D^kf_0\|_\Gamma \leq C_0, \qquad k = 0, \cdots, m,$$

then the solution f to (10) satisfies

$$\|D^kf\|_\Gamma \leq C, \qquad k = 0, \cdots, m, \qquad \forall t > 0,$$

where C_σ, C_0 and C are constants independent of ε.

Theorem 4 (Estimate on $f - \rho M^s$) *With all the assumptions in Theorem 3, and in addition,*

$$\int_{\mathbb{R}^d} \int_{\mathbb{R}^d} (D^k\sigma)v^2 M^s(v)M^s(w)\,dwdv \leq C,$$

$$\left|\int_{\mathbb{R}^d} (D^k\sigma)M^s(w)\,dw\right| \leq C, \quad \|D^k(v \cdot \nabla_x f_0)\|_\Gamma \leq C,$$

for $k = 1, \cdots, m$, then

$$\|D^k(f - \rho M^s)\|_\Gamma^2 \leq e^{-\sigma_{\min}t/2\varepsilon^2} \|D^k(f_0 - \rho_0 M_0^s)\|_\Gamma^2 + C'\varepsilon^2 \leq C\varepsilon^2, \tag{39}$$

for any $t \in (0, T]$ and $0 \leq k \leq m$, where C' and C are constants independent of ε.

Differing from the isotropic scattering, for the anisotropic collision kernel, to obtain the decay rate of $f - \rho M^s$, an exponential decay estimate on $v \cdot \nabla_x f$ is needed [59].

3.3 General Linear Kinetic Equations

While the previous analysis gave decay estimates on the deviation between f and its *local equilibrium*, which can be difficult for more general kinetic equations, the use of *hypocoercivity* to estimate the deviation of f from the *global equilibrium* which is independent of t and x, helps one to deal with more general and even nonlinear equations. For general linear transport equation with *one* conserved quantity:

$$\partial_t f + \frac{1}{\varepsilon}\mathbf{v} \cdot \nabla_x f + \frac{1}{\varepsilon}\nabla_x\phi \cdot \nabla_v f = \frac{1}{\delta\varepsilon}\mathcal{Q}^l(f), \tag{40}$$

where the collision $\mathcal{Q}^l$ includes

- BGK operator $\mathcal{Q}^l = \sigma(x,z)(\Pi f - f)$, where Π is a projection operator onto the local equilibrium;

- Anisotropic scattering operator $\mathcal{Q}^l = \int [\sigma(v \rightarrow v^*, z)f(v^*) - \sigma(v^* \rightarrow v, z)f(v)]\,dv^*$, $\sigma > 0$.

Two regimes will be considered: the high-field regime ($\delta = 1$) and the parabolic regime ($\delta = \varepsilon$).

In [57] the following regularity result was established:

Theorem 5 *Let f be the solution to the kinetic equation (40), and assume the initial data has sufficient regularity with respect to z:* $\|\partial_z^l f_0\| \leq H^l$, *then:*

(1) $\|\partial_z^l f\| \leq Cl! \min\{e^{-\lambda_z t}C(t)^l, e^{(C-\lambda_z)t}2^{l-1}(1+H)^{l+1}\}$, *where C is a constant, C(t) is an algebraic function of t, and* $\lambda_z > 0$ *is uniformly bounded below from zero;*
(2) *f is analytic with uniform convergence radius* $\frac{1}{2(1+H)}$;
(3) *Both the exponential convergence in time and convergence radius are uniform with respect to* ε.

The proof of the results is based on the hypocoercivity property for deterministic equation [21], which gives the exponential decay in time, and a careful analysis of ε-independent decay rate.

3.4 The Vlasov-Poisson-Fokker-Planck System

We now discuss the (*nonlinear*) VPFP system (17)–(18). For simplicity, we only consider $\mathbf{x} = x \in (0, l)$ and $\mathbf{v} = v \in \mathbb{R}$ in one dimension. Define the L^2 space in the measure of

$$d\mu = d\mu(x, v, z) = \omega(z)\,dx\,dv\,dz. \tag{41}$$

With this measure, one has the corresponding Hilbert space with the following inner product and norms:

$$<f, g> = \int_\Omega \int_\mathbb{R} \int_{I_z} fg\,d\mu(x, v, z), \quad \text{or,} \quad <\rho, j> = \int_\Omega \int_{I_z} \rho j\,d\mu(x, z), \tag{42}$$

with norm

$$\|f\|^2 = <f, f>.$$

In order to get the convergence rate of the solution to the global equilibrium, define,

$$h = \frac{f - M^v}{\sqrt{M^v}}, \quad \sigma = \int_\mathbb{R} h\sqrt{M}\,dv, \quad u = \int_\mathbb{R} h v\sqrt{M}\,dv, \tag{43}$$

where h is the (microscopic) *fluctuation* around the equilibrium, σ is the (macroscopic) density fluctuation, and u is the (macroscopic) velocity fluctuation. Then the microscopic quantity h satisfies,

$$\varepsilon \delta \partial_t h + \beta v \partial_x h - \delta \partial_x \phi \partial_v h + \delta \frac{v}{2} \partial_x \phi h + \delta v \sqrt{M} \partial_x \phi = \mathscr{L}^F h, \qquad (44)$$

$$\partial_x^2 \phi = -\sigma, \qquad (45)$$

while the macroscopic quantities σ and u satisfy

$$\delta \partial_t \sigma + \partial_x u = 0, \qquad (46)$$

$$\varepsilon \delta \partial_t u + \varepsilon \partial_x \sigma + \varepsilon \int v^2 \sqrt{M} (1 - \Pi) \partial_x h dv + \delta \partial_x \phi \sigma + u + \delta \partial_x \phi = 0, \quad (47)$$

where $\mathscr{L}^F$ is the so-called linearized Fokker-Planck operator,

$$\mathscr{L}^F h = \frac{1}{\sqrt{M^v}} \mathscr{F} \left(M^v + \sqrt{M^v} h \right) = \frac{1}{\sqrt{M^v}} \partial_v \left(M^v \partial_v \left(\frac{h}{\sqrt{M^v}} \right) \right). \qquad (48)$$

Introduce projections onto $\sqrt{M^v}$ and $v \sqrt{M^v}$,

$$\Pi_1 h = \sigma \sqrt{M^v}, \quad \Pi_2 h = vu \sqrt{M^v}, \quad \Pi h = \Pi_1 h + \Pi_2 h. \qquad (49)$$

Furthermore, we also define the following norms and energies,

- Norms:
  - $\|h\|_{L^2(v)}^2 = \int_{\mathbb{R}} h^2 \, dv$,
  - $\|f\|_{H^m}^2 = \sum_{l=0}^{m} \|\partial_z^l f\|^2$, $\quad \|f\|_{H^1(x,z)}^2 = \|f\|^2 + \|\partial_x f\|^2 + \|\partial_z f\|^2$,
  - $\|h\|_v^2 = \int_{(0,l) \times \mathbb{R} \times I_z} (\partial_v h)^2 + (1 + |v|^2) h^2 \, d\mu(x, v, z)$, $\quad \|h\|_{H_v^m}^2 = \sum_{l=0}^{m} \|\partial_z^l h\|_v^2$;
- Energy terms:
  - $E_h^m = \|h\|_{H^m}^2 + \|\partial_x h\|_{H^{m-1}}^2$, $\quad E_\phi^m = \|\partial_x \phi\|_{H^m}^2 + \|\partial_x^2 \phi\|_{H^{m-1}}^2$;
- Dissipation terms:
  - $D_h^m = \|(1 - \Pi)h\|_{H^m}^2 + \|(1 - \Pi)\partial_x h\|_{H^{m-1}}^2$, $\quad D_\phi^m v = E_\phi^m v$,
  - $D_u^m = \|u\|_{H^m}^2 + \|\partial_x u\|_{H^{m-1}}^2$, $\quad D_\sigma^m = \|\sigma\|_{H^m}^2 + \|\partial_x \sigma\|_{H^{m-1}}^2$.

To get the regularity of the solution in the Hilbert space, one usually uses energy estimates. In order to balance the nonlinear term $\partial_x \phi \partial_v f$, and get a regularity independent of the small parameter ε (or depending on ε in a good way), one needs the hypocoercivity property from the collision operator. The hypocoercivity property one uses most commonly is

$$- \langle \mathscr{L}^F h, h \rangle \geq C \|(1 - \Pi_1)h\|^2, \qquad (50)$$

see [21, 76]. However, this is not enough for the non-linear case. We need the following stronger hypocoercivity (see [22]):

Proposition 1 *For $\mathscr{L}^F$ defined in (48),*

(a) $-\langle \mathscr{L}^F h, h \rangle = -\langle L(1 - \Pi)h, (1 - \Pi)h \rangle + \|u\|^2;$

(b) $-\langle \mathscr{L}^F(1-\Pi)h, (1-\Pi)h \rangle = \|\partial_v(1-\Pi)h\|^2 + \frac{1}{4}\|v(1-\Pi)h\|^2 - \frac{1}{2}\|(1-\Pi)h\|^2;$

(c) $-\langle \mathscr{L}^F(1 - \Pi)h, (1 - \Pi)h \rangle \geq \|(1 - \Pi)h\|^2;$

(d) *There exists a constant $\lambda_0 > 0$, such that the following hypocoercivity holds,*

$$- \langle \mathscr{L}^F h, h \rangle \geq \lambda_0 \|(1 - \Pi)h\|_v^2 + \|u\|^2, \tag{51}$$

and the largest $\lambda_0 = \frac{1}{7}$ in one dimension.

The following results were obtained in [43].

Theorem 6 *For the high field regime ($\delta = 1$), if*

$$E_h^m(0) + \frac{1}{\varepsilon^2} E_\phi^m(0) \leq C_0, \tag{52}$$

then,

$$E_h^m(t) \leq \frac{3}{\lambda_0} e^{-\frac{t}{\varepsilon^2}} \left(E_h^m(0) + \frac{1}{\varepsilon^2} E_\phi^m(0) \right), \qquad E_\phi^m(t) \leq \frac{3}{\lambda_0} e^{-t} \left(\varepsilon^2 E_h^m(0) + E_\phi^m(0) \right); \tag{53}$$

For the parabolic regime ($\delta = \varepsilon$), if

$$E_h^m(0) + \frac{1}{\varepsilon} E_\phi^m(0) \leq \frac{C_0}{\varepsilon}, \tag{54}$$

then,

$$E_h^m(t) \leq \frac{3}{\lambda_0} e^{-\frac{t}{\varepsilon}} \left(E_h^m(0) + \frac{1}{\varepsilon} E_\phi^m(0) \right), \qquad E_\phi^m(t) \leq \frac{3}{\lambda_0} e^{-t} \left(\varepsilon E_h^m(0) + E_\phi^m(0) \right). \tag{55}$$

Here $C_0 = 2\lambda_0^3/(80AC_1)^2, B = 48\sqrt{m} \binom{m}{[m/2]}$ is a constant only depending on m, $[m/2]$ is the smallest integer larger or equal to $\frac{m}{2}$, and C_1 is the Sobolev constant in one dimension, and $m \geq 1$.

These results show that the solution will converge to the global Maxwellian M^v. Since M^v is independent of z, one sees that the impact of the randomness dies out exponentially in time, in both asymptotic regimes.

The above theorem also leads to the following regularity result for the solution to VPFP system:

Theorem 7 *Under the same condition given in Theorem 6, for $x \in [0, l]$, one has*

$$\|f(t)\|_{H_z^m}^2 \leq \frac{3}{\lambda_0} E^m(0) + 2l^2, \tag{56}$$

where $E^m(0) = E_h^m(0) + \frac{1}{\varepsilon^2} E_\phi^m$.

This Theorem shows that the regularity of the initial data in the random space is preserved in time. Furthermore, the bound of the Sobolev norm of the solution is independent of the small parameter ε.

3.5 The Classical Nonlinear Boltzmann Equation

In this subsection, we consider the spatially homogeneous classical Boltzmann equation

$$\partial_t f = \mathcal{Q}^b(f, f) \tag{57}$$

subject to random initial data and random collision kernel

$$f(0, \mathbf{v}, z) = f^0(\mathbf{v}, z), \quad B = B(\mathbf{v}, \mathbf{v}_*, \eta, z), \quad z \in I_z.$$

We define the norms and operators:

$$\|f(t, \cdot, z)\|_{L_\mathbf{v}^p} = \left(\int_{\mathbb{R}^d} |f(t, \mathbf{v}, z)|^p \, d\mathbf{v} \right)^{1/p}, \quad \|f(t, \mathbf{v}, \cdot)\|_{L_z^2} = \left(\int_{I_z} f(t, \mathbf{v}, z)^2 \pi(z) dz \right)^{1/2},$$

$$\||f(t, \cdot, \cdot)\||_k = \sup_{z \in I_z} \left(\sum_{l=0}^k \|\partial_z^l f(t, \mathbf{v}, z)\|_{L_\mathbf{v}^2}^2 \right)^{1/2}.$$

$$\mathcal{Q}^b(g, h)(\mathbf{v}) = \int_{\mathbb{R}^d} \int_{S^{d-1}} B(\mathbf{v}, \mathbf{v}_*, \eta, z) \left[g(\mathbf{v}')h(\mathbf{v}_*') - g(\mathbf{v})h(\mathbf{v}_*) \right] d\eta \, d\mathbf{v}_*,$$

$$\mathcal{Q}_1^b(g, h)(\mathbf{v}) = \int_{\mathbb{R}^d} \int_{S^{d-1}} \partial_z B(\mathbf{v}, \mathbf{v}_*, \eta, z) \left[g(\mathbf{v}')h(\mathbf{v}_*') - g(\mathbf{v})h(\mathbf{v}_*) \right] d\eta \, d\mathbf{v}_*.$$

The regularity, studied in [34], relies on the following estimates of $\mathcal{Q}^b(g, h)$ and $\mathcal{Q}_1^b(g, h)$, which are standard results in the deterministic case [10, 58] and straightforward extension to the uncertain case:

Lemma 1 *Assume the collision kernel B depends on z linearly, B and $\partial_z B$ are locally integrable and bounded in z. If $g, h \in L^1_v \cap L^2_v$, then*

$$\|\mathcal{Q}^b(g,h)\|_{L^2_v}, \ \|\mathcal{Q}^b_1(g,h)\|_{L^2_v} \leq C_B \|g\|_{L^1_v} \|h\|_{L^2_v} , \qquad (58)$$

$$\|\mathcal{Q}^b(g,h)\|_{L^2_v}, \ \|\mathcal{Q}^b_1(g,h)\|_{L^2_v} \leq C_B \|g\|_{L^2_v} \|h\|_{L^2_v} , \qquad (59)$$

where the constant $C_B > 0$ depends only on B and $\partial_z B$.

We state the following theorem proved in [34].

Theorem 8 *Assume that B satisfies the assumption in Lemma 1, and $\sup_{z \in I_z} \|f^0\|_{L^1_v} \leq M$, $\||f^0\||_k < \infty$ for some integer $k \geq 0$. Then there exists a constant $C_k > 0$, depending only on C_B, M, T, and $\||f^0\||_k$ such that*

$$\||f\||_k \leq C_k, \qquad for \ any \quad t \in [0, T] . \qquad (60)$$

This result shows that, even for the nonlinear Boltzmann equation, the regularity of the initial data is preserved in time in the random space.

This result can be easily generalized to the full Boltzmann equation (23) with periodic or vanishing boundary condition in space, we omit the detail. Linear dependence of the collision kernel on the random variable can also be relaxed. See [34] for a general proof.

One should notice that if one considers the Euler regime (by putting an ε^{-1} in front of $\mathcal{Q}^b$, then C_k in (60) will depend on the reciprocal of ε, in addition to being a large k-dependent constant (which is already the case for the deterministic problem [24]). This estimate breaks down in the Euler limit when $\varepsilon \to 0$.

4 Generalized Polynomial Chaos Based Stochastic Galerkin (gPC-sG) Methods for Random Kinetic Equations

In the last two decades, a large variety of numerical methods have been developed in the field of uncertainty quantification (UQ) [27, 32, 55, 77]. Among these methods, the most popular ones are Monte-Carlo methods [64], stochastic collocation methods [4, 5, 78] and stochastic Galerkin methods [3, 5]. The idea of Monte-Carlo methods is to sample randomly in the random space, which results in halfth order convergence. Stochastic collocation methods use sample points on a well-designed grid, and one can evaluate the statistical moments by numerical quadratures. Stochastic Galerkin methods start from an orthonormal basis in the random space, and approximate functions by truncated polynomial chaos expansions. By the Galerkin projection, a deterministic system of the expansion coefficients can be obtained. While Monte-Carlo methods have advantage in very high dimensional random space, the other two methods can achieve spectral accuracy if one adopts the generalized polynomial chaos (gPC) basis [79], which is a great advantage if the dimension of the random space is not too high. In this paper we focus on low dimensional random space, and adopt the stochastic Galerkin (sG) approach.

In the gPC expansion, one approximates the solution of a stochastic problem via an orthogonal polynomial series [79] by seeking an expansion in the following form:

$$f(t, \mathbf{x}, \mathbf{v}, \mathbf{z}) \approx \sum_{|\mathbf{k}|=0}^{M} f_{\mathbf{k}}(t, \mathbf{x}, \mathbf{v}) \Phi_{\mathbf{k}}(\mathbf{z}) := f_M(t, \mathbf{x}, \mathbf{v}, \mathbf{z}), \qquad (61)$$

where $\mathbf{k} = (k_1, \ldots, k_n)$ is a multi-index with $|\mathbf{k}| = k_1 + \cdots + k_n$. $\{\Phi_{\mathbf{k}}(\mathbf{z})\}$ are from $\mathbb{P}_M^n$, the set of all n-variate polynomials of degree up to M and satisfy

$$< \Phi_{\mathbf{k}}, \Phi_{\mathbf{j}} >_\omega = \int_{I_z} \Phi_{\mathbf{k}}(\mathbf{z}) \Phi_{\mathbf{j}}(\mathbf{z}) \omega(\mathbf{z}) \, d\mathbf{z} = \delta_{\mathbf{k}\mathbf{j}}, \quad 0 \le |\mathbf{k}|, |\mathbf{j}| \le M.$$

Here $\delta_{\mathbf{k}\mathbf{j}}$ is the Kronecker delta function. The orthogonality with respect to $\omega(\mathbf{z})$, the probability density function of $\mathbf{z}$, then defines the orthogonal polynomials. For example, the Gaussian distribution defines the Hermite polynomials; the uniform distribution defines the Legendre polynomials, etc. Note that when the random dimension $n > 1$, an ordering scheme for multiple index can be used to re-order the polynomials $\{\Phi_{\mathbf{k}}(\mathbf{z}), 0 \le |\mathbf{k}| \le M\}$ into a single index $\{\Phi_k(\mathbf{z}), 1 \le k \le N_M = \dim(\mathbb{P}_M^n) = \binom{M+n}{M}\}$. Typically, the graded lexicographic order is used, see, for example, Section 5.2 of [77].

Now inserting (61) into a general kinetic equation

$$\begin{cases} \partial_t f + \mathbf{v} \cdot \nabla_{\mathbf{x}} f + \nabla_{\mathbf{x}} \phi \cdot \nabla_{\mathbf{v}} f = \mathcal{Q}(f), & t > 0, \ \mathbf{x} \in \Omega, \ \mathbf{v} \in \mathbb{R}^d, \ \mathbf{z} \in I_z, \\ f(0, \mathbf{x}, \mathbf{v}) = f^0(\mathbf{x}, \mathbf{v}), & \mathbf{x} \in \Omega, \ \mathbf{v} \in \mathbb{R}^d, \ \mathbf{z} \in I_z, \\ f(t, \mathbf{x}, \mathbf{v}) = g(t, \mathbf{x}, \mathbf{v}), & t \ge 0, \ \mathbf{x} \in \partial\Omega, \ \mathbf{v} \in \mathbb{R}^d, \ \mathbf{z} \in I_z. \end{cases} \quad (62)$$

Upon a standard Galerkin projection, one obtains for each $0 \le |\mathbf{k}| \le M$,

$$\begin{cases} \partial_t f_{\mathbf{k}} + \mathbf{v} \cdot \nabla_{\mathbf{x}} f_{\mathbf{k}} + \sum_{|\mathbf{j}|=0}^{M} \nabla_{\mathbf{x}} \phi_{\mathbf{k}\mathbf{j}} \cdot \nabla_{\mathbf{v}} f_{\mathbf{j}} = \mathcal{Q}_{\mathbf{k}}(f_M), & t > 0, \ \mathbf{x} \in \Omega, \ \mathbf{v} \in \mathbb{R}^d, \\ f_{\mathbf{k}}(0, \mathbf{x}, \mathbf{v}) = f_{\mathbf{k}}^0(\mathbf{x}, \mathbf{v}), & \mathbf{x} \in \Omega, \ \mathbf{v} \in \mathbb{R}^d, \\ f_{\mathbf{k}}(t, \mathbf{x}, \mathbf{v}) = g_{\mathbf{k}}(t, \mathbf{x}, \mathbf{v}), & t \ge 0, \ \mathbf{x} \in \partial\Omega, \ \mathbf{v} \in \mathbb{R}^d, \end{cases}$$

$$(63)$$

with

$$\mathcal{Q}_{\mathbf{k}}(f_M) := \int_{I_z} \mathcal{Q}(f_M)(t, \mathbf{x}, \mathbf{v}, \mathbf{z}) \Phi_{\mathbf{k}}(\mathbf{z}) \omega(\mathbf{z}) \, d\mathbf{z}, \quad \phi_{\mathbf{k}\mathbf{j}} := \int_{I_z} \phi(t, \mathbf{x}, \mathbf{z}) \Phi_{\mathbf{k}}(\mathbf{z}) \Phi_{\mathbf{j}}(\mathbf{z}) \omega(\mathbf{z}) \, d\mathbf{z},$$

$$f_{\mathbf{k}}^0 := \int_{I_z} f^0(\mathbf{x}, \mathbf{v}, \mathbf{z}) \Phi_{\mathbf{k}}(\mathbf{z}) \omega(\mathbf{z}) \, d\mathbf{z}, \quad g_{\mathbf{k}} := \int_{I_z} g(t, \mathbf{x}, \mathbf{v}, \mathbf{z}) \Phi_{\mathbf{k}}(\mathbf{z}) \omega(\mathbf{z}) \, d\mathbf{z}.$$

Here the collision operator $\mathscr{Q}(f_M)$ could be either linear or nonlinear depending on the specific problem. We also assume that the potential $\phi(t, \mathbf{x}, \mathbf{z})$ is given a priori for simplicity (the case that it is coupled to a Poisson equation can be treated similarly).

Therefore, one has a system of *deterministic* equations to solve and the unknowns are gPC coefficients $f_\mathbf{k}$, which are independent of $\mathbf{z}$. Mostly importantly, the resulting gPC-sG system is just a vector analogue of its deterministic counterpart, thus allowing straightforward extension of the existing deterministic kinetic solvers (of course special attention is needed for the collision operator which will be discussed later). Once the coefficients $f_\mathbf{k}$ are obtained through some numerical procedure, the statistical information such as the mean, covariance, standard deviation of the true solution f can be approximated as

$$\mathbb{E}[f] \approx f_0, \quad \text{Var}[f] \approx \sum_{|\mathbf{k}|=1}^{M} f_\mathbf{k}^2, \quad \text{Cov}[f] \approx \sum_{|\mathbf{i}|,|\mathbf{j}|=1}^{M} f_\mathbf{i} f_\mathbf{j}.$$

4.1 Property of the Collision Operator Under the gPC-sG Approximation

Due to the truncated approximation (61), the positivity of f is immediately lost. Thus some properties such as the H-theorem no longer holds under the gPC-sG approximation. Yet the conservation of the collision operator, for instance (26), is still valid (whose proof does not require the positivity of f). Normally these need to be analyzed based on the specific collision operator. We give a simple example here (see [40] for the proof).

Lemma 2 *For the semiconductor Boltzmann collision operator (13) with random scattering kernel $\sigma = \sigma(\mathbf{v}, \mathbf{v}_*, \mathbf{z})$, if its gPC-sG approximation $\mathscr{Q}_\mathbf{k}^s = 0$ for every $0 \leq |\mathbf{k}| \leq M$, then it admits a unique solution $f_\mathbf{k} = \rho_\mathbf{k} M^s(\mathbf{v})$, $0 \leq |\mathbf{k}| \leq M$, where $\rho_\mathbf{k} := \int_{\mathbb{R}^d} f_\mathbf{k} \, d\mathbf{v}$.*

This lemma is just a vector analogue of the property (15).

4.2 An Efficient Treatment of the Boltzmann Collision Operator Under the gPC-sG Approximation

As mentioned previously, numerical discretization of the gPC-sG system (63) for most kinetic equations does not present essential difficulties. In principle, any time and spatial discretization used for the deterministic, scalar kinetic equations can be generalized easily to the vectorized form. However, this is not the case for the collision operator, especially when it is nonlinear. To illustrate the idea, we use the classical Boltzmann collision operator as an example.

Under the gPC-sG approximation, the **k**th-mode of the classical Boltzmann collision operator (24) is given by

$$Q_{\mathbf{k}}^b(t, \mathbf{x}, \mathbf{v}) = \int_{I_{\mathbf{z}}} \mathcal{Q}^b(f_M, f_M)(t, \mathbf{x}, \mathbf{v}, \mathbf{z}) \Phi_{\mathbf{k}}(\mathbf{z}) \omega(\mathbf{z}) \, d\mathbf{z}$$

$$= \sum_{|\mathbf{i}|,|\mathbf{j}|=0}^{M} S_{\mathbf{kij}} \int_{\mathbb{R}^d} \int_{S^{d-1}} |\mathbf{v} - \mathbf{v}_*|^{\lambda} \left[f_{\mathbf{i}}(\mathbf{v}') f_{\mathbf{j}}(\mathbf{v}'_*) - f_{\mathbf{i}}(\mathbf{v}) f_{\mathbf{j}}(\mathbf{v}_*) \right] d\eta d\mathbf{v}_*,$$

$$(64)$$

with

$$S_{\mathbf{kij}} := \int_{I_{\mathbf{z}}} b_{\lambda}(\mathbf{z}) \Phi_{\mathbf{k}}(\mathbf{z}) \Phi_{\mathbf{i}}(\mathbf{z}) \Phi_{\mathbf{j}}(\mathbf{z}) \omega(\mathbf{z}) \, d\mathbf{z}, \qquad (65)$$

where we assumed that the collision kernel takes the form (25) with uncertainty in b_{λ}.

Note that the tensor $S_{\mathbf{kij}}$ does not depend on the solution $f_{\mathbf{k}}$, so it can be precomputed and stored for repeated use. But even so, the evaluation of $Q_{\mathbf{k}}^b$ still presents a challenge. A naive, direct computation for each t, $\mathbf{x}$, and $\mathbf{k}$ would result in $O(N_M^2 N_\eta^{d-1} N_{\mathbf{v}}^{2d})$ complexity, where $N_M = \binom{M+n}{M}$ is the dimension of $\mathbb{P}_M^n$, N_η is the number of discrete points in each angular direction, and $N_{\mathbf{v}}$ is the number of points in each velocity dimension. This is, if not impossible, prohibitively expensive.

In [34], we constructed a fast algorithm for evaluating (64). It was shown that the above direct cost $O(N_M^2 N_\eta^{d-1} N_{\mathbf{v}}^{2d})$ can be reduced to $\max\{O(R_{\mathbf{k}} N_\eta^{d-1} N_{\mathbf{v}}^d \log N_{\mathbf{v}}),$ $O(R_{\mathbf{k}} N_M N_{\mathbf{v}}^d)\}$ with $R_{\mathbf{k}} \leq N_M$ by leveraging the singular value decomposition (SVD) and the fast spectral method for the deterministic collision operator [62]. This is achieved in two steps.

First, for each fixed $\mathbf{k}$, decompose the symmetric matrix $(S_{\mathbf{kij}})_{N_M \times N_M}$ as (via a truncated SVD with desired accuracy)

$$S_{\mathbf{kij}} = \sum_{r=1}^{R_{\mathbf{k}}} U_{\mathbf{i}r}^{\mathbf{k}} V_{r\mathbf{j}}^{\mathbf{k}}.$$

Substituting it into (64) and rearranging terms, one gets

$$Q_{\mathbf{k}}^b(\mathbf{v}) = \sum_{r=1}^{R_{\mathbf{k}}} \int_{\mathbb{R}^d} \int_{S^{d-1}} |\mathbf{v} - \mathbf{v}_*|^{\lambda} \left[g_r^{\mathbf{k}}(\mathbf{v}') h_r^{\mathbf{k}}(\mathbf{v}'_*) - g_r^{\mathbf{k}}(\mathbf{v}) h_r^{\mathbf{k}}(\mathbf{v}_*) \right] d\eta d\mathbf{v}_*, \qquad (66)$$

with

$$g_r^{\mathbf{k}}(\mathbf{v}) := \sum_{|\mathbf{i}|=0}^{M} U_{\mathbf{i}r}^{\mathbf{k}} f_{\mathbf{i}}(\mathbf{v}), \qquad h_r^{\mathbf{k}}(\mathbf{v}) := \sum_{|\mathbf{i}|=0}^{M} V_{r\mathbf{i}}^{\mathbf{k}} f_{\mathbf{i}}(\mathbf{v}).$$

Hence one readily reduce the cost from $O(N_M^2 N_\eta^{d-1} N_v^{2d})$ to $\max\{O(R_k N_\eta^{d-1} N_v^{2d})$, $O(R_k N_M N_v^d)\}$, where $R_k \leq N_M$ is the numerical rank of matrix $(S_{kij})_{N_M \times N_M}$.

Next, note that (66) can be formally written as

$$Q_k^b(\mathbf{v}) = \sum_{r=1}^{R_k} \mathcal{Q}^b(g_r^k, h_r^k), \tag{67}$$

and $\mathcal{Q}^b$ is the deterministic collision operator (24) with kernel $B = |\mathbf{v} - \mathbf{v}_*|^\lambda$. In [62], a fast Fourier-spectral method in velocity variable $\mathbf{v}$ was developed for (24) in the case of 2D Maxwell molecule ($\lambda = 0$) and 3D hard-sphere molecule ($\lambda = 1$). Applying this method to (67) with slight modification, one can further reduce the cost from $\max\{O(R_k N_\eta^{d-1} N_v^{2d}), O(R_k N_M N_v^d)\}$ to $\max\{O(R_k N_\eta^{d-1} N_v^d \log N_v), O(R_k N_M N_v^d)\}$, see appendix of [34] for a detailed description (in practice, typically $N_\eta \ll N_v$ [23, 25]).

The above method has been extended to the Fokker-Planck-Landau collision operator in [36]. When the random variable is in high dimension, the problem suffers from the dimension curse. A wavelet based sparse grid method was introduced in [70], in which the matrix $(S_{kij})_{N_M \times N_M}$ is very sparse, and the computational cost can be significantly reduced.

4.3 A Spectral Accuracy Analysis

The regularity results presented previously can be used to establish the spectral convergence of the gPC-sG method. As in Sect. 3.5, we will restrict to the spatially homogeneous Boltzmann equation (57).

Using the orthonormal basis $\{\Phi_k(z)\}$, the solution f to (57) can be represented as

$$f(t, \mathbf{v}, z) = \sum_{k=0}^{\infty} \hat{f}_k(t, \mathbf{v}) \Phi_k(z), \quad \text{where} \quad \hat{f}_k(t, \mathbf{v}) = \int_{I_z} f(t, \mathbf{v}, z) \Phi_k(z) \omega(z) \, dz. \tag{68}$$

Let P_M be the projection operator defined as

$$P_M f(t, \mathbf{v}, z) = \sum_{k=0}^{M} \hat{f}_k(t, \mathbf{v}) \Phi_k(z).$$

Define the norms

$$\|f(t, \mathbf{v}, \cdot)\|_{H_z^k} = \left(\sum_{l=0}^{k} \|\partial_z^l f(t, \mathbf{v}, z)\|_{L_z^2}^2 \right)^{1/2},$$

$$\|f(t, \cdot, \cdot)\|_{L_{\mathbf{v},z}^2} = \left(\int_{I_z} \int_{\mathbb{R}^d} f(t, \mathbf{v}, z)^2 \, d\mathbf{v} \omega(z) \, dz \right)^{1/2}, \tag{69}$$

then one has the following projection error.

Lemma 3 *Assume z obeys uniform distribution, i.e., $z \in I_z = [-1, 1]$ and $\omega(z) = 1/2$ (so $\Phi_k(z)$ are Legendre polynomials). If $\|\|f^0\|\|_m$ is bounded, then*

$$\|f - P_M f\|_{L^2_{v,z}} \le \frac{C}{M^m}, \tag{70}$$

where C is a constant.

Given the gPC approximation of f:

$$f_M(t, \mathbf{v}, z) = \sum_{k=0}^{M} f_k(t, \mathbf{x}, \mathbf{v}) \Phi_k(z), \tag{71}$$

define the error function

$$e_M(t, \mathbf{v}, z) = P_M f(t, \mathbf{v}, z) - f_M(t, \mathbf{v}, z) := \sum_{k=0}^{M} e_k(t, \mathbf{v}) \Phi_k(z),$$

where $e_k = \hat{f}_k - f_k$. Then we have

Theorem 9 ([34]) *Assume the random variable z and initial data f^0 satisfy the assumption in Lemma 3, and the gPC approximation f_M is uniformly bounded in M, then*

$$\|f - f_M\|_{L^2_{v,z}} \le C(t) \left\{ \frac{1}{M^m} + \|e_M(0)\|_{L^2_{v,z}} \right\},$$

where C is a constant depending on t.

Remark 1 Clearly for spectral accuracy, one needs $\|e_M(0)\|_{L^2_{v,z}} \le C/M^m$. In practice, one chooses $f_k(0, \mathbf{v}) = \hat{f}_k(0, \mathbf{v})$, for all $k = 0, \cdots, M$, then $e_M(0) = 0$.

4.4 Numerical Examples

We now show two typical examples of the kinetic equations subject to random inputs. The first one is the classical Boltzmann equation with random boundary condition and the second one is the semiconductor Boltzmann equation with random force field. For simplicity, we assume the random variable z is one-dimensional and obeys uniform distribution.

Example 1 Consider the classical Boltzmann equation (23) with the following boundary condition: the gas is initially in a constant state

$$f^0(x, \mathbf{v}) = \frac{1}{2\pi T^0} e^{-\frac{v^2}{2T^0}}, \quad T^0 = 1, \quad x \in [0, 1].$$

At time $t = 0$, suddenly increase the wall temperature at left boundary by a factor of 2 with a small random perturbation:

$$T_w(z) = 2(T^0 + sz), \quad s = 0.2.$$

The purely diffusive Maxwell boundary condition is assumed at $x = 0$. For other implementation details, see [34].

The deterministic version of this problem has been considered by many authors [1, 23, 26], where they all observed that with the sudden rise of wall temperature, the gas close to the wall is heated and accordingly the pressure there rises sharply, which pushes the gas away from the wall and a shock wave propagates into the domain. The mean of our solution also exhibits a similar behavior, see Fig. 1. Meanwhile, the standard deviation of the solution allows us to predict the propagation of uncertainties quantitatively.

Example 2 Consider the semiconductor Boltzmann equation (10) coupled with a Poisson equation:

$$\beta(z)\partial_{xx}\phi = \rho - h(x, z), \quad \phi(0) = 0, \quad \phi(1) = 5, \quad x \in [0, 1],$$

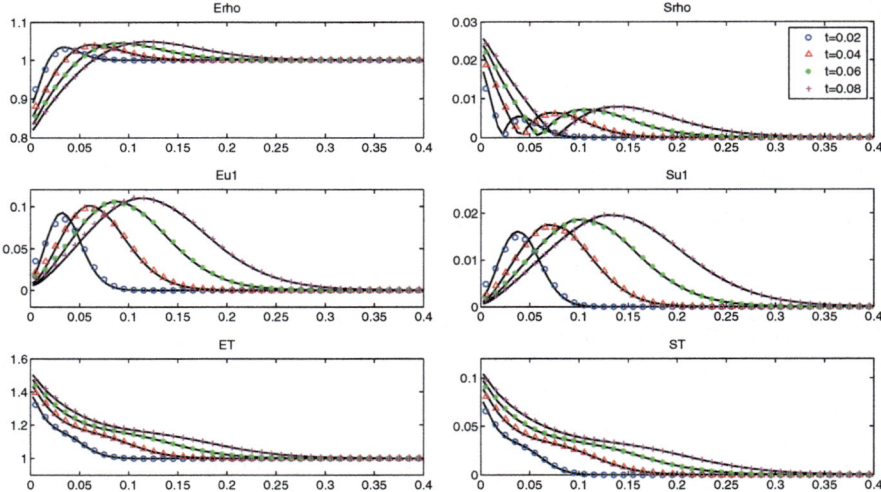

Fig. 1 Example 1. Left column: mean of density, bulk velocity (first component), and temperature. Right column: standard deviation of density, bulk velocity (first component), and temperature. Solid line: stochastic collocation with $N_z = 20, N_v = 64, N_\eta = 8, N_x = 200$. Other legends are the 7-th order gPC-sG solutions at different time with $N_v = 32, N_\eta = 4, N_x = 100$

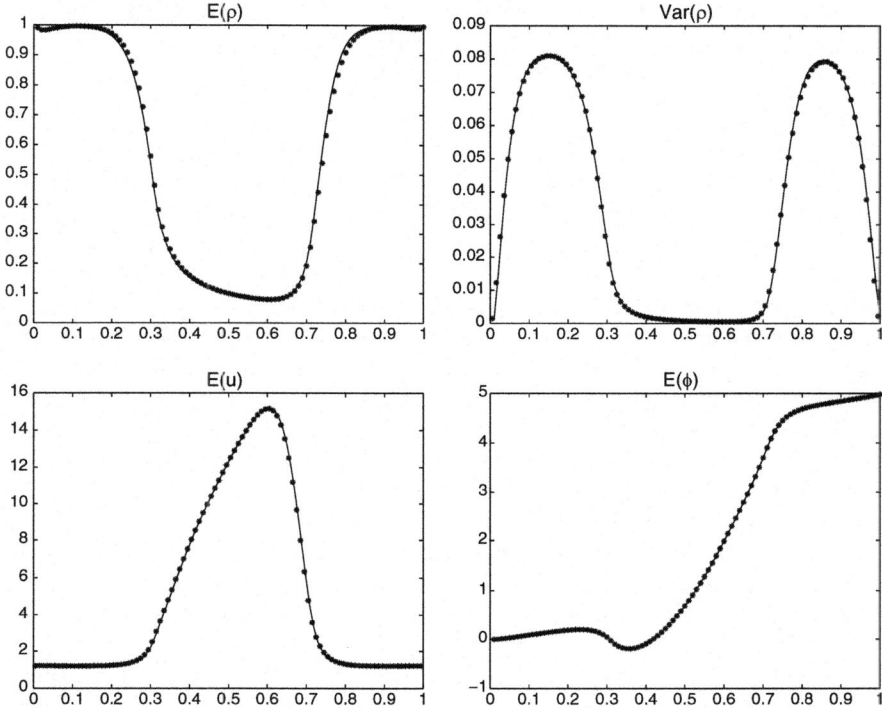

Fig. 2 Example 2. First row: mean and variance of ρ. Second row: mean of velocity u and potential ϕ. Time $t = 0.05$, $\Delta x = 0.01$, $\Delta t = 10^{-5}$, $\varepsilon = 0.001$. Star: 4-th order gPC-sG solutions. Solid line: the reference solutions obtained by stochastic collocation

where we assume the scaled Debye length $\beta(z)$ and the doping profile $h(x, z)$ are subject to uncertainty:

$$\beta(z) = 0.002(1 + 0.2z),$$

$$c(x, z) = \left(1 - (1 - s_0)\rho(0, t = 0)\left[\tanh\left(\frac{x - x_1}{s}\right) - \tanh\left(\frac{x - x_2}{s}\right)\right]\right)(1 + 0.5z),$$

with $s = 0.02$, $s_0 = (1 - 0.001)/2$, $x_1 = 0.3$, $x_2 = 0.7$. For other implementation details, see [40].

The 4-th order gPC solutions and the reference solutions obtained by stochastic collocation are shown in Fig. 2, and they are in good agreement.

5 Stochastic Asymptotic-Preserving (sAP) Schemes for Random Kinetic Equations in Diffusive Scalings

Kinetic equations often have scaling parameters (such as the Knudsen number ε) that asymptotically lead kinetic equations to their hydrodynamic or diffusion limit equations. When ε is small, numerically solving the kinetic equations is challenging since time and spatial discretizations need to resolve ε. Asymptotic-preserving (AP) schemes are those that mimic the asymptotic transitions from kinetic equations to their hydrodynamic/diffusion limits in the discrete setting [35, 37, 38, 53, 54]. Starting from the mid-1990s, the development of AP schemes for such problems has generated many interests, see, for example, [29, 30, 39, 44, 45, 49, 56]. The AP strategy has been proved to be a powerful and robust technique to address multiscale problems in many kinetic problems. The main advantage of AP schemes is that they are very efficient even when ε is small, since they do not need to resolve the small scales numerically, and yet can still capture the macroscopic behavior governed by the limiting macroscopic equations. Indeed, it was proved, in the case of linear transport with a diffusive scaling, an AP scheme converges uniformly with respect to the scaling parameter [29]. This is expected to be true for all AP schemes [38], although specific proofs are needed for specific problems. AP schemes avoid the difficulty of coupling a microscopic solver with a macroscopic one, as the micro solver *automatically* becomes a macro solver as $\varepsilon \to 0$.

Here we are interested in the scenario when the uncertainty (random inputs) and small scaling both present in a kinetic equation. Since the sG method makes the random kinetic equations into deterministic systems which are vector analogue of the original scalar deterministic kinetic equations, one can naturally utilize the deterministic AP machinery to solve the sG system to achieve the desired AP goals. To this aim, the notion of *stochastic asymptotic preserving (sAP)* was introduced in [46]. A scheme is sAP if a sG method for the random kinetic equation becomes a sG approximation for the limiting macroscopic, random (hydrodynamic or diffusion) equation as $\varepsilon \to 0$, with highest gPC degree, mesh size and time step all held fixed. Such schemes guarantee that even for $\varepsilon \to 0$, *all* numerical parameters, including the number of gPC modes, can be chosen only for accuracy requirement and *independent* of ε.

Next we use the linear transport equation (1) as an example to derive a sAP scheme. It has the merit that rigorous convergence and sAP theory can be established, see [47].

5.1 A sAP-sG Method for the Linear Transport Equation

We assume the complete orthogonal polynomial basis in the Hilbert space $H(I_z; \omega(z)\,\mathrm{d}z)$ corresponding to the weight $\omega(z)$ is $\{\phi_i(z), i = 0, 1, \cdots, \}$, where

$\phi_i(z)$ is a polynomial of degree i and satisfies the orthonormal condition:

$$\langle \phi_i, \phi_j \rangle_\omega = \int \phi_i(z)\phi_j(z)\omega(z)\,dz = \delta_{ij}.$$

Here $\phi_0(z) = 1$, and δ_{ij} is the Kronecker delta function. Since the solution $f(t, \cdot, \cdot, \cdot)$ is defined in $L^2([0, 1] \times [-1, 1] \times \mathbb{I}_z; d\mu)$, one has the gPC expansion

$$f(t, x, v, z) = \sum_{i=0}^{\infty} f_i(t, x, v)\,\phi_i(z), \quad \hat{f} = (f_i)_{i=0}^{\infty} := (\bar{f}, \hat{f}_1).$$

The mean and variance of f can be obtained from the expansion coefficients as

$$\bar{f} = E(f) = \int_{I_z} f\omega(z)\,dz = f_0, \quad \text{var}(f) = |\hat{f}_1|^2.$$

Denote the sG solution by

$$f_M = \sum_{i=0}^{M} f_i \phi_i, \quad \hat{f}^M = (f_i)_{i=0}^{M} := (\bar{f}, \hat{f}_1^M), \tag{72}$$

from which one can extract the mean and variance of f_M from the expansion coefficients as

$$E(f_M) = \bar{f}, \quad \text{var}(f_M) = |\hat{f}_1^M|^2 \leq \text{var}(f).$$

Furthermore, we define

$$\sigma_{ij} = \langle \phi_i, \sigma\phi_j \rangle_\omega, \quad \Sigma = (\sigma_{ij})_{M+1,M+1},$$
$$\sigma_{ij}^a = \langle \phi_i, \sigma^a\phi_j \rangle_\omega, \quad \Sigma^a = (\sigma_{ij}^a)_{M+1,M+1},$$

for $0 \leq i, j \leq M$. Let Id be the $(M + 1) \times (M + 1)$ identity matrix. Σ, Σ^a are symmetric positive-definite matrices satisfying [77]

$$\Sigma \geq \sigma_{\min} \text{ Id}.$$

If one applies the gPC ansatz (72) into the transport equation (1), and conduct the Galerkin projection, one obtains

$$\varepsilon \partial_t \hat{f} + v \partial_x \hat{f} = -\frac{1}{\varepsilon}(I - [\cdot])\Sigma \hat{f} - \varepsilon \Sigma^a \hat{f} - \hat{S}, \tag{73}$$

where $\hat{S}$ is defined similarly as (72).

We now use the micro-macro decomposition [56]:

$$\hat{f}(t, x, v, z) = \hat{\rho}(t, x, z) + \varepsilon \hat{g}(t, x, v, z), \tag{74}$$

where $\hat{\rho} = [\hat{f}]$ and $[\hat{g}] = 0$, in (73) to get

$$\partial_t \hat{\rho} + \partial_x [v \hat{g}] = -\Sigma^a \hat{\rho} + \hat{S}, \tag{75a}$$

$$\partial_t \hat{g} + \frac{1}{\varepsilon}(I - [.])(v \partial_x \hat{g}) = -\frac{1}{\varepsilon^2} \Sigma \hat{g} - \Sigma^a \hat{g} - \frac{1}{\varepsilon^2} v \partial_x \hat{\rho}, \tag{75b}$$

with initial data

$$\hat{\rho}(0, x, z) = \hat{\rho}_0(x, z), \quad \hat{g}(0, x, v, z) = \hat{g}_0(x, v, z).$$

It is easy to see that system (75) formally has the diffusion limit as $\varepsilon \to 0$:

$$\partial_t \hat{\rho} = \partial_x(K \partial_x \hat{\rho}) - \Sigma^a \hat{\rho} + \hat{S}, \tag{76}$$

where

$$K = \frac{1}{3} \Sigma^{-1}. \tag{77}$$

This is the sG approximation to the random diffusion equation (8)–(9). Thus the gPC approximation is sAP in the sense of [46].

One can easily derive the following energy estimate for system (75)

$$\int_0^1 \hat{\rho}(t, x)^2 \, dx + \frac{\varepsilon^2}{2} \int_0^1 \int_{-1}^1 \hat{g}(t, x, v)^2 \, dv \, dx$$

$$\leq \int_0^1 \hat{\rho}(0, x)^2 \, dx + \frac{\varepsilon^2}{2} \int_0^1 \int_{-1}^1 \hat{g}(0, x, v)^2 \, dv \, dx.$$

Let f be the solution to the linear transport equation (1)–(2). Use the Mth order projection operator P_M, the error arisen from the gPC-sG can be split into two parts r_N and e_N,

$$f - f_M = f - P_M f + P_M f - f_M := r_M + e_M, \tag{78}$$

where $r_M = f - P_M f$ is the truncation error, and $e_M = P_M f - f_M$ is the projection error.

Here we summarize the results of [47].

Lemma 4 (Truncation Error) *Under all the assumption in Theorems 1 and 2, we have for $t \in (0, T]$ and any integer $k = 0, \ldots, m$,*

$$\|r_M\|_\Gamma \leq \frac{C_1}{M^k}. \tag{79}$$

Moreover,

$$\| [r_M] - r_M \|_\Gamma \leq \frac{C_2}{M^k}\varepsilon, \tag{80}$$

where C_1 and C_2 are independent of ε.

Lemma 5 (Projection Error) *Under all the assumptions in Theorems 1 and 2, we have for $t \in (0, T]$ and any integer $k = 0, \ldots, m$,*

$$\|e_M\|_\Gamma \leq \frac{C(T)}{M^k}, \tag{81}$$

where $C(T)$ is a constant independent of ε.

Combining the above lemmas gives the uniform (in ε) convergence theorem:

Theorem 10 *If for some integer $m \geq 0$,*

$$\|\sigma(z)\|_{H^k} \leq C_\sigma, \quad \|D^k f_0\|_\Gamma \leq C_0, \quad \|D^k(\partial_x f_0)\|_\Gamma \leq C_x, \quad k = 0, \ldots, m, \tag{82}$$

then the error of the sG method is

$$\|f - f_M\|_\Gamma \leq \frac{C(T)}{M^k}, \tag{83}$$

where $C(T)$ is a constant independent of ε.

Theorem 10 gives a uniformly in ε spectral convergence rate, thus one can choose M independent of ε, a very strong sAP property. Such a result is also obtained with the anisotropic scattering case, for the linear semiconductor Boltzmann equation (10) [59].

5.2 A Full Discretization

As pointed out in [46], and also seen in Sect. 4, by using the gPC-sG formulation, one obtains a vector version of the original deterministic transport equation. This enables one to use the deterministic AP methodology. In this paper, we adopt the micro-macro decomposition based AP scheme developed in [56] for the gPC-sG system (75).

We take a uniform grid $x_i = ih, i = 0, 1, \cdots N$, where $h = 1/N$ is the grid size, and time steps $t^n = n\Delta t$. ρ_i^n is the approximation of ρ at the grid point (x_i, t^n) while $g_{i+\frac{1}{2}}^{n+1}$ is defined at a staggered grid $x_{i+1/2} = (i + 1/2)h, i = 0, \cdots N - 1$.

The fully discrete scheme for the gPC system (75) is

$$
\frac{\hat{\rho}_i^{n+1} - \hat{\rho}_i^n}{\Delta t} + \left[v \frac{\hat{g}_{i+\frac{1}{2}}^{n+1} - \hat{g}_{i-\frac{1}{2}}^{n+1}}{\Delta x} \right] = -\Sigma_i^a \hat{\rho}_i^{n+1} + \hat{S}_i, \tag{84a}
$$

$$
\frac{\hat{g}_{i+\frac{1}{2}}^{n+1} - \hat{g}_{i+\frac{1}{2}}^n}{\Delta t} + \frac{1}{\varepsilon \Delta x} (I - [.]) \left(v^+ (\hat{g}_{i+\frac{1}{2}}^n - \hat{g}_{i-\frac{1}{2}}^n) + v^- (\hat{g}_{i+\frac{3}{2}}^n - \hat{g}_{i+\frac{1}{2}}^n) \right) \tag{84b}
$$

$$
= -\frac{1}{\varepsilon^2} \Sigma_i \hat{g}_{i+\frac{1}{2}}^{n+1} - \Sigma^a \hat{g}_{i+\frac{1}{2}}^{n+1} - \frac{1}{\varepsilon^2} v \frac{\hat{\rho}_{i+1}^n - \hat{\rho}_i^n}{\Delta x}.
$$

It has the formal diffusion limit when $\varepsilon \to 0$ given by

$$
\frac{\hat{\rho}_i^{n+1} - \hat{\rho}_i^n}{\Delta t} - K \frac{\hat{\rho}_{i+1}^n - 2\hat{\rho}_i^n + \hat{\rho}_{i-1}^n}{\Delta x^2} = -\Sigma_i^a \hat{\rho}_i^{n+1} + \hat{S}_i, \tag{85}
$$

where $K = \frac{1}{3} \Sigma^{-1}$. This is the fully discrete sG scheme for (76). Thus the fully discrete scheme is sAP.

One important property for an AP scheme is to have a stability condition independent of ε, so one can take $\Delta t \gg O(\varepsilon)$. The next theorem from [47] answers this question.

Theorem 11 *Assume $\sigma^a = S = 0$. If Δt satisfies the following CFL condition*

$$
\Delta t \leq \frac{\sigma_{\min}}{3} \Delta x^2 + \frac{2\varepsilon}{3} \Delta x, \tag{86}
$$

then the sequences $\hat{\rho}^n$ and $\hat{g}^n$ defined by scheme (84) satisfy the energy estimate

$$
\Delta x \sum_{i=0}^{N-1} \left((\hat{\rho}_i^n)^2 + \frac{\varepsilon^2}{2} \int_{-1}^1 \left(\hat{g}_{i+\frac{1}{2}}^n \right)^2 dv \right) \leq \Delta x \sum_{i=0}^{N-1} \left((\hat{\rho}_i^0)^2 + \frac{\varepsilon^2}{2} \int_{-1}^1 \left(\hat{g}_{i+\frac{1}{2}}^0 \right)^2 dv \right)
$$

for every n, and hence the scheme (84) is stable.

Since the right hand side of (86) has a lower bound when $\varepsilon \to 0$ (and the lower bound being that of a stability condition of the discrete diffusion equation (85)), the scheme is asymptotically stable and Δt remains finite even if $\varepsilon \to 0$.

A discontinuous Galerkin method based sAP scheme for the same problem was developed in [17], where uniform stability and rigorous sAP property were also proven.

5.3 Numerical Examples

We now show one example from [47] to illustrate the sAP properties of the scheme. For simplicity, we again assume the random variable z is one-dimensional and obeys uniform distribution.

Example 3 Consider the linear transport equation (1) with $\sigma^a = S = 0$ and random coefficient

$$\sigma(z) = 2 + z,$$

subject to zero initial condition $f(0, x, v, z) = 0$ and boundary condition

$$f(t, 0, v, z) = 1, \quad v \geq 0; \qquad f(t, 1, v, z) = 0, \quad v \leq 0.$$

When $\varepsilon \to 0$, the limiting random diffusion equation is

$$\partial_t \rho = \frac{1}{3\sigma(z)} \partial_{xx}\rho, \tag{87}$$

with initial and boundary conditions:

$$\rho(0, x, z) = 0, \quad \rho(t, 0, z) = 1, \quad \rho(t, 1, z) = 0.$$

The analytical solution for (87) with the given initial and boundary conditions is

$$\rho(t, x, z) = 1 - \text{erf}\left(\frac{x}{\sqrt{\frac{4}{3\sigma(z)}t}}\right). \tag{88}$$

When ε is small, we use this as the reference solution, as it is accurate with an error of $O(\varepsilon^2)$. For other implementation details, see [47].

In Fig. 3, we plot the errors in mean and standard deviation of the gPC numerical solutions at $t = 0.01$ with different gPC orders M. Three sets of results are included: solutions with $\Delta x = 0.04$ (squares), $\Delta x = 0.02$ (circles), $\Delta x = 0.01$ (stars). We always use $\Delta t = 0.0002/3$. One observes that the errors become smaller with finer mesh. One can see that the solutions decay rapidly in M and then saturate where spatial discretization error dominates. It is then obvious that the errors due to gPC expansion can be neglected at order $M = 4$ even for $\varepsilon = 10^{-8}$. From this simple example, we can see that using the properly designed sAP scheme, the time, spatial, and random domain discretizations can be chosen independently of the small parameter ε.

In Fig. 4, we examine the difference between the solution at $t = 0.01$ obtained by the 4th-order gPC method with $\Delta x = 0.01$, $\Delta t = \Delta x^2/12$ and the limiting analytical solution (88). As expected, we observe the differences become smaller as ε is smaller in a quadratic fashion, before the numerical errors become dominant. This, on the other hand, shows the sAP scheme works uniformly for different ε.

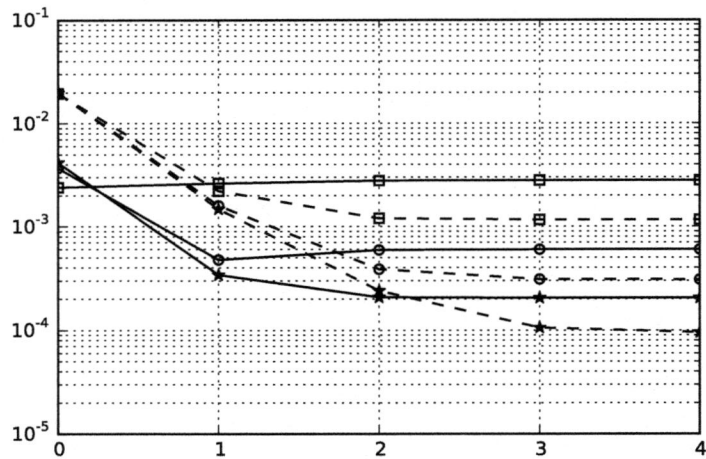

Fig. 3 Example 3. Errors of the mean (solid line) and standard deviation (dash line) of ρ with respect to the gPC order M at $\varepsilon = 10^{-8}$: $\Delta x = 0.04$ (squares), $\Delta x = 0.02$ (circles), $\Delta x = 0.01$ (stars). $\Delta t = 0.0002/3$

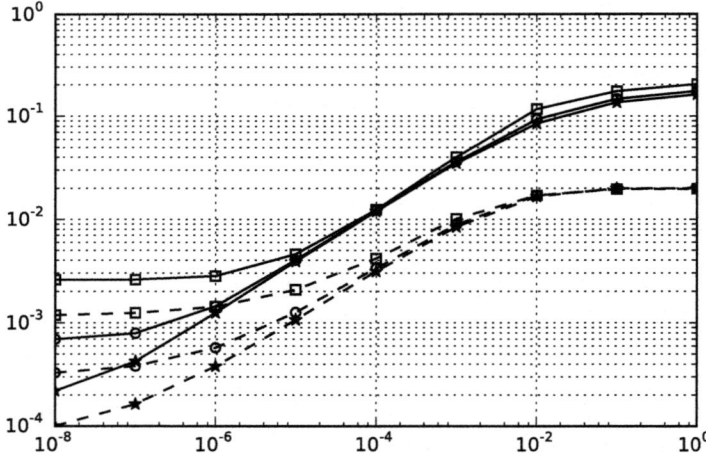

Fig. 4 Example 3. Differences in the mean (solid line) and standard deviation (dash line) of ρ with respect to ε^2, between the limiting analytical solution (88) and the 4th-order gPC solution with $\Delta x = 0.04$ (squares), $\Delta x = 0.02$ (circles) and $\Delta x = 0.01$ (stars)

6 Conclusion and Open Problems

Using the classical Boltzmann equation, linear Boltzmann equations and Vlasov-Poisson-Fokker-Planck system as prototype examples, we have surveyed recent development of uncertainty quantification (UQ) for kinetic equations. The uncertainties for such equations typically come from collision/scattering kernels, boundary data, initial data, forcing terms, among others. We proved the regularity in the random space and then adopted the generalized polynomial chaos based stochastic Galerkin (gPC-sG) approach to handle the random inputs which could yield spectral accuracy, under some regularity assumption on the initial data and random coefficients. Various theoretical and computational issues with respect to the collision operator were studied. When the kinetic equation has diffusive scaling that asymptotically leads to a diffusion equation, we constructed the stochastic Asymptotic-Preserving (sAP) scheme which allows numerical discretization including the gPC order to be chosen independently of the small parameter, hence is highly efficient in diffusive regime.

UQ for kinetic equations is a fairly recent research field, and many interesting problems remain open. We list a few such problems here:

- Nonlinear kinetic equations. Although sG or sAP schemes have been introduced for some nonlinear kinetic equations, for example the Boltzmann equation [34], the Landau equation [36], the radiative heat transfer equations [41], disperse two-phase kinetic-fluid model [42], rigorous analysis—such as regularity, long-time and small ε behavior, spectral convergence, etc.—has been lacking. In particular, for the Boltzmann equation, the behavior of the sG scheme in the Euler regime is not understood.
- High dimensional random space. When the dimension of the random parameter **z** is moderate, sparse grids have been introduced [36, 70] using wavelet approximations. Since wavelet basis does not have high order accuracy, it remains to construct sparse grids with high (or spectral) order of accuracy in the random space. When the random dimension is much higher, new methods need to be introduced to reduce the dimension.
- Study of sampling based methods such as collocation and multi-level Monte-Carlo methods. In practice, sampling based non-intrusive methods are attractive since they are based on the deterministic, or legacy codes. So far there has been no analysis done for the stochastic collocation methods for random kinetic equations. Moreover, multi-level Monte-Carlo method could significantly reduce the cost of sampling based methods [28]. Its application to kinetic equations with uncertainty remains to be investigated.

Despite at its infancy, due to the good regularity and asymptotic behavior in the random space for kinetic equations with uncertain random inputs, the UQ for kinetic equations is a promising research direction that deserves more development in their mathematical theory, efficient numerical methods, and applications. Moreover, since the random parameters in uncertain kinetic equations share some properties of the velocity variable for a kinetic equation, the ideas from kinetic theory can be very useful for UQ [18], and vice versa, thus the marriage of the two fields can be very fruitful.

Acknowledgements J. Hu's research was supported by NSF grant DMS-1620250 and NSF CAREER grant DMS-1654152. S. Jin's research was supported by NSF grants DMS-1522184 and

DMS-1107291: RNMS KI-Net, NSFC grant No. 91330203, and the Office of the Vice Chancellor for Research and Graduate Education at the University of Wisconsin-Madison with funding from the Wisconsin Alumni Research Foundation.

References

1. K. Aoki, Y. Sone, K. Nishino, H. Sugimoto, Numerical analysis of unsteady motion of a rarefied gas caused by sudden changes of wall temperature with special interest in the propagation of a discontinuity in the velocity distribution function, in *Rarefied Gas Dynamics*, ed. by A.E. Beylich (VCH, Weinheim, 1991), pp. 222–231
2. A. Arnold, J.A. Carrillo, I. Gamba, C.-W. Shu, Low and high field scaling limits for the Vlasov- and Wigner-Poisson-Fokker-Planck systems. Transp. Theory Stat. Phys. **30**, 121–153 (2001)
3. I. Babuska, R. Tempone, G.E. Zouraris, Galerkin finite element approximations of stochastic elliptic differential equations. SIAM J. Numer. Anal. **42**, 800–825 (2004)
4. I. Babuska, F. Nobile, R. Tempone, A stochastic collocation method for elliptic partial differential equations with random input data. SIAM J. Numer. Anal. **45**(3), 1005–1034 (2007)
5. J. Back, F. Nobile, L. Tamellini, R. Tempone, Stochastic spectral Galerkin and collocation methods for PDEs with random coefficients: a numerical comparison, in *Spectral and High Order Methods for Partial Differential Equations*, ed. by E.M. Rønquist, J.S. Hesthaven (Springer, Berlin, Heidelberg, 2011)
6. C. Bardos, R. Santos, R. Sentis, Diffusion approximation and computation of the critical size. Trans. Am. Math. Soc. **284**(2), 617–649 (1984)
7. A. Bensoussan, J.-L. Lions, G.C. Papanicolaou, Boundary layers and homogenization of transport processes. Publ. Res. Inst. Math. Sci. **15**(1), 53–157 (1979)
8. P.R. Berman, J.E.M. Haverkort, J.P. Woerdman, Collision kernels and transport coefficients. Phys. Rev. A **34**, 4647–4656 (1986)
9. G.A. Bird, *Molecular Gas Dynamics and the Direct Simulation of Gas Flows* (Clarendon Press, Oxford, 1994)
10. F. Bouchut, L. Desvillettes, A proof of the smoothing properties of the positive part of Boltzmann's kernel. Rev. Mat. Iberoamericana **14**, 47–61 (1998)
11. C. Cercignani, *The Boltzmann Equation and Its Applications* (Springer, New York, 1988)
12. C. Cercignani, *Rarefied Gas Dynamics: From Basic Concepts to Actual Calculations* (Cambridge University Press, Cambridge, 2000)
13. C. Cercignani, R. Illner, M. Pulvirenti, *The Mathematical Theory of Dilute Gases* (Springer, Berlin, 1994)
14. S. Chandrasekhar, Stochastic problems in physics and astronomy. Rev. Mod. Phys. **15**, 1–89 (1943)
15. S. Chandrasekhar, *Radiative Transfer* (Dover, New York, 1960)
16. S. Chapman, T.G. Cowling, *The Mathematical Theory of Non-uniform Gases*, 3rd edn. (Cambridge University Press, Cambridge, 1991)
17. Z. Chen, L. Liu, L. Mu, DG-IMEX stochastic Galerkin schemes for linear transport equation with random inputs and diffusive scalings. J. Sci. Comput. **73**, 566–592 (2017)
18. H. Cho, D. Venturi, G.E. Karniadakis, Numerical methods for high-dimensional probability density function equations. J. Comput. Phys. **305**, 817–837 (2016)
19. P. Degond, *Mathematical Modelling of Microelectronics Semiconductor Devices*. AMS/IP Studies in Advanced Mathematics, vol. 15 (AMS and International Press, Providence, 2000), pp. 77–110
20. G. Dimarco, L. Pareschi, Numerical methods for kinetic equations. Acta Numer. **23**, 369–520 (2014)
21. J. Dolbeault, C. Mouhot, C. Schmeiser, Hypocoercivity for linear kinetic equations conserving mass. Trans. Am. Math. Soc. **367**(6), 3807–3828 (2015)

22. R. Duan, M. Fornasier, G. Toscani, A kinetic flocking model with diffusion. Commun. Math. Phys. **300**(1), 95–145 (2010)
23. F. Filbet, On deterministic approximation of the Boltzmann equation in a bounded domain. Multiscale Model. Simul. **10**, 792–817 (2012)
24. F. Filbet, C. Mouhot, Analysis of spectral methods for the homogeneous Boltzmann equation. Trans. Am. Math. Soc. **363**, 1974–1980 (2011)
25. F. Filbet, C. Mouhot, L. Pareschi, Solving the Boltzmann equation in NlogN. SIAM J. Sci. Comput. **28**, 1029–1053 (2006)
26. I. Gamba, S.H. Tharkabhushanam, Shock and boundary structure formation by spectral-Lagrangian methods for the inhomogeneous Boltzmann transport equation. J. Comput. Math. **28**, 430–460 (2010)
27. R.G. Ghanem, P.D. Spanos, *Stochastic Finite Elements: A Spectral Approach* (Springer, New York, 1991)
28. M.B. Giles, Multilevel Monte Carlo methods. Acta Numer. **24**, 259 (2015)
29. F. Golse, S. Jin, C.D. Levermore, The convergence of numerical transfer schemes in diffusive regimes. I. Discrete-ordinate method. SIAM J. Numer. Anal. **36**(5), 1333–1369 (1999)
30. L. Gosse, G. Toscani, Space localization and well-balanced schemes for discrete kinetic models in diffusive regimes. SIAM J. Numer. Anal. **41**, 641–658 (2004)
31. T. Goudon, J. Nieto, F. Poupaud, J. Soler, Multidimensional high-field limit of the electrostatic Vlasov-Poisson-Fokker-Planck system. J. Differ. Equ. **213**(2), 418–442 (2005)
32. M.D. Gunzburger, C.G. Webster, G. Zhang, Stochastic finite element methods for partial differential equations with random input data. Acta Numer. **23**, 521–650 (2014)
33. J. Hirschfelder, R. Bird, E. Spotz, The transport properties for non-polar gases. J. Chem. Phys. **16**, 968–981 (1948)
34. J. Hu, S. Jin, A stochastic Galerkin method for the Boltzmann equation with uncertainty. J. Comput. Phys. **315**, 150–168 (2016)
35. J. Hu, S. Jin, Q. Li, Asymptotic-preserving schemes for multiscale hyperbolic and kinetic equations, in *Handbook of Numerical Methods for Hyperbolic Problems*, vol. 18, ed. by R. Abgrall, C.-W. Shu (North-Holland, Elsevier, 2017), pp. 103–129
36. J. Hu, S. Jin, R. Shu, A stochastic Galerkin method for the Fokker-Planck-Landau equation with random uncertainties, in *Proceedings of 16th International Conference on Hyperbolic Problems* (to appear)
37. S. Jin, Efficient asymptotic-preserving (AP) schemes for some multiscale kinetic equations. SIAM J. Sci. Comput. **21**, 441–454 (1999)
38. S. Jin, Asymptotic preserving (AP) schemes for multiscale kinetic and hyperbolic equations: a review. Riv. Mat. Univ. Parma **3**, 177–216 (2012)
39. S. Jin, C.D. Levermore, Numerical schemes for hyperbolic conservation laws with stiff relaxation terms. J. Comput. Phys. **126**(2), 449–467 (1996)
40. S. Jin, L. Liu, An asymptotic-preserving stochastic Galerkin method for the semiconductor Boltzmann equation with random inputs and diffusive scalings. Multiscale Model. Simul. **15**(1), 157–183 (2017)
41. S. Jin, H. Lu, An asymptotic-preserving stochastic Galerkin method for the radiative heat transfer equations with random inputs and diffusive scalings. J. Comput. Phys. **334**, 182–206 (2017)
42. S. Jin, R. Shu, A stochastic asymptotic-preserving scheme for a kinetic-fluid model for disperse two-phase flows with uncertainty. J. Comput. Phys. **335**, 905–924 (2017)
43. S. Jin, Y. Zhu, Hypocoercivity and uniform regularity for the Vlasov-Poisson-Fokker-Planck system with uncertainty and multiple scales. SIAM J. Math. Anal., to appear
44. S. Jin, L. Pareschi, G. Toscani, Diffusive relaxation schemes for multiscale discrete-velocity kinetic equations. SIAM J. Numer. Anal. **35**(6), 2405–2439 (1998)
45. S. Jin, L. Pareschi, G. Toscani, Uniformly accurate diffusive relaxation schemes for multiscale transport equations. SIAM J. Numer. Anal. **38**(3), 913–936 (2000)
46. S. Jin, D. Xiu, X. Zhu, Asymptotic-preserving methods for hyperbolic and transport equations with random inputs and diffusive scalings. J. Comput. Phys. **289**, 35–52 (2015)

47. S. Jin, J.-G. Liu, Z. Ma, Uniform spectral convergence of the stochastic Galerkin method for the linear transport equations with random inputs in diffusive regime and a micro-macro decomposition based asymptotic preserving method. Res. Math. Sci. **4**(15), (2017)

48. A. Jüngel, *Transport Equations for Semiconductors*. Lecture Notes in Physics, vol. 773 (Springer, Berlin, 2009)

49. A. Klar, An asymptotic-induced scheme for nonstationary transport equations in the diffusive limit. SIAM J. Numer. Anal. **35**(3), 1073–1094 (1998)

50. K. Koura, H. Matsumoto, Variable soft sphere molecular model for inverse-power-law or Lennard-Jones potential. Phys. Fluids A **3**, 2459–2465 (1991)

51. L.D. Landau, The kinetic equation in the case of Coulomb interaction. Zh. Eksper. Teoret. Fiz. **7**, 203–209 (1937)

52. E.W. Larsen, J.B. Keller, Asymptotic solution of neutron transport problems for small mean free paths. J. Math. Phys. **15**, 75–81 (1974)

53. E.W. Larsen, J.E. Morel, Asymptotic solutions of numerical transport problems in optically thick, diffusive regimes. II. J. Comput. Phys. **83**(1), 212–236 (1989)

54. E.W. Larsen, J.E. Morel, W.F. Miller Jr., Asymptotic solutions of numerical transport problems in optically thick, diffusive regimes. J. Comput. Phys. **69**(2), 283–324 (1987)

55. O.P. Le Maitre, O.M. Knio, *Spectral Methods for Uncertainty Quantification: With Applications to Computational Fluid Dynamics* (Springer, Berlin, 2010)

56. M. Lemou, L. Mieussens, A new asymptotic preserving scheme based on micro-macro formulation for linear kinetic equations in the diffusion limit. SIAM J. Sci. Comput. **31**(1), 334–368 (2008)

57. Q. Li, L. Wang, Uniform regularity for linear kinetic equations with random input based on hypocoercivity (2016). Preprint, arXiv:1612.01219

58. P.-L. Lions, Compactness in Boltzmann's equation via Fourier integral operators and applications. I, II. J. Math. Kyoto Univ. **34**(2), 391–427, 429–461 (1994)

59. L. Liu, Uniform spectral convergence of the stochastic Galerkin method for the linear semiconductor Boltzmann equation with random inputs and diffusive scaling, preprint

60. M. Loève, *Probability Theory*, 4th edn. (Springer, New York, 1977)

61. P.A. Markowich, C. Ringhofer, C. Schmeiser, *Semiconductor Equations* (Springer, Wien, 1990)

62. C. Mouhot, L. Pareschi, Fast algorithms for computing the Boltzmann collision operator. Math. Comp. **75**, 1833–1852 (2006)

63. G. Naldi, L. Pareschi, G. Toscani (eds.), *Mathematical Modeling of Collective Behavior in Socio-Economic and Life Sciences* (Birkhauser, Basel, 2010)

64. H. Niederreiter, P. Hellekalek, G. Larcher, P. Zinterhof, *Monte Carlo and Quasi-Monte Carlo Methods 1996* (Springer, Berlin, 1998)

65. J. Nieto, F. Poupaud, J. Soler, High-field limit for the Vlasov-Poisson-Fokker-Planck system. Arch. Ration. Mech. Anal. **158**(1), 29–59 (2001)

66. M. Per Pettersson, G. Iaccarino, J. Nordstrom, *Polynomial Chaos Methods for Hyperbolic Partial Differential Equations* (Springer, Berlin, 2015)

67. F. Poupaud, Diffusion approximation of the linear semiconductor Boltzmann equation: analysis of boundary layers. Asymptot. Anal. **4**, 293–317 (1991)

68. F. Poupaud, J. Soler, Parabolic limit and stability of the Vlasov-Fokker-Planck system. Math. Models Methods Appl. Sci. **10**(7), 1027–1045 (2000)

69. C. Ringhofer, C. Schmeiser, A. Zwirchmayr, Moment methods for the semiconductor Boltzmann equation in bounded position domains. SIAM J. Numer. Anal. **39**, 1078–1095 (2001)

70. R. Shu, J. Hu, S. Jin, A stochastic Galerkin method for the Boltzmann equation with multidimensional random inputs using sparse wavelet bases. Numer. Math. Theory Methods Appl. **10**, 465–488 (2017)

71. L. Sirovich, Kinetic modeling of gas mixtures. Phys. Fluids **5**, 908–918 (1962)

72. E. Sonnendrucker, Numerical methods for Vlasov equations. Technical report, MPI TU Munich, 2013

73. E.A. Uehling, G.E. Uhlenbeck, Transport phenomena in Einstein-Bose and Fermi-Dirac gases. I. Phys. Rev. **43**, 552–561 (1933)

74. C. Villani, A review of mathematical topics in collisional kinetic theory, in *Handbook of Mathematical Fluid Mechanics*, vol. I, ed. by S. Friedlander, D. Serre (North-Holland, Amsterdam, 2002), pp. 71–305
75. C. Villani, Mathematics of granular materials. Lecture notes, 2005
76. C. Villani, *Hypocoercivity* (American Mathematical Society, Providence, 2009), pp. 949–951
77. D. Xiu, *Numerical Methods for Stochastic Computations* (Princeton University Press, Princeton, 2010)
78. D. Xiu, J.S. Hesthaven, High-order collocation methods for differential equations with random inputs. SIAM J. Sci. Comput. **27**(3), 1118–1139 (2005)
79. D. Xiu, G.E. Karniadakis, The Wiener-Askey polynomial chaos for stochastic differential equations. SIAM J. Sci. Comput. **24**, 619–644 (2002)

Monte-Carlo Finite-Volume Methods in Uncertainty Quantification for Hyperbolic Conservation Laws

Siddhartha Mishra and Christoph Schwab

Abstract We consider hyperbolic systems of conservation laws and review developments in the general area of computational uncertainty quantification (UQ) for these equations. We focus on non-intrusive sampling methods of the Monte-Carlo (MC) and Multi-level Monte-Carlo (MLMC) type. The modeling of uncertainty, within the framework of random fields and random entropy solutions, is discussed. We also describe (ML)MC finite volume methods and present the underlying error bounds and complexity estimates. Based on these bounds, and numerical experiments, we illustrate the gain in efficiency resulting from the use of MLMC methods in this context. Recent progress in the mathematical UQ frameworks of measure-valued and statistical solutions is briefly presented, with comprehensive literature survey.

1 Introduction

Systems of conservation laws are nonlinear partial differential equations of the form

$$\partial_t \mathbf{U} + \nabla_x \cdot \mathbf{F}(\mathbf{c}(x,t), \mathbf{U}) = 0, \tag{1a}$$

$$\mathbf{U}(x, 0) = \mathbf{U}_0(x). \tag{1b}$$

Here, the unknown $\mathbf{U} = \mathbf{U}(x,t) : \mathbb{R}^d \times \mathbb{R}_+ \to \mathbb{R}^N$ is the vector of *conserved variables*, $\mathbf{F} = (\mathbf{F}^1, \ldots, \mathbf{F}^d) : \mathbb{R}^{N \times N} \to \mathbb{R}^{N \times d}$ is the *flux function* and $\mathbf{c} = (\mathbf{c}^1, \ldots, \mathbf{c}^d) : \mathbb{R}^d \times \mathbb{R}_+ \to \mathbb{R}^{N \times d}$ is a spatio-temporal coefficient. We denote $\mathbb{R}_+ := [0, \infty)$. Here, $\mathbf{U}_0$ denotes the prescribed initial data.

In bounded domains, the system (1) needs to be supplemented with suitable boundary conditions.

S. Mishra (✉) · C. Schwab
Seminar for Applied Mathematics, ETH Zürich, Zurich, Switzerland
e-mail: smishra@sam.math.ethz.ch; christoph.schwab@sam.math.ethz.ch

© Springer International Publishing AG, part of Springer Nature 2017
S. Jin, L. Pareschi (eds.), *Uncertainty Quantification for Hyperbolic and Kinetic Equations*, SEMA SIMAI Springer Series 14,
https://doi.org/10.1007/978-3-319-67110-9_7

The system (1) is termed *hyperbolic* if the flux Jacobian matrix has real eigenvalues [11]. Hyperbolic systems of conservation laws arise in a wide variety of models in physics and engineering and we refer to [11] for a wide range of examples. Solutions of (1) can develop discontinuities in finite time, even for smooth initial data (see again [11] and the references there). Therefore, solutions of (1) are weak solutions in that $\mathbf{U} \in (L^1_{loc}(\mathbb{R}^d \times \mathbb{R}_+))^N$ is required to satisfy the integral identity

$$\int_{R_+} \int_{R^d} \left(\mathbf{U}\boldsymbol{\varphi}_t + \sum_{j=1}^d \mathbf{F}^j(\mathbf{c}^j, \mathbf{U})\boldsymbol{\varphi}_{x_j} \right) dxdt + \int_{R^d} \mathbf{U}_0(x)\boldsymbol{\varphi}(x, 0)dx = 0 , \qquad (2)$$

for all test functions $\varphi \in C^1_0(\mathbb{R}_+ \times \mathbb{R}^d)$. It is well known that weak solutions are not necessarily unique [11]. Additional admissibility criteria or *entropy conditions* are necessary to obtain uniqueness. In space dimension $d > 1$, rigorous existence and uniqueness results for conservation (balance) laws and for generic initial data are available only for the scalar case, i.e., in the case $N = 1$.

1.1 Numerical Methods

Numerical methods for the solution of (2) comprise Finite Difference (FD), Finite Volume (FV) and Discontinuous Galerkin (DG) methods. We refer to the textbooks [28, 42] and the references there.

Within the popular FV framework [28], the cell averages of the unknown are updated in time in terms of numerical fluxes across cell interfaces. These numerical fluxes are often obtained by the (approximate) solutions of Riemann problems in the direction normal to the cell interface. Higher order spatial accuracy is obtained by reconstructing cell averages in terms of non-oscillatory piecewise polynomial functions, within the TVD [42], ENO [33] and WENO [55] procedures or using Discontinuous Galerkin methods (see, e.g. [7]). Higher order temporal accuracy is achieved by employing strong stability preserving Runge-Kutta methods [30]. Space-time DG-discretizations can also be employed for High-order spatio-temporal accuracy [34].

1.2 Aims and Scope

Any numerical scheme approximating (1) requires the initial data $\mathbf{U}_0$, the coefficients $\mathbf{c}$ and the flux function $\mathbf{F}$, as well as suitable boundary conditions, as inputs. However, in practice, these inputs are obtained by measurements (observations). Moreover, measurements cannot be precise and always involve some degree of uncertainty. Input uncertainty for (1) implies, upon uncertainty propagation, corresponding uncertainty in the solution.

The modeling, mathematical analysis, numerical approximation and numerical quantification of solution uncertainty, given experimental data, comprise the discipline of uncertainty quantification (UQ).

One aim in this article is to survey computational methods for the efficient, computational UQ for nonlinear, hyperbolic conservation laws with random inputs, and to provide some indications on the numerical analysis of UQ methods for these equations. Our focus in this article is on *non-intrusive computational methods* and their implementation, and on the mathematical analysis of their computational complexity. In our presentation, we emphasize broad applicability for a large class of conservation laws, rather than problem-specific, optimal results.

Our motivation for this focus on non-intrusive methods is as follows: first, non-intrusive methods afford trivial integration of existing, deterministic numerical solvers of forward problems, and are, therefore, popular in computational UQ in science and engineering. Second and as mentioned earlier, nonlinear hyperbolic conservation laws are well known to exhibit *solutions of very low regularity in physical space*, due to *shock formation* even for smooth input data (initial and boundary data, as well as flux functions). Third, *hyperbolicity implies finite speed of propagation* which, in the context of UQ for conservation laws with parametric input uncertainty, implies propagation of singular supports into the domain of parameters that describe the uncertain inputs of the system. The presence of, in general, moving singular supports propagating along characteristics in parametric families of weak solutions precludes high convergence rates of "smooth" computational methods, such as generalized polynomial chaos, PCA etc. for this class of computational UQ problems (we mention, however, that even in the absence of viscosity, there are regularizing effects due to averaging; cases in point are the so-called "transport collapse" regularizations in averaging lemmas (see, e.g. [43, 44] and the references there) or due to statistical ensemble averaging of random entropy solutions (see, e.g. [54])).

We therefore focus in the present survey on *sampling methods* of Monte-Carlo (MC for short) and of Multi-Level MC (MLMC for short) type, as well as on stochastic collocation methods. These methods have in common that their computational realization is based on existing numerical conservation law solvers, for example the finite volume (FV) or discontinuous Galerkin (DG) type, without any modification; this implies, in particular, that existing discretization error bounds for these methods, e.g. from [8, 15, 39, 42] and the references there, can be used for an error analysis of non-intrusive computational UQ for hyperbolic conservation laws.

In contrast, *intrusive computational methods* will require, as a rule, some form of reformulation of the conservation law prior to discretization and entail, usually, significant refactoring resp. redesign of numerical solvers. We refer to, e.g., [24, 59] and references therein for examples of this type, where the so-called *stochastic Galerkin* methodology has been employed and was shown to require significant modifications of numerical schemes as well as of actual, numerical solver.

Given our focus on non-intrusive UQ methods of the MC and Multi-level MC type, we structure this survey as follows: in the first part, we will focus on the very specific problem of UQ for *scalar* conservation laws with random initial data. Here, we describe data and solution uncertainty in terms of random fields and within the framework of random entropy solutions [46]. This is feasible as the underlying deterministic solution operator is well-defined and forms a non-expansive (in time) semi-group on $L^1(\mathbb{R}^d)$. We will formulate both the MC and MLMC methods and combine them with a FV space-time discretization to obtain rigorous convergence rates for the (ML)MC-FV scheme and demonstrate that the MLMC-FV method is significantly more efficient (computationally) than the MC-FV method.

Next, we extend the (ML)MC-FV schemes for UQ of systems of conservation laws with random inputs. Here, the underlying deterministic problem may be ill-posed within the class of entropy solutions [6]. Consequently, the notion of random entropy solutions may not be well-defined. Moreover, there is no rigorous convergence result for the underlying deterministic FV (or any other) discretization frameworks. Hence, we postulate convergence and obtain the corresponding error (and complexity) estimates for the (ML)MC-FV methods. Although this combination is seen to work well in practice, recent results [17, 18, 45] have demonstrated the limitations of this framework. Instead, novel solution concepts such as those of *entropy measure valued solutions* [17, 18] and *statistical solutions* [19] and have been proposed and analyzed. We will conclude with a brief review of these concepts.

2 Preliminaries

2.1 Random Variables in Banach Spaces

Our mathematical formulation of scalar conservation laws with random inputs will use the concept of random fields i.e., random variables taking values in function spaces. We recapitulate basic concepts as presented, for example, in [10, Chap1].

Let E be a Banach space, and let $(\Omega, \mathscr{F})$ be a measurable space, with the set Ω of elementary events, and with $\mathscr{F}$ a corresponding σ-algebra. An *E-valued random variable* (or random variable taking values in E) is any mapping $X : \Omega \to E$ such that the set $\{\omega \in \Omega : X(\omega) \in A\} = \{X \in A\} \in \mathscr{F}$ for any $A \in \mathscr{G}$, i.e. such that X is a $\mathscr{G}$-measurable mapping from Ω into E. Here, $(E, \mathscr{G})$ denotes a measurable space on the Banach space E.

For a Banach space E, we denote the Borel σ-field $\mathscr{B}(E)$. Then, $(E, \mathscr{B}(E))$ is a measurable space and random variables taking values in E i.e. maps $X : \Omega \to E$ are $(\mathscr{F}, \mathscr{B}(E))$ measurable. For a separable Banach space E with norm $\| \circ \|_E$ and (topological) dual E^*, $\mathscr{B}(E)$ is the smallest σ-field of subsets of E containing all sets

$$\{x \in E : \varphi(x) \leq \alpha\}, \ \varphi \in E^*, \ \alpha \in \mathbb{R}.$$

For a separable Banach space, $X : \Omega \to E$ is an E-valued random variable iff for every $\varphi \in E^*$, $\omega \longmapsto \varphi(X(\omega)) \in \mathbb{R}^1$ is an $\mathbb{R}^1$-valued random variable: for any RV $X : \Omega \to E$ on $(\Omega, \mathscr{F})$ which takes values in E, the mapping $\Omega \ni \omega \longmapsto \|X(\omega)\|_E \in \mathbb{R}^1$ is (strongly) measurable. For more details and proofs, we refer to [10] or to [60].

The strongly measurable mapping $X : \Omega \to E$ is *Bochner integrable* if, for any probability measure $\mathbb{P}$ on the measurable space $(\Omega, \mathscr{F})$,

$$\int_\Omega \|X(\omega)\|_E \, \mathbb{P}(d\omega) < \infty. \tag{3}$$

A probability measure $\mathbb{P}$ on $(\Omega, \mathscr{F})$ is a σ-additive set function from Ω into $[0, 1]$ such that $\mathbb{P}(\Omega) = 1$; the triplet $(\Omega, \mathscr{F}, \mathbb{P})$ is called probability space. We shall always assume, unless explicitly stated, that $(\Omega, \mathscr{F}, \mathbb{P})$ is complete.

An E-valued RV is called *simple* if it can assume only finitely many values. A simple RV X, taking values in E, has the explicit form (with χ_A denoting the indicator function of $A \in \mathscr{F}$)

$$X = \sum_{i=1}^N x_i \, \chi_{A_i}, \quad A_i \in \mathscr{F}, \ x_i \in E, \ N < \infty. \tag{4}$$

For simple RVs X taking values in E and for any $\mathscr{B} \in \mathscr{F}$,

$$\int_B X(\omega) \, \mathbb{P}(d\omega) = \int_B X d\mathbb{P} := \sum_{i=1}^N x_i \, \mathbb{P}(A_i \cap B). \tag{5}$$

For such $X(\cdot)$ and for all $B \in \mathscr{F}$,

$$\left\| \int_B X(\omega) \, \mathbb{P}(d\omega) \right\|_E \leq \int_B \|X(\omega)\|_E \, \mathbb{P}(d\omega). \tag{6}$$

For any random variable $X : \Omega \to E$ which is Bochner integrable, there exists a sequence $\{X_m\}_{m \in \mathbb{N}}$ of simple random variables such that, for all $\omega \in \Omega$, $\|X(\omega) - X_m(\omega)\|_E \to 0$ as $m \to \infty$. Therefore, (5) and (6) extend in the usual fashion by continuity to any E-valued random variable. We denote the Bochner-integral

$$\int_\Omega X(\omega) \, \mathbb{P}(d\omega) = \lim_{m \to \infty} \int_\Omega X_m(\omega) \, \mathbb{P}(d\omega) \in E \tag{7}$$

by $\mathbb{E}[X]$ ("expectation" of X).

We introduce for $1 \leq p \leq \infty$ Bochner spaces of p-summable random variables X taking values in the Banach-space E. The set $L^1(\Omega, \mathscr{F}, \mathbb{P}; E)$ comprises all (equivalence classes of) integrable, E-valued random variables X. It is a Banach space if equipped with the norm

$$\|X\|_{L^1(\Omega;E)} := \mathbb{E}(\|X\|_E) = \int_\Omega \|X(\omega)\|_E \, \mathbb{P}(d\omega) . \tag{8}$$

Define $L^p(\Omega, \mathscr{F}, \mathbb{P}; E)$ for $1 \leq p < \infty$ as the set of p-summable random variables taking values E. With the norm

$$\|X\|_{L^p(\Omega;E)} := (\mathbb{E}(\|X\|_E^p))^{1/p}, \ 1 \leq p < \infty \tag{9}$$

$L^p(\Omega, \mathscr{F}, \mathbb{P}; E)$ becomes a Banach space. For $p = \infty$, we denote by $L^\infty(\Omega, \mathscr{F}, \mathbb{P}; E)$ the set of all E-valued random variables which are essentially bounded. This set is a Banach space equipped with the norm

$$\|X\|_{L^\infty(\Omega;E)} := \text{ess} \sup_{\omega \in \Omega} \|X(\omega)\|_E . \tag{10}$$

If $T < \infty$ and $\Omega = [0, T]$, $\mathscr{F} = \mathscr{B}([0, T])$, we write $L^p([0, T]; E)$. Note that for any separable Banach-space E, and for any $r \geq p \geq 1$,

$$L^r(0, T; E), \ C^0([0, T]; E) \in \mathscr{B}(L^p(0, T; E)) . \tag{11}$$

We conclude the section of preliminaries with a criterion for strong measurability.

Lemma 1 ([60, Corollary 1.13]) *Let E_1 and E_2 be Banach spaces, and $(\Omega, \mathscr{F}, \mathbb{P})$ a probability space. If $f : \Omega \to E_1$ is strongly Bochner measurable, and if $\phi : E_1 \to E_2$ is continuous, then the composition $\phi \circ f : \Omega \to E_2$ is strongly Bochner measurable.*

2.2 Monte-Carlo (MC) Sampling in Banach Spaces

Let $\widehat{Y}_i : \Omega \to F$, $i = 1, \ldots, M$, be independent identically distributed ("iid" for short) random variables taking values in the Banach space E. We let

$$E_M[Y^{(k)}] := \frac{1}{M} \sum_{i=1}^M \widehat{Y}_i^{(k)},$$

be the Monte Carlo estimator for $\mathbb{E}[Y^{(k)}]$. A computable estimate $E_M[Y^{(k)}]$ for the kth moment $\mathscr{M}^k(Y)$ of Y will converge in E as $M \to \infty$ at a rate which depends on the integrability of Y. Specifically, at the rate of convergence (in terms of M) of

$$\mathbb{E}\left[\left\|\mathbb{E}[Y^{(k)}] - E_M[Y^{(k)}]\right\|_E^p\right]^{1/p}$$

to zero as $M \to \infty$ for some $1 \leq p < \infty$. If E is a finite dimensional space or a Hilbert space, and if $X \in L^{2k}(\Omega; X)$, the so-called *mean square error* (MSE) is bounded as

$$\mathbb{E}\left[\left\|\mathbb{E}[Y^{(k)}] - E_M[Y^{(k)}]\right\|_E^2\right] \leq \frac{1}{M}\mathrm{Var}[Y^{(k)}] \tag{12}$$

using the independence of the samples $\widehat{Y}_i$.

For general Banach spaces E, the convergence rate depends on the *type* of the Banach space, which is defined as follows [38, p. 246].

Definition 1 Let $1 \leq p \leq \infty$ and $Z_j, j \in \mathbb{N}$ a sequence of Bernoulli-Rademacher random variables. A Banach space E is said to be of *type p* if there is a *type constant* $C > 0$ such that for all finite sequences $(x_j)_{j=1}^N \subset E, N \in \mathbb{N}$,

$$\left\|\sum_{j=1}^N Z_j x_j\right\|_E \leq C \left(\sum_{j=1}^N \|x_j\|_E^p\right)^{1/p}.$$

By the triangle inequality, every Banach space has type 1. Hilbert spaces have type 2. One can show that the L^p-spaces have type $\min\{p, 2\}$ for $1 \leq p < \infty$ [38].

One has the following result from [36], [38, Proposition 9.11] and [9, Section 4] for Banach spaces of type p.

Proposition 1 ([36]) *Let E be a Banach space of type p with a type constant C_t. Then, for every finite sequence $(Y_j)_{j=1}^N$ of independent mean zero random variables in $L^p(\Omega)$, one has the bound*

$$\mathbb{E}\left[\left\|\sum_{j=1}^N Y_j\right\|_E^p\right] \leq (2C_t)^p \sum_{j=1}^N \mathbb{E}\left[\|Y_j\|_E^p\right].$$

Corollary 1 *Let E be a Banach space of type $p \in [1, 2]$ with type constant C_t. Then for every finite sequence $(Y_j)_{j=1}^N$ of iid mean zero random variables with $Y_j(\omega) \sim Y(\omega)$ in $L^p(\Omega)$,*

$$\mathbb{E}\left[\left\|E_M[Y^{(k)}]\right\|_E^p\right] = \mathbb{E}\left[\left\|\frac{1}{N}\sum_{j=1}^N Y_j^{(k)}\right\|_E^p\right] \leq (2C_t)^p N^{1-p} \mathbb{E}\left[\|Y^{(k)}\|_E^p\right].$$

Remark 1 The complexity of MC methods with respect to the type parameter p of the function space E has been investigated in [12, 13] and the references there. The relevance of Proposition 1 for MC methods applied to scalar conservation laws is due to $L^1(\mathbb{R}^d)$ being crucial for the error- and well-posedness analysis of the underlying FV schemes for these problems. The space $L^1(\mathbb{R}^d)$ being a Banach space of type 1, will a priori not allow for convergence rate bounds in MLMC-FV discretizations, as was incorrectly stipulated in [46], and also in related work [51] on combining MLMC discretization with the front-tracking algorithm.

Instead, and as pointed out in [48], lower rates of convergence of FV discretizations in the stronger norms $L^p(\mathbb{R}^d)$ with $p > 1$, with the space $L^p(\mathbb{R}^d)$ being of type $\min\{p, 2\}$, lead to convergence and error vs. work analysis of (ML)MC-FV methods for scalar conservation laws, which will be described in detail subsequently.

3 Scalar Conservation Laws with Random Initial Data

We begin with a review of classical results on deterministic scalar conservation laws (SCLs). We also review random entropy solutions for SCLs with random initial data from [46], and in particular existence and uniqueness of a random entropy solution with finite second moments. Let us mention that SCLs with random input data have received considerable attention in the context of numerical methods for uncertainty quantification; we only mention [22, 63].

We also mention considerable activity on enlarging the class of admissible random flux functions, in particular by so-called "rough path" calculus and the kinetic (re)formulation of the SCL (14)–(15) rather than the Kružkov theory; we refer to [22, 26, 43, 44] and the references there.

3.1 Deterministic Scalar Hyperbolic Conservation Laws

In-order to present the basic ideas in a simple setting, we consider the Cauchy problem for scalar conservation laws (SCL) i.e., (1) with $N = 1$ and with a spatially homogeneous deterministic flux function $f(u)$. Then, (1) can be written as

$$\frac{\partial u}{\partial t} + \mathrm{div}\,(f(u)) = 0 \ \text{ for } (x, t) \in \mathbb{R}^d \times \mathbb{R}_+. \tag{13}$$

with

$$f(u) = (f_1(u), \ldots, f_d(u)) \in C^1(\mathbb{R}; \mathbb{R}^d)\,, \quad \mathrm{div} f(u) = \sum_{j=1}^{d} \frac{\partial}{\partial x_j} f_j(u)\,, \tag{14}$$

We supply the SCL (13) with initial condition

$$u(x, 0) = u_0(x), \ \ x \in \mathbb{R}^d\,, \tag{15}$$

and an entropy admissibility condition, which we choose as the Kružkov entropy condition or an equivalent version of it.

3.2 Entropy Solution

It is well-known that the deterministic Cauchy problem (13), (15) admits, for each $u_0 \in L^1(\mathbb{R}^d) \cap BV(\mathbb{R})$, a unique entropy solution u (see, e.g., [11, 28, 56]). For every $t > 0$, $u(\cdot, t) \in L^1(\mathbb{R}^d)$. We require the (nonlinear) data-to-solution map

$$S : u_0 \longmapsto u(\cdot, t) = S(t) u_0, \quad t > 0 \tag{16}$$

in our subsequent development. To state its properties, we introduce some additional notation: for a Banach-space E with norm $\|\circ\|_E$, and for $0 < T \leq +\infty$, we denote by $C([0, T]; E)$ the space of bounded and continuous functions from $[0, T]$ with values in E, and by $L^p(0, T; E)$, $1 \leq p \leq +\infty$, the space of strongly Bochner measurable functions from $(0, T)$ to E such that for $1 \leq p < +\infty$

$$\|v\|_{L^p(0,T;E)} = \left(\int_0^T \|v(t)\|_E^p \, dt \right)^{\frac{1}{p}}, \tag{17}$$

respectively, if $p = \infty$,

$$\|v\|_{L^\infty(0,T;E)} = \operatorname*{ess\,sup}_{0 \leq t \leq T} \|v(t)\|_E \tag{18}$$

are finite. The following existence result is classical. (see, for example, [35, Thms. 2.13, 2.14, Thm. 4.3] or also [15, 28, 29, 37, 42],

Theorem 1 *Assume that in the SCL (14)–(15) holds $f \in C^1(\mathbb{R}; \mathbb{R}^d)$, and the initial data u_0 satisfies*

$$u_0 \in Z := L^1(\mathbb{R}^d) \cap L^\infty(\mathbb{R}^d) \cap BV(\mathbb{R}^d) . \tag{19}$$

Then there holds:

(1) *The SCL (14)–(15) admits a unique entropy solution $u \in L^\infty(\mathbb{R}^d \times (0, T))$.*
(2) *For every $t > 0$, the (nonlinear) data-to-solution map $S(t)$ given by*

$$u(\cdot, t) = S(t) u_0$$

    satisfies

    (i) *$S(t) : L^1(\mathbb{R}^d) \to L^1(\mathbb{R}^d)$ is a (non-expansive) Lipschitz map, i.e.,*

$$\|S(t)u_0 - S(t)v_0\|_{L^1(\mathbb{R}^d)} \leq \|u_0 - v_0\|_{L^1(\mathbb{R}^d)} . \tag{20}$$

(ii) $S(t)$ maps $(L^1 \cap BV)(\mathbb{R}^d)$ into $(L^1 \cap BV)(\mathbb{R}^d)$ and

$$TV(S(t)u_0) \leq TV(u_0) \quad \forall u_0 \in (L^1 \cap BV)(\mathbb{R}^d) \,. \tag{21}$$

(iii) *There hold the L^∞ and L^1 stability bounds*

$$\|S(t)u_0\|_{L^\infty(\mathbb{R}^d)} \leq \|u_0\|_{L^\infty(\mathbb{R}^d)} \,; \tag{22}$$

$$\|S(t)u_0\|_{L^1(\mathbb{R}^d)} \leq \|u_0\|_{L^1(\mathbb{R}^d)} \,. \tag{23}$$

(iv) *The mapping $S(t)$ is a uniformly continuous mapping from $L^1(\mathbb{R}^d)$ into $C([0, \infty); L^1(\mathbb{R}^d))$, and*

$$\|S(\cdot)u_0\|_{C([0,T];L^1(\mathbb{R}^d))} = \max_{0 \leq t \leq T} \|S(t)u_0\|_{L^1(\mathbb{R}^d)} \leq \|u_0\|_{L^1(\mathbb{R}^d)} \,. \tag{24}$$

Hyperbolic conservation laws exhibit *finite propagation speed* of perturbations. As a consequence, compactly supported initial data gives rise to solutions which are compactly supported for all time, however, with time-dependent supports. We present one version of such a "domain of influence" result, for the SCL (14)–(15).

Proposition 2 *For the Cauchy problem (14)–(15), assume that $f \in C^1(\mathbb{R}; \mathbb{R}^d)$ and that u_0 satisfies (19). Assume moreover that the initial data $u_0 \in Z$ has compact support: there exists a finite, positive constant R such that*

$$\mathrm{supp}(u_0) \subset [-R, R]^d \,. \tag{25}$$

Then, for every $0 < t < \infty$, the unique entropy solution u of the Cauchy problem (14)–(15) is compactly supported as well, and with $\bar{M} := \|u_0\|_{L^\infty(\mathbb{R}^d)} < \infty$, there holds

$$\mathrm{supp}(u(t)) \subset [-(R + tB), R + tB]^d \,, \tag{26}$$

where

$$B := \|\partial_u f\|_{C^0([-\bar{M}, \bar{M}]; \mathbb{R}^d)}, \tag{27}$$

denotes a upper bound on the maximal propagation speed.

3.3 Random Entropy Solution

Based on Theorem 1, we will now formulate the SCL (14)–(15) for random initial data $u_0(\omega; \cdot)$, with deterministic flux. To this end, we denote $(\Omega, \mathscr{F}, \mathbb{P})$ a probability space. We *assume given* a Lipschitz continuous deterministic flux f and random initial data u_0, which satisfies the

Assumption 1 (Assumptions on the Random Input Data)

1. The random initial data u_0 is an $L^1(\mathbb{R}^d)$-valued random variable on $(\Omega, \mathscr{F}, \mathbb{P})$. It is in particular a strongly Bochner measurable map

$$u_0 : (\Omega, \mathscr{F}) \longmapsto \left(L^1(\mathbb{R}^d), \mathscr{B}(L^1(\mathbb{R}^d))\right). \tag{28}$$

2. The map u_0 is also strongly Bochner measurable from $(\Omega, \mathscr{F}, \mathbb{P})$ with values in the space $Z = L^1(\mathbb{R}^d) \cap L^\infty(\mathbb{R}^d) \cap BV(\mathbb{R}^d)$ (taking values in the separable Banach space $L^1(\mathbb{R}^d)$, the random initial data u_0 is in particular separably valued in Z) introduced in (19), so that

$$\Omega \ni \omega \mapsto u_0(\omega; \cdot) \in Z = (L^1 \cap L^\infty \cap BV)(\mathbb{R}^d) \tag{29}$$

is strongly Bochner measurable; here, we equip the space Z in (19), (29) with the norm

$$\|u\|_Z := \|u\|_{L^1(\mathbb{R}^d)} + \|u\|_{L^\infty(\mathbb{R}^d)} + TV(u). \tag{30}$$

3. There holds a uniform bound: for some constant $0 < \bar{M} < \infty$,

$$\|u_0(\omega; \cdot)\|_{L^\infty(\Omega; Z)} \leq \bar{M} < \infty, \tag{31}$$

4. The random initial data u_0 satisfies the bounded support assumption (25) with probability one, i.e. there exists a constant $0 < R < \infty$ such that

$$\operatorname{supp}(u_0(\omega, \cdot)) \subset [-R, R]^d \quad \text{with probability } 1. \tag{32}$$

5. The flux function f is bounded on the set of states: for $\bar{M}$ as in (31), item 3, there holds, with $S = [-\bar{M}, \bar{M}]$, the bound (27).

Since $L^1(\mathbb{R}^d)$ and $C^1(\mathbb{R}^d; \mathbb{R}^d)$ are separable, we may impose on the mapping (28) *kth moment conditions*

$$\|u_0\|_{L^k(\Omega; L^1(\mathbb{R}^d))} < \infty, \quad k \in \mathbb{N}, \tag{33}$$

where the Bochner spaces are defined in Sect. 2. We consider the *random scalar conservation law* (RSCL)

$$\begin{cases} \partial_t u(\omega; x, t) + \operatorname{div}_x(f(\omega; u(\omega; x, t))) = 0, \ t > 0, \\ u(\omega; x, 0) = u_0(\omega; x), \end{cases} \quad x \in \mathbb{R}^d. \tag{34}$$

Definition 2 ([46]) A random field $u : \Omega \ni \omega \rightarrow u(\omega; x, t)$, i.e., a strongly measurable mapping from $(\Omega, \mathscr{F})$ to $C([0, T]; L^1(\mathbb{R}^d))$, is a *random entropy solution* of the SCL (34) with random initial data u_0 satisfying (28)–(33) for some $k \geq 2$ and with a *spatially homogeneous* flux $f(u)$ if it satisfies the following conditions,

(i) Weak solution:
   For $\mathbb{P}$-a.e. $\omega \in \Omega$, $u(\omega; \cdot, \cdot)$ satisfies the following integral identity

$$\int\limits_0^T \int\limits_{\mathbb{R}^d} \left(u(\omega; x, t)\varphi_t(x, t) + \sum_{j=1}^d f_j(\omega; u(\omega; x, t)) \frac{\partial}{\partial x_j} \varphi(x, t) \right) dxdt$$

$$+ \int\limits_{\mathbb{R}^d} u_0(x, \omega))\varphi(x, 0)\, dx = 0, \qquad (35)$$

   for all test functions $\varphi \in C_0^1(\mathbb{R}^d \times [0, T))$.
(ii) Entropy condition: For any pair of (deterministic) entropy η and entropy flux $Q(\cdot)$ i.e., η, Q_j with $j = 1, 2, \ldots, d$ are functions such that η is convex and such that $Q_j'(\cdot) = \eta' f_j'(\cdot)$ for all j, and u satisfies the following inequality

$$\int\limits_0^T \int\limits_{\mathbb{R}^d} \left(\eta(u(\omega; x, t))\varphi_t(x, t) + \sum_{j=1}^d Q_j(u(\omega; x, t)) \frac{\partial}{\partial x_j} \varphi(x, t) \right) dxdt$$

$$+ \int\limits_{\mathbb{R}^d} \eta(u_0(\omega; x)\varphi(x, 0)\, dx \geq 0, \qquad (36)$$

   for all deterministic test functions $0 \leq \varphi \in C_0^1(\mathbb{R}^d \times [0, T))$, $\mathbb{P}$-a.s.

We remark that it suffices to assume that (36) holds for all Kružkov entropy functions $\eta(u) = |u - k|$, where k is any constant, which we assume throughout what follows.

Theorem 2 ([28, Chap. 2, Thms. 5.1,5.2]) *Consider the SCL (14)–(15) with spatially homogeneous, bounded flux $f \in C^1(\mathbb{R}; \mathbb{R}^d)$ with random initial data $u_0 : \Omega \rightarrow L^1(\mathbb{R}^d)$ satisfying Assumption 1 and the kth moment condition (33) for some integer $k \geq 2$. Then there exists a random entropy solution $u : \Omega \rightarrow C([0, T]; L^1(\mathbb{R}^d))$ which is "pathwise" unique, i.e., for $\mathbb{P}$-a.e. $\omega \in \Omega$, described in terms of the deterministic, nonlinear mapping $S(t)$ from Theorem 1 such that*

$$u(\omega; \cdot, t) = S(t)u_0(\omega; \cdot), \quad t > 0, \ \mathbb{P}\text{-}a.e.\ \omega \in \Omega. \qquad (37)$$

Moreover, $u : \Omega \rightarrow C(0, T; L^1(\mathbb{R}^d))$ is $\mathbb{P}$-a.s. separably valued and strongly Bochner measurable.

For every $k \geq 1$, for every $0 \leq t \leq T < \infty$, and for $\mathbb{P}$-a.e. $\omega \in \Omega$ holds

$$\|u\|_{L^k(\Omega; C(0,T; L^1(\mathbb{R}^d)))} \leq \|u_0\|_{L^k(\Omega; L^1(\mathbb{R}^d))}, \tag{38}$$

$$\|S(t)u_0(\omega; \cdot)\|_{(L^1 \cap L^\infty)(\mathbb{R}^d)} \leq \|u_0(\omega; \cdot)\|_{(L^1 \cap L^\infty)(\mathbb{R}^d)} \tag{39}$$

$$TV(S(\omega; t)u_0(\omega; \cdot)) \leq TV(u_0(\omega; \cdot)). \tag{40}$$

There exists $\bar{M} < \infty$ such that

$$\|u_0\|_{L^\infty(\Omega; L^\infty(\mathbb{R}^d))} \leq \bar{M}. \tag{41}$$

and

$$\sup_{0 \leq t \leq T} \|u(\omega; \cdot, t)\|_{L^\infty(\mathbb{R}^d)} \leq \bar{R} \quad \mathbb{P}\text{-a.e. } \omega \in \Omega. \tag{42}$$

Theorem 2 ensures the existence of a unique random entropy solution $u(\omega; x, t)$ with finite kth moments for bounded random flux, provided that $u_0 \in L^k(\Omega, \mathscr{F}, \mathbb{P}; Z)$.

The deterministic maximum principle (22) and (41) imply, in addition, that $\mathbb{P}$-a.e. realization of the random entropy solution u takes values in the state space $S = [-\bar{M}, \bar{M}]$, for *a.e.* $x \in \mathbb{R}^d$ and for all $t > 0$.

4 Monte Carlo and Multi-Level Monte Carlo Finite Volume Methods

We present the Multilevel Monte Carlo Finite Volume Method (MLMC-FVM) for scalar conservation laws. We introduce it in several steps: first, we discuss MC sampling of random initial data, second, FV discretization of the samples on the single, fixed triangulation and, finally, its multi-level extension on a hierarchy of possibly unstructured grids.

4.1 Monte-Carlo Method

The MC method for the SCL with random data $u_0(\omega; x)$ as in (28)–(31) consists in sampling in the probability space. We also assume (33), i.e., the existence of kth moments of u_0 for some $k \in \mathbb{N}$, to be specified. We analyze the error in computable numerical approximations of kth order statistical moments of u. For $k = 1$ we obtain the expected value of the solution random field i.e., $\mathscr{M}^1(u) = \mathbb{E}[u]$. The

MC approximation of $\mathbb{E}[u]$ is defined as the usual statistical sample average: given M independent, identically distributed ("iid" for short) draws of the initial data, $\widehat{u}_0^i$, $i = 1, \ldots, M$, the MC estimate of $\mathbb{E}[u(\cdot; \cdot, t)]$ at time $t > 0$ is the sample average

$$E_M[u(\cdot, t)] := \frac{1}{M} \sum_{i=1}^{M} \widehat{u}^i(\cdot, t) . \tag{43}$$

Here, $\widehat{u}^i(\cdot, t)$ denotes the M unique entropy solutions of the M Cauchy Problems (14)–(15) with iid initial data $\widehat{u}_0^i$. We observe that by

$$\widehat{u}^i(\cdot, t) = \widehat{S}(t) \widehat{u}_0^i \tag{44}$$

we have from (21)–(23) for M MC samples and for any $0 < t < \infty$, for every $1 \leq p \leq \infty$, using the triangle inequality,

$$\|E_M[u(\omega; \cdot, t)]\|_{L^p(\mathbb{R}^d)} = \left\| \frac{1}{M} \sum_{i=1}^{M} \widehat{S}(t) \widehat{u}_0^i(\cdot; \omega) \right\|_{L^p(\mathbb{R}^d)} \leq \frac{1}{M} \sum_{i=1}^{M} \|\widehat{u}_0^i(\omega; \cdot)\|_{L^p(\mathbb{R}^d)} . \tag{45}$$

Using the i.i.d. property of the samples $\{\widehat{u}_0^i\}_{i=1}^{M}$ of the random initial data u_0 and the linearity of the expectation $\mathbb{E}[\cdot]$, we obtain for any $1 \leq p \leq \infty$ from the *assumed strong measurability* of u_0 in the Banach space Z defined in (29), the bound

$$\mathbb{E}\left[\|E_M[u(\cdot; \cdot, t)]\|_{L^p(\mathbb{R}^d)}\right] \leq \mathbb{E}\left[\|u_0\|_{L^p(\mathbb{R}^d)}\right] = \|u_0\|_{L^1(\Omega; L^p(\mathbb{R}^d))} < \infty. \tag{46}$$

As $M \to \infty$, the MC estimates (43) converge in $L^2(\Omega; C([0, T]; L^p(\mathbb{R}^d)))$ and the following convergence rate bound holds.

Theorem 3 *Assume that in the SCL* (14)–(15) *the random initial data u_0 satisfies Assumption 1, items 1–5. In particular, u_0 is with probability one compactly supported in space, i.e., there exists a compact domain $D \subset \mathbb{R}^d$ such that* $\mathrm{supp}(u_0(\omega)) \subset D$ *for almost every $\omega \in \Omega$.*

Assume, moreover, that the random initial data u_0 is strongly Bochner measurable taking values in the space $Z = L^1(\mathbb{R}^d) \cap L^\infty(\mathbb{R}^d) \cap BV(\mathbb{R}^d)$ (cp. (29), (30)), and satisfies

$$u_0 \in L^2(\Omega; L^1(\mathbb{R}^d)) \cap L^2(\Omega; L^\infty(\mathbb{R}^d)) . \tag{47}$$

Assume further that (29), (31) *hold.*
  Then for every $0 < t < \infty$ holds the a priori bound

$$\|u(t)\|_{L^2(\Omega; L^2(\mathbb{R}^d))}^2 \leq \|u_0\|_{L^2(\Omega; L^1(\mathbb{R}^d))} \|u_0\|_{L^2(\Omega; L^\infty(\mathbb{R}^d))} \tag{48}$$

The MC estimates $E_M[u(\cdot, t)]$ in (43) converge, as $M \to \infty$, to $\mathcal{M}^1(u(\cdot, t)) = \mathbb{E}[u(\cdot, t)]$. For $M \in \mathbb{N}$, and for every fixed $0 < t < \infty$, there holds the error bound

$$\|\mathbb{E}[u(\cdot, t)] - E_M[u(\cdot, t)]\|^2_{L^2(\Omega; L^2(\mathbb{R}^d))} \leq M^{-1} \|u_0\|_{L^2(\Omega; L^1(\mathbb{R}^d))} \|u_0\|_{L^2(\Omega; L^\infty(\mathbb{R}^d))} .$$
(49)

Under Assumption 1, item 4, we also have for every $0 < t < T$

$$\|\mathbb{E}[u(\cdot, t)] - E_M[u(\cdot, t)]\|^2_{L^2(\Omega; L^1(\mathbb{R}^d))} \leq C(T) M^{-1} \|u_0\|_{L^2(\Omega; L^1(\mathbb{R}^d))} \|u_0\|_{L^2(\Omega; L^\infty(\mathbb{R}^d))} .$$
(50)

In (50), $C(T)$ is a time dependent constant that also depends on the bounded domain $[-R, R]^d$ on which the random initial data is supported with probability 1.

Proof Under the assumptions of the theorem, by Theorem 2 there exists a unique random entropy solution u.

For any $v \in L^1(\mathbb{R}^d) \cap L^\infty(\mathbb{R}^d)$ holds $\|v\|^2_{L^2(\mathbb{R}^d)} \leq \|v\|_{L^1(\mathbb{R}^d)} \|v\|_{L^\infty(\mathbb{R}^d)}$. For every fixed $0 < t < \infty$ we have from (22), (23),

$$\begin{aligned}
\int_\Omega \|S(t)u_0\|^2_{L^2(\mathbb{R}^d)} &= \int_\Omega \|u(\omega; t)\|^2_{L^2(\mathbb{R}^d)} d\mathbb{P}(\omega) \\
&\leq \int_\Omega \|u_0(\omega)\|_{L^1(\mathbb{R}^d)} \|u_0(\omega)\|_{L^\infty(\mathbb{R}^d)} d\mathbb{P}(\omega) \\
&\leq \|u_0\|_{L^2(\Omega; L^1(\mathbb{R}^d))} \|u_0\|_{L^2(\Omega; L^\infty(\mathbb{R}^d))} .
\end{aligned}$$

Therefore, for every $0 < T < \infty$,

$$\begin{aligned}
\|u\|^2_{L^2(\Omega; C(0, T; L^2(\mathbb{R}^d)))} &= \int_\Omega \sup_{0 < t < T} \|u(\omega; t)\|^2_{L^2(\mathbb{R}^d)} d\mathbb{P}(\omega) \\
&\leq \sup_{0 < t < T} \|S(t)u_0\|^2_{L^2(\Omega; L^2(\mathbb{R}^d))} \\
&\leq \|u_0\|_{L^2(\Omega; L^1(\mathbb{R}^d))} \|u_0\|_{L^2(\Omega; L^\infty(\mathbb{R}^d))} ,
\end{aligned}$$
(51)

which is finite by assumption (47). From this bound follows the MC error bound (49) by referring to the general MC error bound in Corollary 1, with the observation that Hilbert spaces have type $p = 2$, and type constant $2C_t = 1$ or directly from (12).

We show the second part: note that the space $L^1(\mathbb{R}^d)$ is of type 1. From Corollary 1 we cannot expect a MC convergence rate bound in $L^1(\mathbb{R}^d)$ without extra assumptions.

Suppose therefore now that Assumption 1, item 4, holds, i.e. all realizations of u_0 have compact support in a common set $[-R, R]^d$. Then the bound (27) of the flux f in $C^1(S; \mathbb{R}^d)$ and the finite propagation property, Proposition 2, imply that for every $0 < t < \infty$, and $\mathbb{P}$-a.s., that the random entropy solution is likewise compactly supported: from (26) it follows that there holds, for every $t > 0$,

$$\text{supp}(u(\omega; t)) \subset [-(R + tB), R + tB]^d \quad \text{with probability 1} .$$
(52)

Then, for $\mathbb{P}$-a.e. ω,

$$\|u(\omega;t)\|_{L^1(\mathbb{R}^d)} \leq C(t,B,R)\|u(\omega;t)\|_{L^2(\mathbb{R}^d)} \ .$$

Squaring both sides and taking expectations, we find

$$\|u(t)\|^2_{L^2(\Omega;L^1(\mathbb{R}^d))} \leq C(t,B,R)^2 \|u(\omega;t)\|^2_{L^2(\Omega;L^2(\mathbb{R}^d))} \ .$$

Using (48) and reasoning as before, we arrive at (50). □

Remark 2 The bound (51) can be generalized to k-point correlation functions $\mathscr{M}^{(k)}u = \mathbb{E}(u(\cdot,t,x_1)\ldots u(\cdot,t,x_k))$, $x_1,\ldots,x_k \in \mathbb{R}^d$ with $k > 1$ of the random entropy solution: from Jensen's inequality and Fubini's theorem,

$$\begin{aligned}
\|\mathscr{M}^{(k)}u(t)\|^2_{L^2(\mathbb{R}^{kd})} &\leq \int_\Omega \int_{x_1}\ldots\int_{x_k} |u(\cdot,x_1,t)\ldots u(\cdot,x_k,t)|^2 dx_k\ldots dx_1 d\mathbb{P}(\omega) \\
&= \int_\Omega \|u(\cdot,t,\cdot)\|^{2k}_{L^2(\mathbb{R}^d)} d\mathbb{P}(\omega) \\
&= \int_\Omega \|S(t)u_0(\cdot)\|^{2k}_{L^2(\mathbb{R}^d)} d\mathbb{P}(\omega) \\
&\leq \int_\Omega \|S(t)u_0(\cdot)\|^k_{L^1(\mathbb{R}^d)} \|S(t)u_0(\cdot)\|^k_{L^\infty(\mathbb{R}^d)} d\mathbb{P}(\omega) \\
&\leq \int_\Omega \|u_0(\cdot)\|^k_{L^1(\mathbb{R}^d)} \|u_0(\cdot)\|^k_{L^\infty(\mathbb{R}^d)} d\mathbb{P}(\omega) \\
&\leq \|u_0\|^k_{L^{2k}(\Omega;L^1(\mathbb{R}^d))} \|u_0\|^k_{L^{2k}(\Omega;L^\infty(\mathbb{R}^d))} \ .
\end{aligned}$$

From Theorem 3 we see that $L^1(\mathbb{R}^d)$ MC convergence rate bounds can be obtained despite L^1 being a Banach space of type 1; however, as already observed in [13] and the references there, this is only possible by an intermediate bound of samples and, for multilevel MC, for error bounds on FV discretizations in Banach spaces of type $1 < q \leq 2$, as introduced in Definition 1. As observed in [48], Theorem 4.1, the assumption of compactly supported initial data, satisfying (25), and bounded flux (27) which imply (26) at positive time $t > 0$ does afford intermediate $L^2_\omega L^2_x$ bounds which in turn allow MC convergence rate bounds.

The properties (20)–(24) also hold for FV discretizations. Accordingly, we aim at analogous results for MC FV methods for the random SCL. We next introduce FV methods; rather than presenting a particular scheme, we state several properties required in the ensuing error analysis which are satisfied by several popular FV methods.

4.2 Finite Volume Method (FVM)

So far, the MC method was prescribed under the assumption that "pathwise" entropy solutions $\widehat{u}^i(\omega;x,t) = S(\omega;t)\widehat{u}^i_0(\omega;x)$ for the Cauchy problem (14)–(15)

iid initial data samples $\hat{u}_0^i = u_0(\omega_i; x)$ are available exactly. In practice, numerical approximations of $S(t)\hat{u}_0^i$ must be computed and the corresponding discretization errors taken into account.

To analyze MC-FVM approximations we impose sufficient conditions on the FVM to afford the Kuznetsov type error bounds for first order FVM for the *deterministic* SCL (14)–(15); these will be required for the convergence rate bounds of the MLMC FVM as considered in [46, 48] and also for parametric collocation FVM as in [48, Sec.5] in the subsequent chapters. We review the generic first order, explicit FV schemes considered here, as for example in [15, 37, 42].

Denote the time step by $\Delta t > 0$ and a triangulation $\mathcal{T}$ of the spatial domain $D \subset \mathbb{R}^d$ of interest. We assume that $\mathcal{T}$ is a set of open, convex polyhedra $K \subset \mathbb{R}^d$ with plane faces such that standard conditions on shape regularity hold: if $K \in \mathcal{T}$ denotes a generic volume, we define

$$\rho_K = \rho(K) = \max\{\mathrm{diam}(B_r) : B_r \subset \overline{K}\} \tag{53}$$

i.e., the maximum inradius in volume $\overline{K}$ for $K \in \mathcal{T}$ and define, in addition, for a generic mesh $\mathcal{T}$, the *shape regularity constants* (where $\Delta x_K := \mathrm{diam}\, K$)

$$\kappa(\mathcal{T}) := \sup\{\Delta x_K / \rho(K) : K \in \mathcal{T}\}, \quad \mathcal{T} \in \mathfrak{M}. \tag{54}$$

The meshwidth of triangulation $\mathcal{T}$ is

$$\Delta x(\mathcal{T}) = \sup\{\mathrm{diam}(K) : K \in \mathcal{T}\}. \tag{55}$$

For any volume $K \in \mathcal{T}$, we define the *set $\mathcal{N}(K)$ of neighboring volumes*

$$\mathcal{N}(K) := \{K' \in \mathcal{T} : K' \neq K \wedge \mathrm{meas}_{d-1}(\overline{K} \cap \overline{K'}) > 0\}. \tag{56}$$

We assume that the triangulation $\mathcal{T}$ are regular in the sense that the support size of the FV "stencil" at any element $K \in \mathcal{T}$ is uniformly bounded i.e.,

$$\sigma(\mathcal{T}) := \sup_{K \in \mathcal{T}} \#(\mathcal{N}(K)) \leq B < \infty \tag{57}$$

with some bound B which is independent of the particular partition $\mathcal{T}$. The global CFL constant is defined by

$$\lambda := \Delta t / \Delta x(\mathcal{T}). \tag{58}$$

for constant time step Δt; we also set $t_n = n\Delta t$. It is determined by a standard CFL condition (see e.g. [28]) based on the maximum wave speed given by the flux bound (27), see Proposition 2.

We discretize (14)–(15) by an explicit, first order FV scheme on $\mathscr{T}$:

$$v_K^{n+1} = H(\{v_{K'}^n : K' \in \mathscr{N}(K) \cup K\}), \quad K \in \mathscr{T} \tag{59}$$

where $H : \mathbb{R}^{(2k+1)^d} \to \mathbb{R}$, with k denoting the size of the stencil of the FV scheme, is continuous and where v_K^n denotes an approximation to the cell average of u at time $t_n = n\Delta t$.

In our subsequent developments, we write the FVM in operator form. To this end, we introduce the operator $H_{\mathscr{T}}(v)$ which maps a sequence $\underline{v} = (v_K)_{K \in \mathscr{T}}$ into $H_{\mathscr{T}}((v_K)_{K \in \mathscr{T}})$. The time explicit FVM (59) takes the abstract form

$$\underline{v}^{n+1} = H_{\mathscr{T}}(\underline{v}^n), \quad n = 0, 1, 2, \ldots \tag{60}$$

For the ensuing convergence analysis, we shall assume and use several properties of the FV scheme (60); these properties are satisfied by many commonly used FVM of the form (60), on regular or irregular meshes $\mathscr{T}$ in $\mathbb{R}^d$.

To state the assumptions, we introduce further notation: for any initial data $u_0(x) \in L^1(\mathbb{R}^d)$, we define the FVM approximation at time $t = 0$, $(v_K^0)_{K \in \mathscr{T}}$ by the *cell averages*

$$v_K^0 = \frac{1}{|K|} \int_K u_0(x)\, dx, \text{ where } K \in \mathscr{T}. \tag{61}$$

Interpreting the vector $\underline{v} = (v_K)_{K \in \mathscr{T}} \in \mathbb{R}^{\#\mathscr{T}}$ as cell averages, we associate with $\underline{v}$ the piecewise constant function $v_{\mathscr{T}}(x, t)$ defined a.e. in $\mathbb{R}^d \times (0, \infty)$ by

$$v_{\mathscr{T}}(x, t)\big|_K := v_K^n, \quad K \in \mathscr{T}, \quad t \in [t_n, t_{n+1}). \tag{62}$$

We denote space of all piecewise constant functions on $\mathscr{T}$ (i.e., the "simple" or "step" functions on $\mathscr{T}$) by $S(\mathscr{T})$. Given any $v_{\mathscr{T}} \in S(\mathscr{T})$, we define the (mesh-dependent) norms:

$$\|\underline{v}\|_{L^1(\mathscr{T})} = \sum_{K \in \mathscr{T}} |K|\, |v_K| = \|v_{\mathscr{T}}\|_{L^1(\mathbb{R}^d)}, \quad \|\underline{v}\|_{L^\infty(\mathscr{T})} = \sup_{K \in \mathscr{T}} |v_K| = \|v_{\mathscr{T}}\|_{L^\infty(\mathbb{R}^d)}.$$

As in [46, 48], we consider FVM schemes in the MC-FVM algorithms which satisfy the following standard assumptions which are analogous to (22)–(24).

Assumption 2 *The abstract FV scheme* (60) *satisfies*

1. Stability: $\forall t \geq 0$

$$\|v_{\mathscr{T}}(\cdot, t)\|_{L^\infty(\mathbb{R}^d)} \leq \|v_{\mathscr{T}}(\cdot, 0)\|_{L^\infty(\mathbb{R}^d)}, \tag{63}$$

$$\|v_{\mathscr{T}}(\cdot, t)\|_{L^1(\mathbb{R}^d)} \leq \|v_{\mathscr{T}}(\cdot, 0)\|_{L^1(\mathbb{R}^d)}, \tag{64}$$

$$TV(v_{\mathscr{T}}(\cdot, t)) \leq TV(v_{\mathscr{T}}(\cdot, 0)), \tag{65}$$

2. **Lipschitz continuity:** *For any two sequences* $\underline{v} = (v_K)_{K \in \mathcal{T}}$, $\underline{w} = (w_K)_{K \in \mathcal{T}}$ *we have*

$$\|H_{\mathcal{T}}(\underline{v}) - H_{\mathcal{T}}(\underline{w})\|_{L^1(\mathcal{T})} \leq \|\underline{v} - \underline{w}\|_{L^1(\mathcal{T})} \tag{66}$$

or, equivalently,

$$\|H_{\mathcal{T}}(v_{\mathcal{T}}) - H_{\mathcal{T}}(w_{\mathcal{T}})\|_{L^1(\mathbb{R}^d)} \leq \|v_{\mathcal{T}} - w_{\mathcal{T}}\|_{L^1(\mathbb{R}^d)} . \tag{67}$$

3. **Convergence:** *If in the CFL bound* (58) *the CFL constant* $\lambda = \Delta t / \Delta x(\mathcal{T})$ *is kept constant as* $\Delta x(\mathcal{T}) \to 0$, *the approximate solution* $v_{\mathcal{T}}(x, t)$ *generated by* (59)–(62) *converges to the unique entropy solution* u *of the scalar conservation laws* (14)–(15) *at* $L^1(\mathbb{R}^d)$*-rate* $0 < s \leq 1$*: there exists* $C > 0$ *independent of* Δx *such that, as* $\Delta x \to 0$, *for every* $\bar{t}$ *and for* $(\Delta t)^s \leq \bar{t} \leq T$, *it holds*

$$\|u(\cdot, \bar{t}) - v_{\mathcal{T}}(\cdot, \bar{t})\|_{L^1(\mathbb{R}^d)} \leq \|u_0 - v_{\mathcal{T}}^0\|_{L^1(\mathbb{R}^d)} + C \bar{t} \, TV(u_0) \, \Delta x^s . \tag{68}$$

Let us mention that (63)–(65) hold in particular for monotone schemes on Cartesian meshes, see [28, 37]. The classical error analysis of Kuzsnetsov, see e.g. [15], imply the convergence rate $s = 1/2$ in (68). In case of monotone schemes on general FV meshes, one might lose the bound on the total variation of the approximations, and the convergence rate, i.e., the rate s in (68) is correspondingly reduced, see [8].

The error bound (68) contains an initial data approximation error $\|u_0 - v_{\mathcal{T}}^0\|_{L^1(\mathbb{R}^d)}$. This error vanishes for step function initial data on $\mathcal{T}$ (as, e.g., in the solution of Riemann problems). More generally, this error can be bounded by Δx^s *provided* that u_0 has appropriate regularity: under Assumption 2, for $u_0 \in BV(\mathbb{R}^d)$ and for the cell-average projection $v_{\mathcal{T}}^0$ in (61), we obtain

$$\|u_0 - v_{\mathcal{T}}^0\|_{L^1(\mathbb{R}^d)} \leq C(\kappa, \sigma) \Delta x TV(u_0) \leq C(\kappa, \sigma) \Delta x^s TV(u_0), \tag{69}$$

as $s \leq 1$ in (68). Here, the constant $C(\kappa, \sigma)$ depends on the stencil size constant σ and the shape regularity constant κ in (57), and (54), respectively.

Explicit FV schemes (59), (60) subject to the CFL stability condition (58) exhibit a discrete finite domain of dependence result analogous to Proposition 2.

Proposition 3 (Discrete Finite Dependence Domain) *Under Assumption 2 and the assumptions of Proposition 2, in particular under the compact support assumption* (25) *on the random initial data* u_0, *for the explicit FV scheme* (53), (61) *there holds:*

1. *the projection of the initial data on triangulation* $\mathcal{T}$, $v_{\mathcal{T}}^0$, *defined in* (61), (62), *has compact support independently of* $\mathcal{T}$*: there exists* $c > 0$ *such that, for all* $0 < h(\mathcal{T}) \leq 1$, *and with probability* 1, *holds*

$$\text{supp}(v_{\mathcal{T}}^0) \subset [-(1+c)R, (1+c)R]^d . \tag{70}$$

2. *the discrete solutions satisfy a dependence domain result: with probability* 1 *and with the constant* $c > 1$ *from* (70) *for every* $t > 0$ *holds*

$$\text{supp}(v_{\mathscr{T}}(\cdot, t)) \subset [-(1 + c)(R + tB), (1 + c)(R + tB)]^d \,, \tag{71}$$

where B denotes the bound (27) *on the flux, and where* $c > 1$ *is as in* (70).

We refer to [42, Chapter 3.6] for a detailed discussion.

Let us finally mention that the work for the realization of scheme (59)–(62) on a bounded domain $D \subset \mathbb{R}^d$ scales as (using the CFL stability condition (58), i.e., $\Delta t / \Delta x(\mathscr{T}) \leq \lambda = \text{const.}$)

$$\text{Work}_{\mathscr{T}} = \mathcal{O}\left(\Delta t^{-1} \Delta x^{-d}\right) = \mathcal{O}\left(\Delta x^{-(d+1)}\right) \,, \tag{72}$$

with the constant implied in $\mathcal{O}(\cdot)$ depending on on the support domain D of the solution.

4.3 MC-FVM

In the Monte Carlo Finite Volume Methods (MC-FVMs), we combine MC sampling of the random initial data with the FVM (60). In the convergence analysis of these schemes, we shall require the application of the FVM (60) to random initial data $u_0 \in L^\infty(\Omega; (L^1 \cap L^\infty \cap BV)(\mathbb{R}^d))$: given a draw $u_0(x; \omega)$ of u_0, the FVM (60)–(62) produces a family $v_{\mathscr{T}}(x, t; \omega)$ of random step functions on $\mathscr{T}$.

Proposition 4 *Consider the FVM* (60)–(62) *for the approximation of the entropy solution corresponding to a draw* $u_0(\omega; x)$ *of the random initial data, satisfying Assumption 1.*

Then, if the FVM satisfies Assumption 2, and provided that $u_0 \in L^k(\Omega; Z)$, *the random grid functions* $\Omega \ni \omega \mapsto v_{\mathscr{T}}(\omega; x, t)$ *defined by* (58)–(62) *satisfy, for every* $0 < \bar{t} < \infty$, $0 < \Delta x < 1$, *and every* $k \in \mathbb{N} \cup \{\infty\}$, *the stability bounds*

$$\|v_{\mathscr{T}}(\cdot; \cdot, \bar{t})\|_{L^k(\Omega; L^\infty(\mathbb{R}^d))} \leq \|u_0\|_{L^k(\Omega; L^\infty(\mathbb{R}^d))} \,, \tag{73}$$

$$\|v_{\mathscr{T}}(\cdot; \cdot, \bar{t})\|_{L^k(\Omega; L^1(\mathbb{R}^d))} \leq \|u_0\|_{L^k(\Omega; L^1(\mathbb{R}^d))} \,. \tag{74}$$

We also have error bounds

$$\|u(\cdot; \cdot, \bar{t}) - v_{\mathscr{T}}(\cdot; \cdot, \bar{t})\|_{L^k(\Omega; L^1(\mathbb{R}^d))}$$
$$\leq \|u_0(\cdot; \cdot) - v_{\mathscr{T}}^0(\cdot; \cdot)\|_{L^k(\Omega; L^1(\mathbb{R}^d))} + C\bar{t}\Delta x^s \|TV(u_0(\cdot; \cdot))\|_{L^k(\Omega)} \,. \tag{75}$$

Remark 3

1. In order for $\|TV(u_0(\cdot;\cdot))\|_{L^k(\Omega)}$ in (75) to be meaningful, a sufficient condition is that $u_0 : \Omega \to BV(\mathbb{R}^d)$ be strongly measurable, which we assumed in (31).
2. The initial data approximation error term $\|u_0(\cdot;\cdot) - v_{\mathscr{T}}^0(\cdot;\cdot)\|_{L^k(\Omega;L^1(\mathbb{R}^d))}$ in (75) can be bounded as in (69) provided that the random initial data u_0 has sufficient regularity: if $u_0 : \Omega \to Z$ is strongly measurable, and if $u_0 \in L^k(\Omega;Z)$, then (69) implies

$$\|u_0 - v_{\mathscr{T}}^0\|_{L^k(\Omega;L^1(\mathbb{R}^d))} \leq C\Delta x . \tag{76}$$

This approximation error bound holds without assumption of bounded support on u_0.

4.3.1 Definition of the MC-FVM Scheme

We consider once more the SCL (14)–(15) with random data u_0 and with flux f satisfying (28)–(33). We assume the moment condition $u_0 \in L^k(\Omega;Z)$ for sufficiently large $k \in \mathbb{N}$. The MC-FVM scheme for the MC estimation of the mean (i.e., the ensemble average) of the random entropy solution is as follows.

Definition 3 (MC-FVM Scheme) Generate a sample of $M \in \mathbb{N}$ i.i.d. realizations $\{\widehat{u}_0^i\}_{i=1}^M$ of initial data, approximated on the triangulation $\mathscr{T}$ by cell average projections (61).

$$\widehat{u}^i(\cdot, t) = S(t)\widehat{u}_0^i(\cdot), \quad i = 1, \ldots, M. \tag{77}$$

Let $H_{\mathscr{T}}(\cdot)$ be a FVM scheme (59)–(62) satisfying Assumption 2. Then the MC-FVM approximations of $\mathscr{M}^k(u(\cdot, t))$ are defined as statistical estimates from the ensemble

$$\{\widehat{v}_{\mathscr{T}}^i(\cdot, t)\}_{i=1}^M \tag{78}$$

obtained by (60) from the FV approximations $\widehat{v}_{\mathscr{T}}^i(\cdot, 0)$ of the M i.i.d initial data samples $\{\widehat{u}_0^i(x)\}_{i=1}^M$ by (61): specifically, the first moment of the random solution $u(\omega; \cdot, t)$ at time $t > 0$, is approximated by the sample average of FV solutions,

$$\mathscr{M}^1(u(\cdot, t)) \approx E_M[v_{\mathscr{T}}(\cdot, t)] := \frac{1}{M} \sum_{i=1}^M \widehat{v}_{\mathscr{T}}^i(\cdot, t). \tag{79}$$

4.3.2 Convergence Rates for MC-FVM

We next address the convergence of $E_M[v_{\mathscr{T}}]$ to the mean $\mathbb{E}(u)$.

Theorem 4 *Assume that all realizations of the random initial data u_0 are supported on one common, bounded domain $[-R, R]^d \subset \mathbb{R}^d$ for some $0 < R < \infty$ and satisfy (28)–(32). Assume further given a FVM (59)–(62) such that (58) holds and such that Assumption 2 is satisfied; in particular, assume that the deterministic FVM scheme converges at rate $s > 0$ in $C([0, T]; L^1(\mathbb{R}^d))$ for every $0 < T < \infty$, i.e. (68) holds.*

Then, the MC estimate $E_M[v_{\mathscr{T}}(\cdot, t)]$ defined in (79) satisfies, for every M, the error bound

$$\|\mathbb{E}[u(\cdot, t)] - E_M[v_{\mathscr{T}}(\omega; \cdot, t)]\|_{L^2(\Omega; L^1(\mathbb{R}^d))}$$

$$\leq C(D, T)\Big[M^{-\frac{1}{2}} \|u_0\|_{L^2(\Omega; L^1(\mathbb{R}^d))} \tag{80}$$

$$+ \|u_0 - v_{\mathscr{T}}^0\|_{L^\infty(\Omega; L^1(\mathbb{R}^d))} + t\Delta x^s \|TV(u_0(\omega; \cdot))\|_{L^\infty(\Omega)}\Big]$$

where $C > 0$ depends on the final time T and the domain D, in which the initial data is supported $\mathbb{P}$-a.s. but is independent of M and of Δx as $M \to \infty$ and as $\lambda \Delta x = \Delta t \downarrow 0$.

Proof The proof of the above theorem proceeds along the lines of the proof of [46, Thm. 4.6]. However, we point out that there was an error in the argument of the proof of [46, Theorem 4.6] due to the incorrect derivation of a direct MC sampling convergence rate in the type one Banach space $L^1(\mathbb{R}^d)$. On the other hand and as mentioned in the previous section, we may use the local support assumptions on the initial data, and the finite speed of propagation implied by hyperbolicity, to obtain FV convergence rate bounds in the type 2 space $L^2(\mathbb{R}^d)$ from which follows the claimed convergence rate.

For fixed $t > 0$, we have

$$\|\mathbb{E}[u(\cdot, t)] - E_M[v_{\mathscr{T}}(\cdot, t)]\|_{L^2(\Omega; L^1(\mathbb{R}^d))} \leq \|\mathbb{E}[u(\cdot, t)] - E_M[u(\cdot, t)]\|_{L^2(\Omega; L^1(\mathbb{R}^d))}$$
$$+ \|E_M[u(\cdot, t)] - E_M[v_{\mathscr{T}}(\cdot, t)]\|_{L^2(\Omega; L^1(\mathbb{R}^d))}$$
$$=: \mathrm{I} + \mathrm{II} \quad .$$

Term I is a MC error which can be bounded by (50).

Term II is essentially a discretization error bound. By the (pathwise) FV error bounds (68) and by (21)–(24) and Assumption 2 by the triangle inequality that

$$\mathrm{II} = \|E_M[u(\cdot, t; \omega) - v_{\mathscr{T}}(\cdot, t)]\|_{L^2(\Omega; L^1(\mathbb{R}^d))}$$

$$\leq \frac{1}{M} \sum_{j=1}^{M} \|\hat{u}^j(\cdot, t; \omega) - \hat{v}_{\mathscr{T}}^j(\cdot, t; \omega)]\|_{L^2(\Omega; L^1(\mathbb{R}^d))}$$

$$\leq \|u(\cdot, t; \omega) - v_{\mathscr{T}}(\cdot, t; w)]\|_{L^2(\Omega; L^1(\mathbb{R}^d))}$$

$$\leq C\{\|u_0 - v_{\mathscr{T}}^0\|_{L^2(\Omega; L^1(\mathbb{R}^d))} + t\Delta x^s \|TV(u_0(\cdot, w))\|_{L^2(\Omega)}\} .$$

$\square$

The initial data approximation error $\|u_0 - v_{\mathscr{T}}^0\|_{L^2(\Omega;L^1(\mathbb{R}^d))}$ can be bounded by Δx as indicated in Remark 3, item 2.

4.3.3 Work Estimates

To calculate the error versus computational work, we estimate the asymptotic complexity of computing the estimators along the lines of [48]. In doing this, we *assume that the computational domain $D \subset \mathbb{R}^d$ is bounded and suitable boundary conditions are specified on ∂D.* In this *bounded, computational domain D*, the work for one time step (59), (60) is of order $\mathscr{O}\left(\Delta x^{-d}\right)$ (with $\mathscr{O}(\cdot)$ depending on the size of the domain and on the size of stencil employed in the FV scheme) we find from the CFL condition (58) that the total computational work to obtain $\{v_{\mathscr{T}}(\cdot,t)\}_{0<t\leq T}$ in D is by (72)

$$\text{Work}(\mathscr{T}) = \mathscr{O}\left(\Delta x^{-d-1}\right), \quad \text{as } \lambda\Delta x = \Delta t \downarrow 0. \tag{81}$$

The work for the computation of the MC estimate $E_M[v_{\mathscr{T}}(\cdot,t)]$ is assumed to scale as

$$\text{Work}(M, \mathscr{T}) = \mathscr{O}\left(M\Delta x^{-d-1}\right), \quad \text{as } \Delta t = \lambda\Delta x \downarrow 0. \tag{82}$$

The bound (80) allows to infer convergence order estimates in terms of work. To derive these, we choose $M^{-1/2} \sim \Delta x^s$ in (80). Setting the implied constant equal to one results in $M = \lceil \Delta t^{-2s} \rceil$. Inserting in (82) yields

$$\text{Work}(\mathscr{T}) = \mathscr{O}\left(\Delta t^{-2s} \Delta x^{-(d+1)}\right) \overset{(58)}{=} \mathscr{O}\left(\Delta x^{-(d+1)-2s}\right) \tag{83}$$

so that we obtain from (80)

$$\|\mathbb{E}[u(\cdot,t)] - E_M[v_{\mathscr{T}}(\cdot,t)]\|_{L^2(\Omega;L^1(\mathbb{R}^d))} \leq C\Delta t^s \leq C(\text{Work}(\mathscr{T}))^{-s/(d+1+2s)}. \tag{84}$$

Regarding the convergence rate in the estimate (84), for the *deterministic FV scheme* holds

$$\text{Work}(\mathscr{T}) = \mathscr{O}\left(\Delta t^{-1} \Delta x^{-d}\right) \overset{(58)}{=} \mathscr{O}\left(\Delta x^{-(d+1)}\right).$$

The bound on the FV error, (68), becomes when written in terms of work, equal to

$$\|u(\cdot,\bar{t}) - v_{\mathscr{T}}(\cdot,\bar{t})\|_{L^1(\mathbb{R}^d)}$$
$$\leq \|u_0 - v_{\mathscr{T}}^0\|_{L^1(\mathbb{R}^d)} + C\bar{t}\, TV(u_0)\, (\text{Work}(\mathscr{T}))^{-s/(d+1)}. \tag{85}$$

Ignoring initial data approximation errors, which are negligible in comparison to the computational work for the time marching, the exponent $-s/(d+1)$ for the deterministic FVM as compared to $-s/(d+1+2s)$ for the MC-FVM. For low order FV schemes (i.e., for small values of the convergence rate s) and in space dimensions $d = 2, 3$, we observe a considerably reduced rate of convergence of the MC-FVM. For high order FV schemes, we recover in (84) the MC rate $1/2$ of the error in terms of work.

4.4 Multilevel MC-FVM

Next, we present and analyze a scheme that allows us to achieve a better accuracy versus work bound for the random initial data u_0, compared to the standard MC-FVM error bound (84). The Multilevel Monte Carlo Finite Volume (MLMC-FVM) scheme is based on MC sampling with level dependent sample sizes M_ℓ on different levels ℓ of resolution of the FVM. Throughout, we assume the explicit FV scheme satisfies Assumption 2 and the CFL stability condition (58). To define the MLMC-FVM, we start by reviewing notation as used in [48].

4.4.1 Notation

The MLMC-FVM is defined as a *multilevel discretization* in x and t with level dependent numbers M_ℓ of samples. To this end, we assume we are given a family $\{\mathcal{T}_\ell\}_{\ell=0}^\infty$ of *nested triangulations* of $\mathbb{R}^d$ such that the mesh width

$$\Delta x_\ell = \Delta x(\mathcal{T}_\ell) = \sup\{\mathrm{diam}(K) : K \in \mathcal{T}_\ell\} = \mathcal{O}\left(2^{-\ell}\Delta x_0\right), \quad \ell \in \mathbb{N}_0, \qquad (86)$$

where K denotes a generic FV cell $K \in \mathcal{T}$. We also assume the family $\mathfrak{M} = \{\mathcal{T}_\ell\}_{\ell=0}^\infty$ of meshes to be shape regular; if $K \in \mathcal{T}_\ell$ denotes a generic cell, we recall, for a generic mesh $\mathcal{T} \in \mathfrak{M}$, the *shape regularity constants* $\kappa(\mathcal{T})$ defined in (54). We say that *the family $\mathfrak{M}$ of meshes is κ-shape regular*, if there exists a constant $\kappa(\mathfrak{M}) < \infty$ such that with ρ_K denoting the diameter of the largest ball inscribed into K

$$\kappa(\mathfrak{M}) = \sup_{\mathcal{T} \in \mathfrak{M}} \kappa(\mathcal{T}) = \sup_{\mathcal{T} \in \mathfrak{M}} \sup_{K \in \mathcal{T}} \frac{\mathrm{diam}(K)}{\rho_K}. \qquad (87)$$

For a mesh hierarchy $\mathfrak{M} = \{\mathcal{T}_\ell\}_{\ell=0}^\infty$, we denote

$$\Delta x_\ell := \Delta x(\mathcal{T}_\ell), \qquad \mathcal{T}_\ell \in \mathfrak{M}, \quad \ell = 0, 1, \dots . \qquad (88)$$

4.4.2   MLMC-FVM

The MLMC FVM consists in estimates of $\mathbb{E}[u(\cdot, t)]$ obtained by replacing $u(\cdot, t)$ by a FV discretization, on a sequence $\mathfrak{M}$ of discretizations $\{\mathscr{T}_\ell\}_{\ell \in \mathbb{N}_0}$ which we assume to be nested. We denote in the present section the FV approximation $v_\mathscr{T}$ on triangulation $\mathscr{T} \in \mathfrak{M}$ by $v_\ell(\cdot, t)$. On $\mathscr{T}_\ell \in \mathfrak{M}$, the CFL condition (58) takes the form

$$\Delta t_\ell \leq \lambda \Delta x_\ell, \quad \ell = 0, 1, 2, \ldots, . \tag{89}$$

We assume the CFL constant $\lambda > 0$ to be independent of ℓ and of the input realization ω; this will allow for deterministic error vs. work bounds; we refer to [49] for a discussion of error vs. work of MLMC for nonuniform (log-gaussian) random inputs.

As the FV scheme is CFL stable, we may generate a sequence $\{v_\ell(\cdot, t)\}_{\ell=0}^\infty$ of stable FV approximations on triangulation $\mathscr{T}_\ell$ for time steps of sizes Δt_ℓ which satisfy the CFL condition (89) with respect to mesh $\mathscr{T}_\ell \in \mathfrak{M}$. We set in what follows $v_{-1}(\cdot, t) := 0$. Then, given a target (finest) level $L \in \mathbb{N}$ of spatial resolution, we may use the linearity of the expectation operator to write, as is customary in MLMC analysis (see, e.g., [27])

$$\mathbb{E}[v_L(\cdot, t)] = \sum_{\ell=0}^{L} \mathbb{E}\left[v_\ell(\cdot, t) - v_{\ell-1}(\cdot, t)\right]. \tag{90}$$

We next estimate each term in the sum (90) by a Monte-Carlo method with a level-dependent number of samples, M_ℓ, to obtain the MLMC-FVM estimator,

$$E^L[u(\cdot, t)] = \sum_{\ell=0}^{L} E_{M_\ell}\left[v_\ell(\cdot, t) - v_{\ell-1}(\cdot, t)\right]. \tag{91}$$

Here, $E_M[v_\mathscr{T}(\cdot, t)]$ is the standard MC estimator defined in (79), and $v_\ell(\cdot, t)$ denotes the FV solution on $\mathscr{T}_\ell$, computed under the CFL assumption (89), with $\Delta t_\ell \leq \lambda \Delta x_\ell$ where $\Delta x_\ell := \Delta x(\mathscr{T}_\ell)$ denotes the meshwidth at mesh level ℓ (see (55)) and where the CFL constant $\lambda > 0$ is independent of ℓ. We emphasize that the form of the estimator (90) implies that the same draw of the random initial data should be approximated on two successive meshes in the hierarchy.

4.4.3   Convergence Analysis

The MLMC-FVM mean field error

$$\left\|\mathbb{E}[u(\cdot, t)] - E^L[u(\cdot, t)]\right\|_{L^2(\Omega; L^1(\mathbb{R}^d))} \tag{92}$$

for $0 < t < \infty$ and $L \in \mathbb{N}$ was analyzed in [46] for the SCL (13) with random initial data and deterministic flux. Analogous results for the more general SCL with random fluxes was shown in [48]. The choice of the sample sizes $\{M_\ell\}_{\ell=0}^\infty$ is such that, for every $L \in \mathbb{N}$, the MLMC error (92) is of order $(\Delta x_L)^s$, where s is the order of convergence in the Kuznetsov type error bound (68). The design of MLMC-FVM is based on a judicious choice of MC sample numbers $\{M_\ell\}_{\ell=0}^\infty$ at the discretization levels ℓ. To derive it, we observe that for each L, the error bound (92) holds with work bounded by

$$\text{Work}_L = \sum_{\ell=0}^L M_\ell \mathcal{O}\left(\Delta x_\ell^{-d-1}\right) = \mathcal{O}\left(\sum_{\ell=0}^L M_\ell \Delta x_\ell^{-d-1}\right). \tag{93}$$

The MLMC convergence analysis in [46] used incorrectly a MC convergence estimate for the space $L^1(\mathbb{R}^d)$ which is a Banach space of type 1. This error was rectified in a recent paper [48], where the following bound on the variance of the FV details was shown:

$$\|(v_\ell - v_{\ell-1})(\cdot, t)\|^2_{L^2(\Omega; L^1(\mathbb{R}^d))} \leq C(D, T) \Delta x_\ell^s \|u_0\|^2_{L^2(\Omega; W^{s,2}(\mathbb{R}^d))} . \tag{94}$$

Theorem 5 ([48, Thm. 4.7]) *Suppose that Assumption 1, items 1–5 hold, and that, moreover, (28)–(32) and (87)–(89) are valid. Then, for any sequence $\{M_\ell\}_{\ell=0}^\infty$ of MC sample numbers at mesh level ℓ, we have for the MLMC-FVM estimate $E^L[u(\cdot, t)]$ in (91) the error bound*

$$\left\| \mathbb{E}[u(\cdot, t)] - E^L[u(\cdot, t)] \right\|_{L^2(\Omega; L^1(\mathbb{R}^d))}$$
$$\leq C\left(D, T, \|u_0\|_{L^\infty(\Omega; Z)}\right) \left[\Delta x_L^s + \sum_{\ell=1}^L M_\ell^{-\frac{1}{2}} \Delta x_\ell^{\frac{s}{2}} + M_0^{-\frac{1}{2}} \right] \tag{95}$$

where C is a constant that depends on the final time $T < \infty$, the initial data and on the bounded domain D which contains, according to (52), the support of $u(t)$ with probability one, but is independent of L.

Proof We calculate for any $t \in [0, T]$

$$\left\| \mathbb{E}[u(\cdot, t)] - E^L[u(\cdot, t)] \right\|_{L^2(\Omega; L^1(\mathbb{R}^d))} \leq \underbrace{\left\| \mathbb{E}[u(\cdot, t)] - \mathbb{E}[v_L(\cdot, t)] \right\|_{L^2(\Omega; L^1(\mathbb{R}^d))}}_{I}$$
$$+ \underbrace{\left\| \mathbb{E}[v_L(\cdot, t)] - E^L[u(\cdot, t)] \right\|_{L^2(\Omega; L^1(\mathbb{R}^d))}}_{II}$$

In complete analogy with the estimation of term *II* in proof of Theorem 4, the term *I* in the above estimate can be readily estimated in terms of the spatio-temporal discretization error of the FV scheme at the finest mesh resolution with diameter

Δx_L as follows,

$$I \leq C(T, \|u_0\|_{L^\infty(\Omega;Z)}) \Delta x_L^s.$$

To estimate term II in the above estimate, we use the discrete finite dependence domain result, Proposition 3, and the bounded support assumption (32) on the random initial data and proceed as follows,

$$II = \left\| \mathbb{E}[v_L(\cdot,t)] - E^L[u(\cdot,t)] \right\|_{L^2(\Omega;L^1(\mathbb{R}^d))}$$

$$= \left\| \sum_{l=0}^{L} \left[\mathbb{E}[v_l(\cdot,t) - v_{l-1}(\cdot,t)] - E_{M_\ell} \left[v_\ell(\cdot,t) - v_{\ell-1}(\cdot,t) \right] \right] \right\|_{L^2(\Omega;L^1(\mathbb{R}^d))}$$

$$\leq \sum_{l=0}^{L} \left\| \mathbb{E}[v_l(\cdot,t) - v_{l-1}(\cdot,t)] - E_{M_\ell} \left[v_\ell(\cdot,t) - v_{\ell-1}(\cdot,t) \right] \right\|_{L^2(\Omega;L^1(\mathbb{R}^d))}$$

$$\leq C(D,T) \sum_{l=0}^{L} \left\| \mathbb{E}[v_l(\cdot,t) - v_{l-1}(\cdot,t)] - E_{M_\ell} \left[v_\ell(\cdot,t) - v_{\ell-1}(\cdot,t) \right] \right\|_{L^2(\Omega;L^2(\mathbb{R}^d))}$$

$$\leq C(D,T) \left\{ \frac{\|v_0(\cdot,t)\|_{L^2(\Omega;L^2(\mathbb{R}^d))}}{M_0^{\frac{1}{2}}} + \sum_{l=1}^{L} \left(\frac{\|(v_\ell - v_{\ell-1})(\cdot,t)\|_{L^2(\Omega;L^2(\mathbb{R}^d))}}{M_l^{\frac{1}{2}}} \right) \right\}$$

In the final estimate, we used the compact support assumption (50) and in the final step, we used the standard Hilbert space MC estimate (12) for the *detail* $v_\ell - v_{\ell-1}$. Accordingly, we need to bound the variance of the details $v_\ell - v_{\ell-1}$ in the (Hilbert) space $L^2(\mathbb{R}^d)$ according to

$$\|(v_\ell - v_{\ell-1})(\cdot,t)\|_{L^2(\Omega;L^2(\mathbb{R}^d))} \leq C(D,T) \|u_0\|_{L^2(\Omega;Z)} \Delta x_\ell^{\frac{s}{2}}. \tag{96}$$

Note that the use of the (Hilbertian) $L^2(\mathbb{R}^d)$ norm in the error bound entails the convergence rate $s/2$ of the FV detail $v_\ell - v_{\ell-1}$, where $0 < s \leq 1$ denotes the $L^1(\mathbb{R}^d)$ convergence rate of the (deterministic) FV approximation. Substituting the above in the estimate for term I and using the stability of the numerical solution at the coarsest level of discretization Δx_0, we arrive at (95). □

The error bound (95) is then used to select MC sample numbers M_ℓ at discretization level ℓ to achieve a prescribed tolerance $\varepsilon \sim \Delta x_L^s$, with minimal computational work.

A (standard in MLMC by now) Lagrange multiplier argument (see, e.g., Giles [27] and references therein) allows to solve the corresponding constrained minimization problems results in sample number choices obtained, for example, in [48] for $0 < s < d + 1$,

$$
M_\ell \sim \frac{\Delta x_\ell^{\frac{(s+d+1)}{2}}}{\Delta x_L^{2s}} \sum_{k=0}^{L} \Delta x_k^{\frac{(s-(d+1))}{2}} \sim \frac{\Delta x_\ell^{\frac{(s+d+1)}{2}}}{\Delta x_L^{\frac{(3s+d+1)}{2}}} \tag{97}
$$

with $\sim$ denoting equivalence uniform with respect to L and ℓ.

As in [46], we use the sample numbers M_ℓ in (97) to obtain the following error estimate in terms of work

$$
\left\| \mathbb{E}[u(\cdot, t)] - E^L[u(\cdot, t)] \right\|_{L^2(\Omega; L^1(\mathbb{R}^d))} \leq C \left(\text{Work}(\{M_\ell\}_{\ell=0}^L; \mathscr{T}_L) \right)^{-s/(d+1+s)}. \tag{98}
$$

The complexity estimate (98) shows that the MLMC FVM can be more efficient than the MC FVM (84), in terms of computational work that needs to be performed for obtaining the same error. However, to achieve a comparable error in $L^2(\Omega; L^1(\mathbb{R}^d))$, the MLMC method is more expensive than a single deterministic solve.

Remark 4 The above discussion on random entropy solutions and (ML)MC methods considered the simplest case of a scalar conservation law with random initial data. These notions and methods were extended to the case where the flux function in a scalar conservation law is random. In a recent paper [48], where the appropriate notion of random entropy solutions were defined and shown to exist, provided that the uncertainty in the flux satisfied certain assumptions, which ensure the random flux to be Bochner measurable and $\mathbb{P}$-a.s. separably valued in the space of Lipschitz continuous flux functions. Both the MC-FV and MLMC-FV methods were analyzed in this context and the MLMC-FV method was shown to satisfy the same error vs computational work i.e., (98) as in the case of deterministic fluxes and random data. Consequently, the MLMC method is significantly more efficient than the corresponding MC-FV method. We refer to [48] for details.

5 Statistical Moments

The error bounds for the MLMC-FVM obtained up to this point addressed the numerical estimation of the "ensemble average", or mean-field. Here, we briefly comment on efficient numerical approximation of 2- and k-point correlation functions of random solutions. When the physical problem is posed in d spatial variables, *spatial k-point correlation functions* of random or of statistical solutions can be represented (under the provision of sufficient regularity) as deterministic functions of kd variables.

The "natural", FV approximation of k-point spatial correlations of random entropy solutions as well as of correlation margins of statistical EMV solutions introduced in [17, 18], will involve k-fold (algebraic) tensorisation of (finite dimensional, by the bounded support assumption (32) on the random initial data, and by the corresponding bounded support property (71) of FV approximations implied by the uniform hyperbolicity of the SCL). This can increase computational complexity due to the low convergence rate $1/2$ of MC sampling, and to the so-called "curse of dimensionality", which entails complexity $\mathscr{O}\left(\Delta x^{-kd-1}\right)$ in k-point correlation estimation. This can be prohibitive, in particular in space dimension $d = 3$, even for two-point correlations where $k = 2$. Two algorithmic strategies are next presented which allow, to some extent and under appropriate regularity conditions, to reduce the computational complexity: first, the MC statistical estimation of kth central statistical moments in the physical dimension D as analyzed recently in [3], and second, k-*fold sparse tensor products* of FV solutions in the physical domain $D \subset \mathbb{R}^d$ as proposed in [46]. They render k-point correlations in $D^k \subset \mathbb{R}^{kd}$ numerically accessible in $\mathscr{O}\left(\Delta x^{-d-1} |\log \Delta x|^{k-1}\right)$ operations.

5.1 Estimation of kth Order Central Statistical Moments

Given a random entropy solution u, a "natural" MC estimator based on M iid samples for the kth central moment of $u(\cdot; x, t)$ reads

$$S_M^k[u] := \frac{1}{M} \sum_{i=1}^{M} (u^i - E_M[u])^k , \qquad (99)$$

assuming availability of exact solution samples u^i of the random entropy solution. The estimator (99) is to be interpreted pointwise w.r. to $x \in D$ and $t \in [0, T]$. It is, however, well-known to be a statistically biased estimator. Unbiased estimators are known. For example,

$$\tilde{S}_M^2[u] := \frac{M}{M-1} S_M^2[u] , \quad \tilde{S}_M^3[u] := \frac{M^2}{(M-1)(M-2)} S_M^3[u] \qquad (100)$$

are unbiased estimators of $\mathscr{M}^2(u)$ and $\mathscr{M}^3(u)$. For $k \geq 4$, unbiased estimators $\tilde{S}_M^k[u]$ can be obtained as polynomial expressions of $S_M^r[u]$ for $r = 1, \ldots, K$ which are, however, not unique. We refer to [3, Lemma 3] for details.

A MLMC estimator of $\mathscr{M}^k(u)(x, \ldots, x; t)$ is introduced and analyzed in [3, Theorem 1].

As the corresponding algorithms only access the FV solver through iid samples of the random initial data and random flux, respectively, they are nonintrusive and embarrasingly parallel. Operating only in the physical domain $D \subseteq \mathbb{R}^d$, they do not require additional data structures for tensorization of FV solutions. Central statistical moments $\mathscr{M}^k(u)(x, \ldots, x; t)$ can, however, also be numerically approximated as "diagonals" of statistical k-point correlations $\mathscr{M}^k(u)$. We discuss this next.

5.2 Estimation of k-Point Spatial Correlations

The work to form a single tensor product over a bounded computational domain $D \subset \mathbb{R}^d$ (such as, e.g., the support domain of the solution at time t in (26)) grows, ignoring timestepping, as $O(\Delta x^{-kd})$ which may entail a computational effort that is, for moment orders $k \geq 2$, prohibitive. To reduce the complexity of k-point correlation function estimation, a so-called "sparse grid" or "sparse tensor product" approach was proposed in [2, 46, 62]. We refer to [31, 53, 58] and the references there for general presentations of sparse tensor product spaces.

5.2.1 k-Point Correlation Estimation by Sparse Tensorization of FV Solutions

As is standard in multi-resolution analysis, the cell-average projections P_ℓ : $L^1(\mathbb{R}^d) \to S_\ell$, defined in (88), (104), allow us to introduce spaces of *increments* or *details* in the FV mesh hierarchy $\mathfrak{M} = \{\mathscr{T}_\ell\}_{\ell=0}^\infty$:

$$W_\ell := (P_\ell - P_{\ell-1})S_\ell, \qquad \ell \geq 0 \tag{101}$$

where $P_{-1} := 0$ so that $W_0 = S_0$. Then, for any $L \in \mathbb{N}_0$, we have the multilevel decomposition

$$S_L = W_0 \oplus \ldots \oplus W_L = \bigoplus_{\ell=0}^{L} W_\ell . \tag{102}$$

The k-point correlations $(v_L(\cdot, t))^{(k)}$ of FV solutions at time $t > 0$ belong to the (algebraic) tensor product space

$$(S_L)^{(k)} := \underbrace{S_L \otimes \ldots \otimes S_L}_{k \text{ times}} = \sum_{|\underline{\ell}|_\infty \leq L} S_{\ell_1} \otimes \ldots \otimes S_{\ell_k} = \bigoplus_{|\underline{\ell}|_\infty \leq L} \bigotimes_{j=1}^{k} W_{\ell_j} . \tag{103}$$

In the numerical realization of 2- and k point correlation functions and correlation margins of measure valued solutions, the computational realization of approximations in the tensor product space $(S_L)^{(k)}$ is necessary. To formulate it, we introduce the k-fold algebraic tensor products of the FV cell-average projections by

$$P_L^{(k)}v := \underbrace{P_L \otimes \ldots \otimes P_L}_{k \text{ times}} : L^1(\mathbb{R}^{kd}) \to (S_L)^{(k)} . \tag{104}$$

In the case that $N_L := \dim S_L < \infty$ (as e.g. when the spaces S_ℓ are only defined on a bounded domain $D \subset \mathbb{R}^d$ such as the "support box" (26) in Proposition 2) then $\dim((S_L)^{(k)}) = N_L^k$ which is prohibitive. *Sparse Tensor approximations of k-point correlation functions*, $(v(\cdot, t))^{(k)}$ are "compressed" FV approximations which involve FV spaces of piecewise constant functions on coarser meshes. They are defined in terms of the *increment space* W_ℓ in (101) by

$$\widehat{(S_L)}^{(k)} := \bigoplus_{|\underline{\ell}|_1 \leq L} \bigotimes_{j=1}^{k} W_{\ell_j} \tag{105}$$

where $|\underline{\ell}|_1 := \ell_1 + \ldots + \ell_k$ and where algebraic tensor products are implied. Note that a realization of the sparse tensor product space $\widehat{(S_L)}^{(k)}$ according to its definition (105) requires construction of explicit bases for W_ℓ in (101); on unstructured, simplicial triangulations $\mathscr{T}$ as in our Assumption 2 such bases can be numerically constructed by agglomeration, see, e.g. [1]. A one-scale FV solution on the finest mesh $\mathscr{T}_L$ can be converted to a ML representation in $O(N_L)$ operations by the so-called *pyramid scheme* (see, e.g., [5, pp. 225–294]). If the mesh family $\mathfrak{M}$ used in the pathwise FV approximation (see Assumption 2) is generated by recursive dyadic refinements of the initial triangulation $\mathscr{T}_0$, when $N_L = \dim S_L < \infty$ (as is the case e.g. on bounded domains $D \subset \mathbb{R}^d$) it holds (see, e.g. [53, 58])

$$\dim \widehat{(S_L)}^{(k)} = O(N_L(\log_2 N_L)^{k-1}) . \tag{106}$$

Having at hand the sparse tensor product space $\widehat{(S_L)}^{(k)}$ in (105), we also define the *sparse tensor projection*

$$\widehat{(P_L)}^{(k)} := \bigoplus_{|\underline{\ell}|_1 \leq L} \bigotimes_{j=1}^{k} (P_{\ell_j} - P_{\ell_j-1}) : L^1(\mathbb{R}^{kd}) \to \widehat{(S_L)}^{(k)} . \tag{107}$$

We refer to [31, 58] and the references there, from where we briefly recapitulate approximation properties of sparse tensor product projections: for any function $U(x_1, \ldots, x_k)$ which belongs to $(W^{s,1}(\mathbb{R}^d))^{(k)}$ being the space of functions of k variables $x_1, \ldots, x_d \in \mathbb{R}^d$ which are, with respect to each variable, in $W^{s,1}(\mathbb{R}^d)$, it holds

$$\|U - \widehat{(P_L)}^{(k)} U\|_{L^1(\mathbb{R}^{kd})} \leq C(\Delta x_L)^s |\log \Delta x_L|^{k-1} \|U\|_{(W^{s,1}(\mathbb{R}^d))^{(k)}} \tag{108}$$

where $C > 0$ is independent of Δx_L (it depends only on k, d and the shape regularity of the family $\mathfrak{M}$ of triangulations, but is independent of Δx_L).

5.2.2 Sparse MLMC-FVM Estimator

The MLMC sparse FV estimator is based on a sparse tensor product FV approximation for each MC sample. To define it, we recall that $E_M[\cdot]$ denotes the MC estimate based on M samples): for a given sequence $\{M_\ell\}_{\ell=0}^L$ of MC sample numbers at level $\ell = 0, \ldots, L$, the sparse tensor MLMC estimate of $\mathscr{M}^k[u(\cdot, t)]$ is, for $0 < t < \infty$, defined by

$$\widehat{E^{L,(k)}}[u(\cdot, t)] := \sum_{\ell=0}^L E_{M_\ell}[\widehat{P_\ell}^{(k)}(v_\ell(\cdot, t))^{(k)} - \widehat{P_{\ell-1}}^{(k)}(v_{\ell-1}(\cdot, t))^{(k)}]. \tag{109}$$

We observe that (109) is identical to (103) except for the sparse formation of the k-point correlation functions of the FV solutions corresponding to the initial data samples $\hat{u}_0^i$. In bounded domains, this reduces the work for the formation of the k-point correlation function from N_L^k to $O(N_L(\log_2 N_L)^{k-1})$ per sample at mesh level L. As is well known (see, e.g. [31, 58]) use of sparse rather than full tensor products essentially preserves (i.e., up to logarithmic w.r. to Δx terms) the order s of FV convergence of sparse tensor product k point correlation function approximations.

5.2.3 Combination Formula

Sparse tensor products are particularly easy to realize when the FV scheme already produces FV solutions to SCLs in MRA format. Such schemes are nonstandard, but available even on unstructured meshes as we admit in Assumption 2, in a development [25, 50] originating in the seminal work of A. Harten [1, 4]. Often, however, only single-level (one-scale) numerical FV approximations on triangulations $\mathscr{T}_\ell$, $\ell = 0, \ldots, L$ are available. In order to realize the MLMC-FV estimator (109), for each realization the approximations must be converted to a MR representation. This can be achieved also on unstructured meshes in linear complexity by the so-called *pyramid scheme* (see, e.g., [52] for a definition and an algorithm).

An alternative approach which obviates MR based numerical methods is the so-called *combination formula*, as proposed for this purpose (in a different context) in [32, Lemma 12, Thm. 13]. For the projector $\widehat{(P_L)}^{(k)}$ in (107), the *combination identity*

$$\widehat{(P_L)}^{(k)} = \sum_{i=0}^{k-1} (-1)^i \binom{k-1}{i} \sum_{|\underline{\ell}|_1 = k-i} P_{\underline{\ell}}^{(k)}, \quad \text{where } P_{\underline{\ell}}^{(k)} := \bigotimes_{j=1}^k P_{\ell_j} \tag{110}$$

holds. The combination identity (110) implies that sparse tensor MC-FV approximations of k-point correlation functions can be numerically built from (pointwise) products of standard, one-scale FV approximations of the SCL (13)–(15) with iid samples of the random initial data u_0 on mesh levels $\underline{\ell} := (\ell_1, \ldots, \ell_k)$. When inserted into (109), the combination identity (110) provides an explicit realization of the MLMC-FVM estimator of k-point correlations of random entropy solutions, based exclusively on (parallel) standard FV solves on all mesh levels.

5.2.4 Error Bounds and Complexity Analysis

We can now generalize Theorems 4 and 5 to sparse tensor MLMC-FV estimates for the kth moments of random entropy solutions; it also applies to estimates of correlation measures of entropy statistical solutions as introduced in [19].

Theorem 6 *Assume that the random initial data u_0 satisfies Assumption 1, items 1–4 and that the FV scheme satisfies Assumption 2, items 1–3. In particular, assume that the deterministic FV scheme converges at rate $s > 0$ according to (68) and that for u_0, the bounded support Assumption 2, item 4, (32) holds.*

Assume further that the FV scheme satisfies the CFL condition (58) and that the random initial data u_0 is Bochner-integrable of order $2k$ in the spaces $Z = (L^1 \cap L^\infty \cap BV)(\mathbb{R}^d)$ and in $W^{s,1}(\mathbb{R}^d)$, i.e.

$$u_0 \in L^{2k}(\Omega; Z) \cap L^{2k}(\Omega; W^{s,1}(\mathbb{R}^d)) . \tag{111}$$

Then, MLMC-FVM estimator $\widehat{E^{L,(k)}}[u(\cdot, t)]$ defined in (109) satisfies, for every sequence $\{M_\ell\}_{\ell=0}^L$ of MC samples, the error bound

$$\|\mathscr{M}^k u(\cdot, t) - \widehat{E^{L,(k)}}[u(\cdot, t; \omega)]\|^2_{L^2(\Omega; L^1(\mathbb{R}^{kd}))}$$

$$\lesssim (1 \vee t)\Delta x_L^{2s} |\log \Delta x_L|^{2(k-1)} \left\{ \|TV(u_0(\cdot, \omega))\|^{2k}_{L^k(\Omega; d\mathbb{P})} + \|u_0(\cdot; \omega)\|^{2k}_{L^\infty(\Omega; W^{s,1}(\mathbb{R}^d))} \right\}$$

$$+ \left\{ \sum_{\ell=0}^L \frac{\Delta x_\ell^s |\log \Delta x_\ell|^{k-1}}{M_\ell} \right\} \left\{ \|u_0(\cdot; \omega)\|^{2k}_{L^{2k}(\Omega; W^{s,1}(\mathbb{R}^d))} + t^2 \|TV(u_0(\cdot; \omega))\|^{2k}_{L^{2k}(\Omega; d\mathbb{P})} \right\} .$$

Here, the constant implied in $\lesssim$ depends on the order k of the moment to be estimated, on the physical space dimension d, and on the support size constant $R > 0$ in the bounded support assumption (32).

Then, the total work to compute the MLMC estimates $\widehat{E^{L,(k)}}[u(\cdot; t)]$ is bounded by (with $O(\cdot)$ depending on the size of D)

$$\widehat{\text{Work}_L^{MLMC}} = O\left(\sum_{\ell=0}^L M_\ell \Delta x_\ell^{-(d+1)} |\log \Delta x|^{k-1} \right) . \tag{112}$$

Based on Theorem 6, we infer that the choice (97) of numbers M_ℓ of MC samples at level ℓ should also be used in the MLMC estimation of k-point correlation functions for $k > 1$, provided the order s of the underlying deterministic FVM scheme (59)–(61) satisfies

$$0 \leq s < d + 1 . \tag{113}$$

In particular, in a bounded domain $D \subset \mathbb{R}^d$ containing the bounded support in (52),

$$\| \mathscr{M}^k u(\cdot, t) - \widehat{E^{L,(k)}}[u(\cdot, t; \omega)] \|_{L^2(\Omega; L^1(D^k))} \leq C(\widehat{\text{Work}_L^{MLMC}})^{-s'/(d+1+s)} \qquad (114)$$

for any $0 < s' < s$ (with constant growing as $0 < s' \to s \leq 1$). The computational domain D can, in particular, contain the bounded support domains of the exact and discrete solutions at time $t > 0$, as indicated in (71), (52).

6 Monte Carlo and Multi-Level Monte Carlo Methods for Systems of Conservation Laws

6.1 General Considerations

We consider the general system of conservation laws (1), with random initial data in (115b) as well as possibly random coefficients and random flux functions in (115a). A notion of random entropy solutions can be defined for this general case, analogous to Definition 2 for scalar conservation laws. We refer to [47] for details. However, there are no well-posedness results for random entropy solutions for systems as even the underlying deterministic problem may not be well-posed, particularly in several space dimensions [6]. One approach to developing numerical approximations in this case is to *assume existence of random entropy solutions* and to design efficient methods for numerical approximations of their solutions:

It is fairly straightforward to extend the MCFV scheme, given in Sect. 4.3.1, to general, nonlinear hyperbolic systems of conservation laws with random inputs. A convergence rate estimate, analogous to (80) can be proved, *provided* that one postulates a convergence rate, analogous to (68), for the underlying spatio-temporal FV discretization, see [47]. Similarly, the MLMC method, as described in Sect. 4.4 can also be readily extended to this general case and a convergence rate estimate, similar to (95), once the underlying spatio-temporal discretization converges like in (68) or at least a estimate of the type (96) holds. Consequently, a complexity estimate as (98) can be shown under these assumptions, demonstrating that the MLMC-FV method is more efficient than the MC-FV method.

6.2 Numerical Experiments

We present a few numerical experiments involving systems of conservation laws to illustrate the robustness of the MLMC method and its comparison with the Monte Carlo method.

6.2.1 Uncertain Orszag-Tang Vortex

This test is taken from [47], section 6.2. The system of conservation laws under consideration are the ideal magnetohydrodynamics (MHD) equations of plasma physics. We consider the ideal MHD equations on the two-dimensional domain $[0, 1]^2$ with periodic boundary conditions. The random initial data is an uncertain version of the well-known Orszag-Tang benchmark test problem. We consider a random initial data with 8 sources of uncertainty, namely in the amplitudes of the initial density and pressure, phases of the initial sinusoidal velocity fields and the phases and amplitude of the initial solenoidal magnetic fields. The mean and the variance of the density, computed at time $T = 1.0$ with an MLMC-FV method, with eight levels of resolution, a finest mesh of 4096^2 and with four samples at the finest level, with the underlying FV method using a HLLC Riemann solver, a second-order WENO reconstruction and upwind treatment of the Godunov-Powell source term [23], are shown in Fig. 1. In this case, one computes a reference solution with the above configuration and calculates the $L^2(\Omega, L^1(D))$ error for both the mean and variance, with MC and MLMC methods (of both first and second order spatio-temporal discretizations). The errors for the mean and variance are plotted in Figs. 2 and 3, respectively. They show that in this example, the MLMC FV method is at least 50–60 times faster than the single-level MC FV method for the mean and 10–20 times faster for the variance, to achieve a prescribed error tolerance in the engineering range of accuracy. Thus, this justifies the complexity estimates, described here, at least for this example.

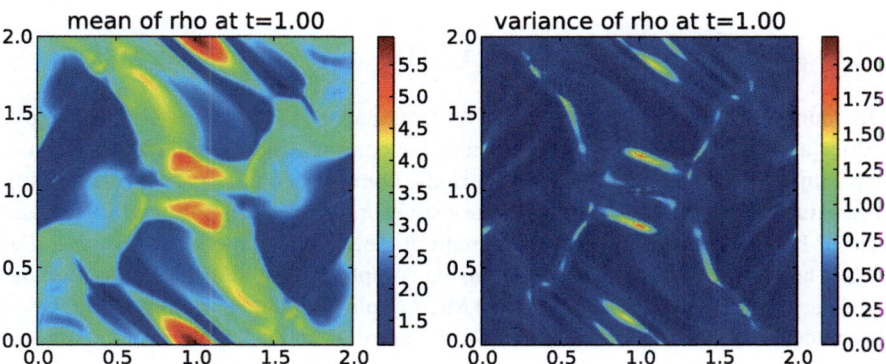

Fig. 1 Uncertain Orszag-Tang vortex solution at $t = 1.0$ using MLMC-FVM (8 sources of uncertainty). Left: Convergence of the sample mean of random density. Right: Convergence of the sample variance of the random density. Reproduced from [57]

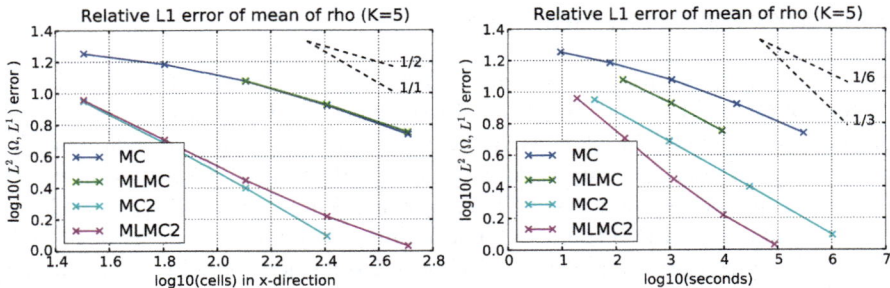

Fig. 2 Convergence of sample mean in the uncertain Orszag-Tang vortex simulation (8 sources of uncertainty). Left: Error vs. Mesh resolution. Right: Error vs. Runtime. Reproduced from [57]

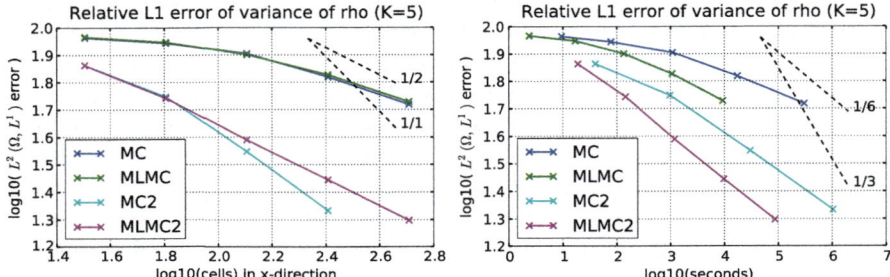

Fig. 3 Convergence of sample variance in the uncertain Orszag-Tang vortex simulation (8 sources of uncertainty). Left: Error vs. Mesh resolution. Right: Error vs. Runtime. Reproduced from [57]

6.2.2 A Random Kelvin-Helmholtz Problem

This numerical example is taken from a recent paper [18] and the MLMC computations are presented in [45]. We consider the compressible Euler equations in the
two-dimensional domain $[0, 1]^2$ with periodic boundary conditions. The uncertainty
arises due to the initial data being a (very small) random perturbation of the classic
Kelvin-Helmholtz problem, see [18], with 20 sources of uncertainty in the initial
data. The mean and variance of the density, computed with a Monte Carlo method,
on a Cartesian 1024^2 grid and with 400 MC samples are shown in Fig. 4 (Top Row).
The underlying FV scheme is the third-order entropy stable TeCNO scheme of [16].
Surprisingly for this test case, the variance of the solution is at least three orders of
magnitude higher than the variance of the initial data. This amplification of variance
is due to the generation of structures at smaller and smaller scales, when the shear
flow interacts with the contact discontinuity. In this particular case, the MLMC
method provides *no gain* in computational efficiency over the standard Monte Carlo
method. This is depicted in Fig. 4 (Bottom row, Left), where the L^1 difference in
the mean of the density, computed with the MLMC and MC methods, at the same
grid resolution for the finest grid and the same number of samples at the finest grid
resolution of MLMC, with respect to an MC reference solution computed with 1024

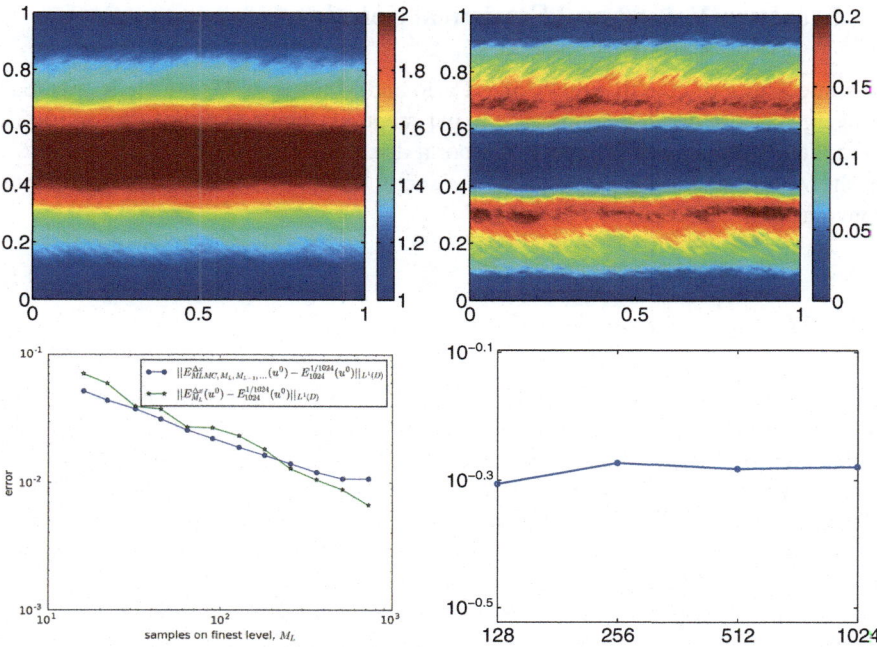

Fig. 4 UQ for the random Kelvin-Helmholtz problem. Top Row: Density at time $T = 2$ computed with MC method and TeCNO3 scheme with the algorithm proposed in [18] Left: Convergence of the sample mean, Right: Convergence of the sample variance. Bottom Row Left: Comparison between MC and MLMC method with respect to error in mean of the density with respect to a reference solution (Reproduced from [45]). Bottom Right: Lack of convergence for a single sample for the density (reproduced from [45])

samples, is compared. The results show that error due to MLMC is comparable to the error due to the MC calculation, provided that the number of samples at the finest grid level of the MLMC calculation is the same as the number of MC samples. Thus, in this case, the coarse levels of the MLMC method do not increase the accuracy of the computation and are redundant. Given this observation, it is clear that an error estimate of the form (95) cannot hold for this particular example. In fact, even an error estimate of the form (68) for the underlying spatio-temporal discretization, does not hold for this example. We see this from Fig. 4 (Bottom row, Right) where the difference in L^1 between two successive mesh resolutions for a single sample is shown. This figure show that the error remains constant with respect to resolution and the underlying FV scheme does not converge for this particular test case. It was also shown in [45] that even weaker convergence bounds, such as the variance of the difference between successive resolutions (96), does not hold for this particular problem, as structures at even smaller scales are generated upon mesh refinement.

7 Measure-Valued and Statistical Solutions

It is interesting to note that the Monte Carlo and Multi-level Monte Carlo methods converge for the previous numerical experiment, as shown in Fig. 4 (bottom left), even though the underlying spatio-temporal discretization may not converge in L^1, as shown in Fig. 4 (bottom right). What exactly do the (ML)MC-FV computations converge to in this case?

7.1 Measure-Valued Solutions

This question was partially answered in recent papers [17, 18] and references therein. There, the authors proved that a Monte Carlo based algorithm, together with an entropy stable scheme such as the TeCNO scheme of [16], converges to an *entropy measure valued solution* of the underlying system of conservation laws (1). Measure valued solutions were first proposed by DiPerna [14] and are in fact *Young measures* i.e., space-time parametrized probability measures on the phase space $\mathbb{R}^N$ of the system (1). Let $D \subset \mathbb{R}^d$ be the domain and $D_t := D \times \mathbb{R}_+$, we define *Young measure* from D_t to $\mathbb{R}^N$ as a map which associates to each point $(x, t) \in D_t$ a probability measure on $\mathbb{R}^N$. More precisely, a Young measure is a weak* measurable map $\nu : D_t \to \mathscr{P}(\mathbb{R}^N)$, meaning that the mapping

$$(x, t) \mapsto \langle \nu_{x,t}, g \rangle = \int_{\mathbb{R}^N} g(\xi) d\nu_{x,t}(\xi) \text{ is Borel measurable for every } g \in C_0(\mathbb{R}^N).$$

The set of all Young measures from D_t into $\mathbb{R}^N$ is denoted by $\mathbf{Y}(D_t, \mathbb{R}^N)$. Given this notation for Young measures, one can rewrite the following system of N-conservation laws,

$$\partial_t u + \nabla_x \cdot f(u) = 0, \quad (x, t) \in D_t, \tag{115a}$$

$$u(x, 0) = \bar{u}(x), \quad x \in D. \tag{115b}$$

in terms of the following *measure-valued Cauchy problem*,

$$\partial_t \langle \nu_{x,t}, \xi \rangle + \nabla_x \cdot \langle \nu_{x,t}, f(\xi) \rangle = 0, \quad (x, t) \in D_t,$$
$$\langle \nu_{x,0}, \xi \rangle = \langle \sigma_x, \xi \rangle, \quad x \in D, \tag{116}$$

with possibly Young measure-valued initial data σ_x. The above system (116) has to hold in the sense of distributions. Entropy (admissibility) conditions can be imposed by interpreting an associated entropy inequality in the Young measure sense [14]

Global existence of measure-valued solutions was shown recently in [17, 18] and references therein, by proving convergence of the following Monte Carlo based ensemble averaging algorithm.

Algorithm 1.
Let the initial data for an underlying time-dependent PDE (116) be given as a Young measure $\sigma \in \mathbf{Y}(D, \mathbb{R}^N)$ i.e., a Young measure $D \mapsto \mathscr{P}(\mathbb{R}^N)$.

- **Step 1:** Let $u_0 : \Omega \mapsto L^p(\mathbb{R}^d)$ be a random field on a probability space $(\Omega, \Sigma, \mathbb{P})$ with law σ, i.e., $\sigma(E) = \mathbb{P}(u_0(\omega) \in E)$.
- **Step 2:** Evolve the initial random field by applying a suitable numerical scheme, with solution map $\mathscr{S}_t^\Delta$, to the initial data $u_0(\omega)$ for every $\omega \in \Omega$, obtaining an approximate random field $u^\Delta(\omega; \cdot, t) := \mathscr{S}_t^\Delta u_0(\omega; \cdot)$.
- **Step 3:** Define the *approximate measure-valued solution* ν^Δ as the law of u^Δ with respect to $\mathbb{P}$, i.e. for all Borel sets $E \subset \mathbb{R}^N$,

$$\nu_{x,t}^\Delta(E) = \mathbb{P}(u(\omega; x, t) \in E).$$

It was shown in [18, Appendix A.3.1] that ν^Δ are indeed Young measures. The existence of a random field u_0 with a given law σ, as required in Step 1, is guaranteed by proposition A.3 of [18].

The numerical method in Step 2 of Algorithm 1, can be appropriate structure preserving Finite Volume Methods, such as the arbitrary high-order entropy stable TeCNO schemes of [16]. The last ingredient in our numerical approximation of measure-valued solutions is to find, and approximate, the random field $u_0(\omega; x)$ which appears in Algorithm 1, resulting in the following algorithm.

Algorithm 2.
Let $\Delta = (\Delta x_1, \ldots, \Delta x_d)$ denote the grid size parameter and let $M \in \mathbb{N}$. Let further $\sigma^\Delta \in \mathbf{Y}(\mathbb{R}^d, \mathbb{R}^N)$ denote the initial Young measure.

- **Step 1:** For some probability space $(\Omega, \Sigma, \mathbb{P})$, draw M i.i.d. random fields $u_0^{\Delta,1}, \ldots, u_0^{\Delta,M} : \Omega \times \mathbb{R}^d \to \mathbb{R}^N$, with [the same] law σ^Δ.
- **Step 2:** For each k and *for a fixed* $\omega \in \Omega$, use a suitable numerical scheme to numerically approximate the conservation law (115a) with initial data $u_0^{\Delta,k}(\omega)$. Denote $u^{\Delta,k}(\omega; \cdot, t) = \mathscr{S}_t^\Delta u_0^{\Delta,k}(\omega; \cdot)$ as the computed solutions.
- **Step 3:** Define the approximate measure-valued solution

$$\nu_{x,t}^{\Delta,M} := \frac{1}{M} \sum_{k=1}^{M} \delta_{u^{\Delta,k}(\omega; x, t)}. \tag{117}$$

Note that, as in any Monte Carlo method, the approximation $\nu^{\Delta,M}$ depends on the choice of $\omega \in \Omega$, i.e. on the choice of seed in the random number generator. However, one can prove that the convergence rate of approximation is independent of this choice, P-almost surely.

The approximate solutions $\nu^{\Delta,M}$ were proved to converge to an entropy measure-valued solution of (116) as $(\Delta, M) \to (0, \infty)$ in [18]. This convergence is in the weak-* topology on Young measures that amounts to convergence of one-point statistical quantities of interest such as the mean, variance, point statistics and probability density functions etc. Thus, the results of Fig. 4 are justified, mathematically. Furthermore, the computations of [18] also demonstrated that *the measure-valued solution may not be atomic even if the initial data is an atomic young measure* concentrated on a L^1 function. *Thus, there seems to be no fully deterministic version of multi-dimensional systems of conservation laws and the initial value problem for* (115), *hitherto considered deterministically, should in fact be considered as a problem in uncertainty quantification (UQ).*

Measure-valued solutions for (116) exist for all times ([17] and references therein) and are able to capture limits of numerical approximation of systems of conservation laws (115). They also serve as an UQ framework within which the random initial data is represented as a Young measure (or point probability distribution). However, measure-valued solutions are not necessarily unique, particularly when the initial data is non-atomic. This holds true even for scalar conservation laws, see for instance example 9.1 in [17], even for the one-dimensional Burgers' equation. Thus, measure-valued solutions need to be augmented with additional constraints in order to recover uniqueness. This is consistent with the observations, reported in [17, 18], that the computed measure valued solution when realized as a limit of the MC-FV algorithm, is stable with respect to the choice of numerical method and with respect to perturbations of the underlying random initial data. We refer to [40] for numerical approximation of EMV solutions and convergence analysis for a combined MC FV method for the velocity formulation of the incompressible Euler equations.

7.2   Statistical Solutions

An attempt to constrain measure-valued solutions in order to recover uniqueness has been made recently in [19]. In this paper, the authors propose a concept of *statistical solutions* of systems of conservation laws as a suitable solution paradigm as well as computational UQ framework. Statistical solutions of the Navier-Stokes equations in the sense of Foias and Prodi [20] are time-parametrized families of probability measures on a L^p function space. We refer to the surveys [21, 41] for their mathematical theory for the incompressible Navier-Stokes equations.

In these references, statistical solutions are time-dependent probability measures on divergence free L^2 functions that evolve based on either *Cylindrical moments* (Liouville equation) or *Characteristic functionals* (Hopf equation) resulting in a functional differential equation on an infinite-dimensional space.

Although a viable concept for viscous flows such as the incompressible Navier-Stokes equations, it is unclear how the statistical solutions in the sense of [21] can be extended to inviscid problems such as systems of conservation laws (115). Moreover, probability measures on L^p spaces are *non-local* and local statistical quantities such as one-point statistics or multi-point correlations are hard to interpret in this setting. Hence, the linkage between statistical solutions (in the sense of [21] or of the closely related notion of statistical solutions introduced by Vishik and Fursikov in [61]) and measure-valued solutions is unclear.

These difficult issues were tackled in a recent paper [19] in which the authors were able to *localize* probability measures on infinite-dimensional function spaces by relating them to Young measures as described below.

7.2.1   Correlation Measures

Let $\mathbf{U} = \mathbb{R}^N$ and $D^k := D \times \cdots \times D$ denote the k-fold cartesian product of D. In [19], the authors defined correlation measures as a collection $\nu = (\nu^1, \nu^2, \ldots)$ of maps $\nu^k : D^k \to \mathscr{P}(\mathbf{U}^k)$ satisfying the following properties:

(i) *Weak-* *measurability:* Each map $\nu^k : D^k \to \mathscr{P}(\mathbf{U}^k)$ is weak-* measurable, in the sense that the map $x \mapsto \langle \nu_x^k, f \rangle$ from $x \in D^k$ into $\mathbb{R}$ is Borel measurable for all $f \in C_0(\mathbf{U}^k)$ and $k \in \mathbb{N}$. In other words, ν^k is a Young measure from D^k to $\mathbf{U}^k$.

(ii) L^p-*boundedness:* ν is L^p-bounded, in the sense that

$$\int_D \langle \nu_x^1, |\xi|^p \rangle \, dx < +\infty.$$

(iii) *Symmetry:* If σ is a permutation of $\{1, \ldots, k\}$ and $f \in C_0(\mathbb{R}^k)$ then $\langle \nu_{\sigma(x)}^k, f(\sigma(\xi)) \rangle = \langle \nu_x^k, f(\xi) \rangle$ for a.e. $x \in D^k$. Here, we denote $\sigma(x) = \sigma(x_1, x_2, \ldots, x_k) = (x_{\sigma_1}, x_{\sigma_2}, \ldots, x_{\sigma_k})$.

(iv) *Consistency:* If $f \in C_0(\mathbf{U}^k)$ is of the form $f(\xi_1, \ldots, \xi_k) = g(\xi_1, \ldots, \xi_{k-1})$ for some $g \in C_0(\mathbf{U}^{k-1})$, then $\langle \nu_{x_1, \ldots, x_k}^k, f \rangle = \langle \nu_{x_1, \ldots, x_{k-1}}^{k-1}, g \rangle$ for almost every $(x_1, \ldots, x_k) \in D^k$.

(v) *Diagonal continuity (DC):* If $B_r(x) := \{ y \in D : |x - y| < r \}$ then

$$\lim_{r \to 0} \int_D \frac{1}{|B_r(0)|} \int_{B_r(x)} \langle \nu_{x,y}^2, |\xi_1 - \xi_2|^p \rangle \, dy \, dx = 0. \tag{118}$$

Each element ν^k is called a *correlation marginal*. The consistency property implies that the kth correlation marginal ν^k determines all $\nu^l, l \leq k$. Thus, the family of correlation marginals is a *hierarchy of young measures*. The *equivalence between correlation measures and probability measures on L^p* is described by the following result, which is [19, Thm. 2.7]:

For every correlation measure ν defined as above, there exists a unique probability measure $\mu \in \mathscr{P}(L^p(D))$ satisfying

$$\int_{L^p} \|u\|_{L^p}^p \, d\mu(u) < \infty \tag{119}$$

such that, for all $k \in \mathbb{N}$ and for every $g \in L^1(D^k : C_0(\mathbf{U}^k))$: there holds

$$\int_{D^k} \int_{\mathbf{U}^k} g(x, \xi) \, d\nu_x^k(\xi) dx = \int_{L^p} \int_{D^k} g(x, u(x)) \, dx d\mu(u) \qquad (120)$$

(where $u(x)$ denotes the vector $(u(x_1), \ldots, u(x_k))$). Conversely, for every probability measure $\mu \in \mathscr{P}(L^p(D))$ with finite L^p bound, there exists a unique correlation measure ν satisfying (120). The relation (120) is also valid for any measurable $g :$ $D \times \mathbf{U} \to \mathbb{R}$ such that $|g(x, \xi)| \le C|\xi|^p$ for a.e. $x \in D$.

Moreover, it was also shown in [19] (Theorem 2.20) that the probability measure μ (equivalently the associated correlation measure ν) was uniquely determined in terms of *moments* or *correlation functions* of the correlation measure ν given by

$$m^k : D^k \to \mathbf{U}^k, \qquad m^k(x) := \int_{\mathbf{U}^k} \xi_1 \otimes \cdots \otimes \xi_k \, d\nu_x^k(\xi), \qquad k \in \mathbb{N}. \qquad (121)$$

Here, $\mathbf{U}^{\otimes k}$ refers to the tensor product space $\mathbf{U} \otimes \cdots \otimes \mathbf{U} \simeq \mathbb{R}^{kN}$ (repeated k times), and $\xi_1 \otimes \cdots \otimes \xi_k$ is a functional defined by its action on the dual space $(\mathbf{U}^{\otimes k})^* = \mathbf{U}^{\otimes k}$ through

$$(\xi_1 \otimes \cdots \otimes \xi_k) : (\zeta_1 \otimes \cdots \otimes \zeta_k) = (\xi_1 \cdot \zeta_1) \cdots (\xi_k \cdot \zeta_k).$$

7.2.2 Definition of Statistical Solutions

Once it is established that probability measures on L^p are completely characterized by moments (correlation functions) of the associated correlation measure, one can evolve an initial probability measure on $L^p(D)$ in time by writing evolution equations for these correlation functions. Following [19], we define statistical solution of (115a) with an initial data $\bar{\mu} \in \mathscr{P}(L^1(\mathbb{R}^d, \mathbb{R}^N))$ as a weak*-measurable mapping $t \mapsto \mu_t \in \mathscr{P}(L^1(\mathbb{R}^d, \mathbb{R}^N))$ such that the corresponding correlation measures $(\nu_t^k)_{k \in \mathbb{N}}$ satisfy the following equations in the sense of distributions,

$$\partial_t \langle \nu_{t,x}^k, \xi_1 \otimes \cdots \otimes \xi_k \rangle + \sum_{i=1}^k \nabla_{x_i} \cdot \langle \nu_{t,x}^k, \xi_1 \otimes \cdots \otimes f(\xi_i) \otimes \cdots \otimes \xi_k \rangle = 0, \quad \forall k \in \mathbb{N}.$$
$$(122)$$

Note that the first equation in the hierarchy for $k = 1$ precisely agrees with the definition of measure-valued solutions (116). Thus, *a statistical solution is a measure-valued solution that includes information about the evolution of all possible multi-point correlations* in the underlying functions. Hence, statistical solutions are considerably more constrained than measure-valued solutions providing some hope for uniqueness. Moreover, statistical solutions reduce to standard weak solutions as long as the initial data and the resulting statistical solution are *atomic* i.e., $\bar{\mu} = \delta_{\bar{u}}, \mu_t = \delta_{u(t)}$ with $\bar{u}, u(t) \in L^p(D)$. On the other hand, a non-atomic initial probability measure $\bar{\mu}$ can be used to model input uncertainty. Hence, statistical solutions provide a UQ framework [19].

Currently, well-posedness results for statistical solutions are only available for scalar conservation laws. In [19], a concept of entropy statistical solutions was proposed for scalar conservation laws, Definition 4.3 therein. This concept generalizes Kružkhov entropies to probability measures on L^1. Well-posedness of *entropy statistical solutions* for scalar conservation laws was shown in [19, Thm. 4.7]. These entropy statistical solutions were also shown to satisfy a non-expansive property with respect to the 1-Wasserstein metric on probability measures on L^1. The mathematical analysis of the convergence of the MC-FV algorithms 1, 2, in the sense of [18], for scalar conservation laws is currently in progress. Some (preliminary) findings are as follows: the same Monte Carlo ensemble averaging algorithm 2, proposed in [18], also converges, under additional assumptions, to a statistical solution of the underlying nonlinear, hyperbolic system as will be shown in a forthcoming paper. Thus, statistical solutions may provide a suitable mathematical and numerical solution framework for multi-dimensional systems of conservation laws as well as of computational uncertainty quantification for them.

The algorithms proposed in [17, 18] are Monte Carlo based. A MLMC version of this algorithm was designed and shown to converge to an entropy measure valued solution in a recent paper [45]. Moreover, it was shown in [45] that if the variance of the details, similar to (96), converge at an algebraic rate i.e. if $s > 0$ in (96), then the MLMC algorithm for approximating entropy measure-valued solutions will be more efficient than the Monte Carlo version. However, such an estimate may not hold as shown in Fig. 4 and the MC and MLMC versions will be comparable (see Fig. 4 bottom left).

Currently, computation of k-point correlation functions within the framework of statistical solutions uses a full tensor format. This can be prohibitively expensive for even moderate k.

The adaptation of the sparse-tensor algorithms from [46] as described in Sect. 4.4 to this framework in order to accelerate the computations of multi-point statistical quantities of interest is currently under development.

Acknowledgements The work of CS was supported in part by ERC grant no. 247277. SM acknowledges partial support from ERC STG NN. 306279 SPARCCLE. CS and SM acknowledge also partial support from ETH grant no. CH1-03 10-1.

References

1. R. Abgrall, A. Harten, Multiresolution representation in unstructured meshes. SIAM J. Numer. Anal. **35**(6), 2128–2146 (electronic) (1998). https://doi.org/10.1137/S0036142997315056
2. A. Barth, C. Schwab, N. Zollinger, Multi-level Monte Carlo finite element method for elliptic PDEs with stochastic coefficients. Numer. Math. **119**(1), 123–161 (2011). https://doi.org/10.1007/s00211-011-0377-0
3. C. Bierig, A. Chernov, Estimation of arbitrary order central statistical moments by the multilevel Monte Carlo method. Stoch. Partial Differ. Equ. Anal. Comput. **4**(1), 3–40 (2016). https://doi.org/10.1007/s40072-015-0063-9
4. B.L. Bihari, A. Harten, Multiresolution schemes for the numerical solution of 2-D conservation laws. I. SIAM J. Sci. Comput. **18**(2), 315–354 (1997). https://doi.org/10.1137/S1064827594278848
5. H. Bijl, D. Lucor, S. Mishra, C. Schwab, *Uncertainty Quantification in Computational Fluid Dynamics*. Lecture Notes in Computational Science and Engineering, vol. 92 (Springer, Cham, 2013). https://doi.org/10.1007/978-3-319-00885-1
6. E. Chiodaroli, C. De Lellis, O. Kreml, Global ill-posedness of the isentropic system of gas dynamics. Commun. Pure Appl. Math. **68**, 1157–1190 (2015)
7. B. Cockburn, C.W. Shu, TVB Runge-Kutta local projection discontinuous Galerkin finite element method for conservation laws. II. General framework. Math. Comput. **52**, 411–435 (1989)
8. B. Cockburn, F. Coquel, P.G. LeFloch, Convergence of the finite volume method for multidimensional conservation laws. SIAM J. Numer. Anal. **32**(3), 687–705 (1995). https://doi.org/10.1137/0732032
9. S. Cox, M. Hutzenthaler, A. Jentzen, J. van Neerven, T. Welti, Convergence in Hölder norms with applications to Monte Carlo methods in infinite dimensions. Technical Report 2016-28, Seminar for Applied Mathematics, ETH Zürich, 2016
10. G. Da Prato, J. Zabczyk, *Stochastic Equations in Infinite Dimensions*. Encyclopedia of Mathematics and its Applications, vol. 44 (Cambridge University Press, Cambridge, 1992). https://doi.org/10.1017/CBO9780511666223
11. C.M. Dafermos, *Hyperbolic Conservation Laws in Continuum Physics*. Grundlehren der Mathematischen Wissenschaften [Fundamental Principles of Mathematical Sciences], vol. 325, 3rd edn. (Springer, Berlin, 2010). https://doi.org/10.1007/978-3-642-04048-1
12. T. Daun, S. Heinrich, Complexity of Banach space valued and parametric integration, in *Monte Carlo and Quasi-Monte Carlo Methods 2012*. Springer Proceedings in Mathematics & Statistics, vol. 65 (Springer, Heidelberg, 2013), pp. 297–316. https://doi.org/10.1007/978-3-642-41095-6_12
13. T. Daun, S. Heinrich, Complexity of parametric initial value problems in Banach spaces. J. Complexity **30**(4), 392–429 (2014). https://doi.org/10.1016/j.jco.2014.01.002
14. R.J. DiPerna, Measure-valued solutions to conservation laws. Arch. Ration. Mech. Anal. **88**, 223–270 (1985)
15. R. Eymard, T. Gallouët, R. Herbin, Finite volume methods, in *Handbook of Numerical Analysis*, vol. VII (North-Holland, Amsterdam, 2000), pp. 713–1020
16. U. Fjordholm, S. Mishra, E. Tadmor, Arbitrarily high-order accurate entropy stable essentially nonoscillatory schemes for systems of conservation laws. SIAM J. Numer. Anal. **50**(2), 544–573 (2012). https://doi.org/10.1137/110836961
17. U.S. Fjordholm, S. Mishra, E. Tadmor, On the computation of measure-valued solutions. Acta Numer. **25**, 567–679 (2016). https://doi.org/10.1017/S0962492916000088
18. U. Fjordholm, R. Kappeli, S. Mishra, E. Tadmor, Construction of approximate entropy measure valued solutions for systems of conservation laws. J. Found. Comput. Math **17**(3), 763–827 (2017). https://doi.org/10.1007/s10208-015-9299-z
19. U.S. Fjordholm, S. Lanthaler, S. Mishra, Statistical solutions of hyperbolic conservation laws I: Foundations. Arch. Ration. Mech. Anal. **226**, 809–849 (2017)

20. C. Foiaş, G. Prodi, Sur les solutions statistiques des équations de Navier-Stokes. Ann. Mat. Pura Appl. (4) **111**, 307–330 (1976)
21. C. Foias, O. Manley, R. Rosa, R. Temam, *Navier-Stokes Equations and Turbulence* (Cambridge University Press, Cambridge, 2001)
22. P.K. Friz, B. Gess, Stochastic scalar conservation laws driven by rough paths. Ann. Inst. H. Poincaré Anal. Non Linéaire **33**(4), 933–963 (2016). https://doi.org/10.1016/j.anihpc.2015.01.009
23. F. Fuchs, A. McMurry, S. Mishra, N.H. Risebro, K. Waagan, Approximate Riemann solver based high-order finite volume schemes for the MHD equations in multi-dimensions. Commun. Comput. Phys. **9**, 324–362 (2011). https://doi.org/10.4208/cicp.171109.070510a
24. G. Geraci, P.M. Congedo, R. Abgrall, G. Iaccarino, A novel weakly-intrusive non-linear multiresolution framework for uncertainty quantification in hyperbolic partial differential equations. J. Sci. Comput. **66**(1), 358–405 (2016). https://doi.org/10.1007/s10915-015-0026-3
25. N. Gerhard, S. Müller, Adaptive multiresolution discontinuous Galerkin schemes for conservation laws: multi-dimensional case. Comput. Appl. Math. **35**(2), 321–349 (2016). https://doi.org/10.1007/s40314-014-0134-y
26. B. Gess, B. Perthame, P.E. Souganidis, Semi-discretization for stochastic scalar conservation laws with multiple rough fluxes. SIAM J. Numer. Anal. **54**(4), 2187–2209 (2016). https://doi.org/10.1137/15M1053670
27. M.B. Giles, Multilevel Monte Carlo methods. Acta Numer. **24**, 259–328 (2015). https://doi.org/10.1017/S096249291500001X
28. E. Godlewski, P.A. Raviart, *Hyperbolic Systems of Conservation Laws*. Mathématiques & Applications (Paris) [Mathematics and Applications], vol. 3/4 (Ellipses, Paris, 1991)
29. E. Godlewski, P.A. Raviart, *Numerical Approximation of Hyperbolic Systems of Conservation Laws*. Applied Mathematical Sciences, vol. 118 (Springer, New York, 1996)
30. S. Gottlieb, C. Shu, E. Tadmor, High order time discretizations with strong stability property. SIAM Rev. **43**, 89–112 (2001)
31. M. Griebel, H. Harbrecht, A note on the construction of L-fold sparse tensor product spaces. Constr. Approx. **38**(2), 235–251 (2013). https://doi.org/10.1007/s00365-012-9178-7
32. H. Harbrecht, M. Peters, M. Siebenmorgen, Combination technique based k-th moment analysis of elliptic problems with random diffusion. J. Comput. Phys. **252**, 128–141 (2013). https://doi.org/10.1016/j.jcp.2013.06.013
33. A. Harten, B. Engquist, S. Osher, S.R. Chakravarty, Uniformly high order accurate essentially non-oscillatory schemes. J. Comput. Phys. **71**, 231–303 (1987)
34. A. Hiltebrand, S. Mishra, Entropy stable shock capturing streamline diffusion space-time discontinuous Galerkin (DG) methods for systems of conservation laws. Numer. Math. **126**(1), 103–151 (2014)
35. H. Holden, N.H. Risebro, *Front Tracking for Hyperbolic Conservation Laws*. Applied Mathematical Sciences, vol. 152 (Springer, New York, 2011). https://doi.org/10.1007/978-3-642-23911-3. First softcover corrected printing of the 2002 original
36. U. Koley, N.H. Risebro, C. Schwab, F. Weber, A multilevel monte carlo finite difference method for random scalar degenerate convection diffusion equations. J. Hyperbolic Diff. Equ. **14**(3), 415–454 (2017)
37. D. Kröner, *Numerical Schemes for Conservation Laws*. Wiley-Teubner Series Advances in Numerical Mathematics (Wiley, Chichester, 1997)
38. M. Ledoux, M. Talagrand, *Probability in Banach Spaces*. Classics in Mathematics (Springer, Berlin, 2011). Isoperimetry and processes, Reprint of the 1991 edition
39. P.G. LeFloch, *Hyperbolic Systems of Conservation Laws*. Lectures in Mathematics ETH Zürich (Birkhäuser, Basel, 2002). https://doi.org/10.1007/978-3-0348-8150-0. The theory of classical and nonclassical shock waves
40. F. Leonardi, A projection method for the computation of admissible measure valued solutions of the incompressible Euler equations. Technical Report 2017-06, Seminar for Applied Mathematics, ETH Zürich, 2017

41. F. Leonardi, S. Mishra, C. Schwab, Numerical approximation of statistical solutions of planar, incompressible flows. Math. Models Methods Appl. Sci. **26**(13), 2471–2524 (2016). https://doi.org/10.1142/S0218202516500597
42. R.J. LeVeque, *Finite Volume Methods for Hyperbolic Problems.* Cambridge Texts in Applied Mathematics (Cambridge University Press, Cambridge, 2002). https://doi.org/10.1017/CBO9780511791253
43. P.L. Lions, B. Perthame, P.E. Souganidis, Scalar conservation laws with rough (stochastic) fluxes. Stoch. Partial Differ. Equ. Anal. Comput. **1**(4), 664–686 (2013)
44. P.L. Lions, B. Perthame, P.E. Souganidis, Scalar conservation laws with rough (stochastic) fluxes: the spatially dependent case. Stoch. Partial Differ. Equ. Anal. Comput. **2**(4), 517–538 (2014). https://doi.org/10.1007/s40072-014-0038-2
45. K. Lye, Multilevel Monte-Carlo for measure valued solutions. Technical Report 2016-51 (revised), Seminar for Applied Mathematics, ETH Zürich, 2016
46. S. Mishra, C. Schwab, Sparse tensor multi-level Monte Carlo finite volume Methods for hyperbolic conservation laws with random initial data. Math. Comput. **81**, 1979–2018 (2012). https://doi.org/10.1090/S0025-5718-2012-02574-9
47. S. Mishra, C. Schwab, J. Sukys, Multi-level monte carlo finite volume methods for nonlinear systems of conservation laws in multi-dimensions. J. Comput. Phys. **231**(8), 3365–3388 (2012). https://doi.org/10.1016/j.jcp.2012.01.011
48. S. Mishra, N.H. Risebro, C. Schwab, S. Tokareva, Numerical solution of scalar conservation laws with random flux functions. SIAM/ASA J. Uncertain. Quantif. **4**(1), 552–591 (2016). https://doi.org/10.1137/120896967
49. S. Mishra, C. Schwab, J. Šukys, Multi-level Monte Carlo finite volume methods for uncertainty quantification of acoustic wave propagation in random heterogeneous layered medium. J. Comput. Phys. **312**, 192–217 (2016). https://doi.org/10.1016/j.jcp.2016.02.014.
50. S. Müller, Multiresolution schemes for conservation laws, in *Multiscale, Nonlinear and Adaptive Approximation* (Springer, Berlin, 2009), pp. 379–408. https://doi.org/10.1007/978-3-642-03413-8_11
51. N.H. Risebro, C. Schwab, F. Weber, Multilevel monte carlo front-tracking for random scalar conservation laws. BIT Numer. Math. (2016). https://doi.org/10.1007/s10543-015-0550-4
52. G. Schmidlin, C. Schwab, Wavelet Galerkin BEM on unstructured meshes by aggregation, in *Multiscale and Multiresolution Methods*. Lecture Notes in Computational Science and Engineering, vol. 20 (Springer, Berlin, 2002), pp. 359–378 https://doi.org/10.1007/978-3-642-56205-1_12
53. C. Schwab, C.J. Gittelson, Sparse tensor discretizations of high-dimensional parametric and stochastic PDEs. Acta Numer. **20**, 291–467 (2011). https://doi.org/10.1017/S0962492911000055.
54. C. Schwab, S. Tokareva, High order approximation of probabilistic shock profiles in hyperbolic conservation laws with uncertain initial data. ESAIM: Math. Model. Numer. Anal. **47**(3), 807–835 (2013). https://doi.org/10.1051/m2an/2012060
55. C.W. Shu, S. Osher, Efficient implementation of essentially non-oscillatory schemes - II. J. Comput. Phys. **83**(1), 32–78 (1989)
56. J. Smoller, *Shock Waves and Reaction-Diffusion Equations.* Grundlehren der Mathematischen Wissenschaften [Fundamental Principles of Mathematical Sciences], vol. 258, 2nd edn. (Springer, New York, 1994)
57. J. Sukys, Robust multi-level monte carlo finite volume methods for systems of conservation laws with random input data. Ph.D. thesis, ETH Zürich, 2014
58. R.A. Todor, A new approach to energy-based sparse finite-element spaces. IMA J. Numer. Anal. **29**(1), 72–85 (2009). https://doi.org/10.1093/imanum/drm041
59. J. Tryoen, O. Le Maître, M. Ndjinga, A. Ern, Intrusive Galerkin methods with upwinding for uncertain nonlinear hyperbolic systems. J. Comput. Phys. **229**(18), 6485–6511 (2010). https://doi.org/10.1016/j.jcp.2010.05.007.
60. J. van Neerven, Stochastic evolution equations. Lecture Notes, ISEM (2007/8)

61. M.I. Višik, A.V. Fursikov, Solutions statistiques homogènes des systèmes differentiels paraboliques et du système de Navier-Stokes. Ann. Scuola Norm. Sup. Pisa Cl. Sci. (4) **4**(3), 531–576 (1977)
62. T. von Petersdorff, C. Schwab, Sparse finite element methods for operator equations with stochastic data. Appl. Math. **51**(2), 145–180 (2006). https://doi.org/10.1007/s10492-006-0010-1
63. J. Wehr, J. Xin, Front speed in the Burgers equation with a random flux. J. Stat. Phys. **88**(3–4), 843–871 (1997). https://doi.org/10.1023/B:JOSS.0000015175.70862.77

Printed by Printforce, the Netherlands